Airport
Planning and
Management

About the Authors

Seth B. Young, Ph.D., is an associate professor of Aviation in the College of Engineering at The Ohio State University in Columbus, Ohio, and the President of the International Aviation Management Group, Inc. Dr. Young has extensive experience in airport management consulting and is a certified member of the American Association of Airport Executives. Dr. Young is the chair of the National Academies' Transportation Research Board Committee on Aviation System Planning and holds an instrument-rated FAA commercial pilot and certified flight instructor certificate. He is the co-author of *Planning and Design of Airports*, Fifth Edition, also from McGraw-Hill.

Alexander T. Wells, Ed.D., recently retired professor at the College of Business at Embry-Riddle Aeronautical University in Daytona Beach, Florida, is a consultant to airport management with over 25 years of experience. He is the author of such leading texts as *Commercial Aviation Safety*, Second Edition, also from McGraw-Hill.

Airport Planning and Management

Sixth Edition

Seth B. Young
Alexander T. Wells

New York   Chicago   San Francisco   Lisbon   London
Madrid   Mexico City   Milan   New Delhi   San Juan
Seoul   Singapore   Sydney   Toronto

The McGraw·Hill Companies

Library of Congress Cataloging-in-Publication Data
Young, Seth B.
Airport planning and management / Seth B. Young, Alexander T. Wells. — 6th ed.
    p. cm.
Previous ed. by Alexander T. Wells and Seth B. Young.
ISBN 978-0-07-175024-0 (pbk.)
1. Airports—Planning.   2. Airports—Management.   I. Wells, Alexander.   II. Title.
TL725.3.P5W452 2011
387.7'36068—dc22   2011007247

McGraw-Hill books are available at special quantity discounts to use as premiums and sales promotions, or for use in corporate training programs. To contact a representative please e-mail us at bulksales@mcgraw-hill.com.

Airport Planning and Management, Sixth Edition

    5 6 7 8 9 0   QFR/QFR   1 6 5 4 3

ISBN 978-0-07-175024-0
MHID 0-07-175024-X

The pages within this book were printed on acid-free paper.

Sponsoring Editor
   Larry Hager

Acquisitions Coordinator
   Michael Mulcahy

Editorial Supervisor
   David E. Fogarty

Project Manager
   Aloysius Raj, Newgen Publishing and
   Data Services

Copy Editor
   Anupama Gopinath, Newgen Publishing
   and Data Services

Proofreader
   Kathrin Immanuel, Newgen Publishing
   and Data Services

Indexer
   Ken Hassman

Production Supervisor
   Pamela A. Pelton

Composition
   Newgen Publishing and Data Services

Art Director, Cover
   Jeff Weeks

Contents

Preface

In 1986, the first edition of *Airport Planning and Management* pioneered an innovative structure for a basic airport principles course designed for several similar, yet distinct, markets: the college student enrolled in an aviation program, as well as someone in the field of airport management or operations who is seeking further education. Since that time, five editions of the text were published, each edition reflecting updates that have occurred in the constantly evolving aviation industry. The response of both professors and students over the years has been gratifying. *Airport Planning and Management* and its accompanying test bank have been more widely used than any other teaching material for an airport course.

In the twenty-five years since the first edition of this text was published, the world of civil aviation, including airport management, has witnessed tremendous changes in technology, structure, and political environments. The aviation industry adjusted to major regulatory change, experienced economic woes, experienced record economic prosperity, adapted to a new world of enhanced security, and most recently, struggled through a worldwide economic downturn, and is poised for an entirely new paradigm of technology. In addition, the world of airport management has continued to evolve into more of an analytic and business-oriented discipline, applying theories of operations, economics, finance, and public administration to adapt to ever-changing environments.

We have made our best attempt to bring the sixth edition of *Airport Planning and Management* to a new standard of quality as a resource for current and future airport managers. We have worked hard to enhance the best and proven elements of the earlier editions while adding new perspectives, theories, and information gained from our respective teaching, research, and aviation experiences. The entire text has been critically revised, updated, and reorganized. In addition, significant text has been added and rewritten. Clear and interesting communication has been a priority, as in past editions.

Recognizing that a course in airport planning and management is normally a student's first exposure to the field, this text provides a significant amount of introductory material. While no one text can be the exhaustive source on any particular topic, this text attempts to provide a body of information that will allow students to gain knowledge of the various facets of airport planning and management at a fundamental, yet also comprehensively rich, level. The focus of this text is to build a solid foundation of understanding of all the elements that are of concern to airport management. Influenced by our combined experience of more than 50 years in teaching aviation management at the college level, we believe that the information contained in this text is commensurate with university level study.

It is recognized that instructors will supplement the material found in this text with current case studies, examples drawn from their own experiences, timely news and Internet sources, and industry and academic journals. Students are encouraged to explore and keep abreast of current periodicals, such as *Airport, Airport Business, Air Transport World,* and *Aviation Week.* It is hoped that the ability to reason accurately and objectively about problems facing airports and the development of a lasting interest in airport planning and management will be two valuable byproducts of the text's basic objectives.

Organization of the Sixth Edition

The nearly eight years since the publication of the fifth edition of this text has seen some of the most dramatic changes in civil aviation, and particularly airport management, and the implementation of an entirely new paradigm for the National Aviation System is on the horizon. Not coincidentally, the sixth edition of *Airport Planning and Management* includes significant revisions, and also includes new content that will hopefully be current for several years.

The text has been reorganized into three parts: Airports and Airport Systems, Airport Operations Management, and Airport Administrative Management. Each part is designed to address airport planning and management from specific perspectives.

Part I: Airports and airport systems

Part I provides an overview of airports from a systems perspective and provides background and historical information regarding the development of airports and the rules that airport management must adhere to. Within this part are three chapters.

> **Chapter 1:** *Airports and airport systems: An introduction* provides a comprehensive overview of airports in the United States, the national administrative structure of airports, and basic definitions that describe airports and types of airport activity.
>
> **Chapter 2:** *Airports and airport systems: Organization and administration* describes the public and private ownership and administrative

structures that exist for civil use airports in the United States and internationally. A comprehensive sample of employment positions that exist at airports is presented, as are descriptions of the duties of the airport manager, and an introduction to the public relations issues facing airport management.

Chapter 3: *Airports and airport systems: A historical and legislative perspective* includes an account of the development of airports within the civil aviation system that has been thoroughly reviewed and updated through early 2010, including the latest legislation debate regarding airport funding, and, of course, airport security.

Part II: Airport operations management

Part II has been written to provide the airport management student, as well as the new airport management employee, with a comprehensive information source describing the facilities and operations that exist within an airport's property including the airfield, airspace, terminals, and ground access systems. This part may be valuable not only as a text but also as a reference guide for those not in academic study. Within this part are five chapters.

Chapter 4: *The airfield* describes the facilities that exist on an airport to facilitate the operation of aircraft, including a full description of runways, taxiways, and navigational aids, along with associated signage, lighting, and markings. Much of the information contained in this chapter is sourced directly from the Federal Aviation Administration's *Airman's Information Manual,* a guide designed to provide pilots of civil aircraft with full descriptions of the aviation environment.

Chapter 5: *Airspace and air traffic management* provides a fundamental, yet detailed, description of the national airspace and air traffic control system, as it relates to airport management. A brief history of air traffic control is provided, as is a description of the management structure of the current air traffic control system. The basics of air traffic control are described, including the various classes of airspace and the rules by which they are operated. In addition, a description of the current and future planned enhancements to the air traffic control system is provided, to allow the airport manager to best prepare for the future of air traffic management.

Chapter 6: *Airport operations management under 14 CFR Part 139* has been moved immediately following the airfield and air traffic management chapters. This chapter discusses how the facilities described in Chapters 4 and 5 must be managed at airports certified to accommodate commercial air service under FAR Part 139—Certification of Airports. This edition has been updated to reflect the major revisions to FAR Part 139 in 2004.

Chapter 7: *Airport terminals and ground access* describes the infrastructure used to facilitate the transfer of passengers and cargo between aircraft and their ultimate origins and destinations within a metropolitan area. The chapter includes a historical account of the development of airport terminals, a description of the various airport terminal geometries that have been constructed, the components of the airport terminal, including aircraft aprons and gates, passenger processing facilities, and vehicle access facilities, such as roadways, curbsides, parking lots, and public transit systems.

Chapter 8: *Airport security* has been updated to describe the historical, current, and possible future of the operation of an airport from security perspectives. Historical accounts of airport security–related events are described, as is a comprehensive analysis of the events of September 11, 2001. The Transportation Security Administration and the associated regulations that affect airport management are discussed. In addition, current and future technologies that may be used to enhance airport security are described.

Part III: Airport administrative management

Part III has been designed to provide the airport management student with fundamental concepts and regulations that govern airport planning and management. This part focuses on the financial, administrative, and planning aspects of airport management. This part contains five chapters.

Chapter 9: *Airport financial management* presents the various strategies that exist to account and pay for the land, labor, and capital required to maintain financially stable airport operations and development. Airport accounting strategies are described, as are issues concerning airport insurance, revenue generating strategies, airport budgeting, and airport funding and financing strategies.

Chapter 10: *The economic, political, and social role of airports* describes the impacts that airports have on their surrounding communities, including the economic benefits of additional transportation service and associated economic activity and the environmental impacts such as noise, air and water quality, and industrialization. In addition, the political role of airport management when dealing with tenants of the airport and the outside community is described.

Chapter 11: *Airport planning* describes the strategies employed on local, regional, and national levels to prepare airports for future aviation activity. The chapter describes system planning on national and regional levels, and focuses on airport master planning, including demand forecasting, airport layout plans, runway orientation, land use planning, obstruction clearances, terminal area plans, and economic evaluation of planning alternatives. This chapter is designed to prepare the university level student for more advanced study in airport planning and design.

Chapter 12: *Airport capacity and delay* has been enhanced from previous editions by adding updated information regarding the latest developments in regulations and technologies that affect airport capacity and delay. In addition, this chapter introduces fundamental concepts that govern the laws of airport capacity and delay.

Chapter 13: *The future of airport management* concludes the text by presenting issues that may potentially have significant impacts on the future of airport planning and management. Included in this chapter are descriptions of new aircraft technologies, ranging from super-jumbo aircraft to small aircraft transportation systems. The text concludes with a brief discussion regarding the needs of future airport managers to further educate themselves in the many facets of management, particularly from a business perspective, as airports further develop as efficient business focused operating systems.

Learning tools

The purpose of this book is to help students learn the basic ingredients in the process of planning and managing an airport and also to provide a reference for those currently in the business of airport management. Towards these ends, we have employed various learning tools that recur throughout the text, including:

- *Chapter outlines:* Each chapter opens with an outline of the major topics to be covered.
- *Chapter objectives:* After the outline, each chapter includes the broad objectives that the student should be able to accomplish upon completing the chapter.
- *Figures, tables, and pictures:* Within each chapter are graphical representations of the material to compliment the text.
- *Logical organization and frequent headings:* The material covered has been put in a systematic framework so that the reader can find continuity and logic in the flow of the text.
- *Key terms:* Each chapter concludes with a list of key terms and other references used in the text. The terms may also be found in a glossary at the end of the text.
- *Review questions:* A series of questions posed for review and discussion follow at the end of each chapter. These questions are intended to encourage the student to summarize and further discuss the information learned from reading the chapter material.

- *Suggested readings:* A list of suggested reading is included after the end of each chapter for those who wish to pursue the material covered in more depth.
- *Glossary:* All key terms appearing at the end of each chapter, as well as many other terms used in the text and other of significance in airport planning and management, are included in the glossary.
- *Complete index:* The text includes a complete index to help the reader find needed information.

Supplemental materials

The material contained in this text is supplemented for instructors with effective teaching tools, including a test bank with over 1,000 questions in true/false, multiple choice, and fill-in-the-blank format, covering all chapters of the text; outlines of each chapter; as well as color graphics of many images found with in the text, in Microsoft Power Point format. These materials may be requested by instructors by contacting the publisher or authors.

It is hoped that this latest edition of *Airport Planning and Management* continues to meet the needs of students, instructors, and those already in the airport management industry as they seek fundamental knowledge of concern to airport planners and managers. As always, we welcome any feedback from our readers. Learning about the exciting world of airport planning and management should be educational and enjoyable. As university professors, industry professionals, and authors, we hope that we have contributed to this mission with this text.

Seth Young & Alex Wells

Acknowledgments

This sixth edition of *Airport Planning and Management* is in fact my second opportunity to revise this text. In the nearly eight years since Dr. Wells invited me to co-author this text, I have been blessed to see students who were the first users of the last edition become leaders in the airport industry, many of whom provided guidance in creating this latest edition. I hope that yet a new generation of airport planners and managers may benefit from this text in the years to come. I continue to thank my students and colleagues, current and past, at The Ohio State University, Embry-Riddle Aeronautical University, the American Association of Airport Executives, the Transportation Research Board, the University of California at Berkeley, Leigh Fisher, the Federal Aviation Administration, and airport managers throughout the world for their support and professional relationships. I would like to specifically thank Mr. Jeff Price, for his contributions to the text's chapter on airport security; Ms. Keri Spencer, for her contributions to the text's chapter on managing airports under FAR Part 139; and Dr. Kim Kenville, Dr. David Byers, and Dr. Ted Syme, for their general contributions and support for this latest edition. It is a pleasure to be associated with such worthy colleagues. My sincere thanks go out to all of my airport and aviation industry partners. Special thanks, of course, to Dr. Alex Wells, who honored me with the opportunity to again co-author the latest edition of this worthy text. I hope that the material found within our text is aptly able to communicate the great body of information I've gained from their valuable sources of knowledge.

Finally, most special thanks to my friends and family, especially my parents, Rosalie and Dennis Young, whose emphasis on the importance of education has penned an indelible mark on my personal and professional life, and have supported me through all that life throws at us. As with our aviation industry and life in general, the short-term impacts of the events that surround us merely support long-term growth.

Seth Young

I am sincerely appreciative of the many public and private institutions that have provided resource material from which I was able to shape this text. In this regard, I am particularly indebted to the Federal Aviation Administration for their numerous publications.

Faculty and students at University Aviation Association institutions who have reviewed material in the previous four editions have significantly shaped this book. To them I owe a special thanks because they represent the true constituency of any textbook author.

I am also indebted to many practicing airport planners and managers for their ideas and to the American Association of Airport Executives (AAAE) who adopted this book in their certification program for a number of years before developing their own material.

Finally, I must thank my wife, Mary, for considerable patience and encouragement throughout the process.

Alex Wells

Airport Planning and Management

Part I

Airports and airport systems

1

Airports and airport systems:
An introduction

Outline

- Introduction
  - Airports in the United States—An overview
  - The national administrative structure of airports
- Airport management on an international level
- The National Plan of Integrated Airport Systems
  - Commercial service airports
  - General aviation airports
  - Reliever airports
- The rules that govern airport management
- Organizations that influence airport regulatory policies

Objectives

The objectives of this section are to educate the reader with information to:

- Discuss the ownership characteristics of airports in the United States and internationally.
- Describe the National Plan of Integrated Airport Systems (NPIAS) and its application to categorizing public-use airports in the United States.
- Describe the governmental administrative organizations in the United States that oversee airports.
- Identify federal regulations and advisory circulars that influence airport operations.

Introduction

It is often said that managing an airport is like being mayor of a city. Similar to a city, an airport is comprised of a huge variety of facilities, systems, users, workers, rules, and regulations. Also, just as cities thrive on trade and commerce with other cities, airports are successful in part by their ability to successfully be the location where passengers and cargo travel to and from other airports. Furthermore, just as cities find their place as part of its county's, state's, and country's economy, airports, too, must operate successfully as part of the nation's system of airports. In this chapter, the airport system in the United States will be described in a number of ways. First, the national airport system, as a whole, will be described. Next, the various facilities that make up the airport system will be described. Finally, the various rules and regulations that govern the airport system will be described.

Airports in the United States—An overview

The United States has by far the greatest number of airports in the world. More than half the world's airports and more than two-thirds of the world's 400 busiest airports are located in the United States. There are more than 19,000 civil landing areas in the United States, including heliports, seaplane bases, and "fixed-wing" landing facilities. Most of these facilities are privately owned, and for private use only. Such facilities include helipads operated at hospitals and office buildings, private lakes for seaplane operations, and, most common, small private airstrips that accommodate the local owners of small aircraft operations. Many of these facilities are nothing more than a cleared area known as a "grass strip." Nevertheless, they are recognized and registered as civil-use landing areas and are, at least, operationally part of the United States system of airports.

There are approximately 5,200 airports that are open for use to the general public, nearly all of which have at least one lighted and/or paved runway. Of the 5,200 public-use airports in the United States, approximately 4,200 are publicly owned, either by the local municipality, county, state, or by an "authority" made up of municipal, county, and/or state officials. The remaining 1,000 are privately owned, by individuals, corporations, or private airport management companies (Fig. 1-1).

A few states, notably Alaska, Hawaii, and Rhode Island, own all the airports within the state, operating as a broad airport system. The federal government used to operate airports, including Ronald Reagan Washington National Airport and Washington Dulles International Airport, but ownership has since been transferred to an independent public authority known as the Metropolitan Washington Airports Authority (MWAA). Many airports in the United States were originally owned by the federal government, specifically the military, as they were created for military use during World Wars I and II. Since then, many

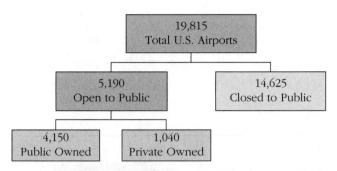

Figure 1-1. *Number of existing and proposed airports by ownership and use (January 2008).* (Figure courtesy FAA)

such airports were transferred to local municipal ownership. The transfers of most of these airports were made with provisions that permit the federal government to recapture its interest under certain conditions and also to review and approve any transfer of formal federal properties destined for nonairport use. Approximately 600 civil airports have these encumbrances. In addition, Army, Air Force Reserve, and National Guard units operate out of many civil airports, usually under some type of lease arrangements. These airports are known as **joint-use civil-military airports.**

The vast majority of the public-use civil airports in the United States, whether publicly or privately owned, are actually quite small, each serving a very small portion of the nation's number of aircraft operations (takeoffs and landings) and even a smaller portion of the total number of commercial air transportation passengers. Much of the activity that occurs at these airports includes operations in small aircraft for recreational purposes, flight training, and transportation by individuals and small private groups. Although most of the flying public rarely, if ever, utilizes many of these airports, the smaller airport facilities play a vital role in the United States system of airports (Fig. 1-2).

Airports are often described by their levels of activity. The activity, services, and investment levels vary greatly among the nation's airports. The most common measures used to describe the level of activity at an airport are the number of passengers served, the amount of cargo carried, and the number of operations performed at the airport.

The number of passengers served at an airport is typically used to measure the level of activity at airports that predominately serve commercial passengers traveling on the world's air carriers. Measuring passenger activity provides airport management with information that will allow for the proper planning and management for facilities used by passengers, including passenger terminals, parking garages, gate areas, and concessions.

Figure 1-2. *Many airports are no more than private grass strips.*
(Photo by Seth Young)

Specifically, the term **enplanements** (or *enplaned passengers*) is used to describe the number of passengers that board an aircraft at an airport. Annual enplanements are often used to measure the amount of airport activity, and even evaluate the amount of funding to be provided for improvement projects. The term **deplanements** (or *deplaned passengers*) is used to describe the number of passengers that deplane an aircraft at an airport.

The term *total passengers* is used to describe the number of passengers that either board or deplane an aircraft at an airport. At many airports, the number of total passengers is roughly double the number of annual enplanements. However, at airports where the majority of passengers are **transfer passengers,** the number of passengers is more than double the number of enplanements. This is because transfer passengers are counted twice, once when deplaning their arriving flight, and then again when boarding their next flight. Because of this distortion, passenger volumes are not often used to estimate passenger activity at an airport, although the largest airports serving as airline hubs often use the passenger volumes to advertise their grandeur. To remove this bias, most official measures of airport passenger activity are given in terms of enplanements.

Cargo activity is typically used to measure the level of activity at airports that handle freight and mail. Airports located near major seaports, railroad hubs, and large metropolitan areas, as well as airports served by the nation's cargo carriers (such as FedEx and UPS) accommodate thousands of tons of cargo annually.

The number of **aircraft operations** is used as a measure of activity at all airports, but is the primary measure of activity at general aviation (GA) airports. An aircraft operation is defined as a takeoff or a landing. When an aircraft makes a landing and then immediately takes off again, it is known as a "touch and go" and is counted as two operations. This activity is common at many GA airports where there is a significant amount of flight training. When an aircraft takes off and lands at an airport without landing at any other airport, the aircraft is said to be performing **local operations.** An **itinerant operation** is a flight that takes off from one airport and lands at another.

Another, albeit, indirect measure of airport activity is identified by the number of aircraft "based" at the airport. A **based aircraft** is an aircraft that is registered as a "resident" of the airport. Typically, the owner of such an aircraft will pay a monthly or annual fee to park the aircraft at the airport, either outside in a designated aircraft parking area or in an indoor hangar facility. The number of based aircraft is used to indirectly measure activity primarily at smaller airports where private "general" aviation is dominant. At airports that primarily handle the air carriers, relatively few aircraft are actually based.

Operations and based aircraft are measures of activity that influence the planning and management primarily of the **airside** of airports, such as the planning and management of runways, taxiways, navigational aids, gates, and aircraft parking areas.

In general, airport management measure the activity levels of their airports on the basis of all levels of passenger, cargo, operations, and based aircraft activity; virtually all airports, especially the largest airports in the nation, accommodate passengers and cargo, as well as air carrier and private aircraft operations.

The national administrative structure of airports

All civil-use airports, large and small, in one way or another, utilize the United States' Civil Aviation System. The civil aviation system is an integral part of the United States' transportation infrastructure. This vital infrastructure is administered through the United States **Department of Transportation (DOT),** led by the secretary of transportation (Fig. 1-3).

The DOT is divided into several administrations that oversee the various modes of national and regional transportation in the United States. Such administrations include:

FHWA—The Federal Highway Administration

FMCSA—The Federal Motor Carrier Safety Administration

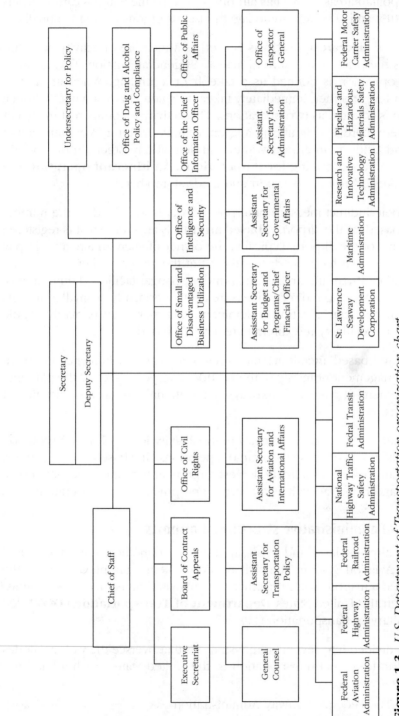

Figure 1-3. *U.S. Department of Transportation organization chart.*

FRA—The Federal Railroad Administration

FTA—The Federal Transit Administration

MARAD—The Maritime Administration

NHTSA—The National Highway Traffic Safety Administration

The administration that oversees civil aviation is the **Federal Aviation Administration (FAA).** The FAA's primary mission is to oversee the safety of civil aviation. The FAA is responsible for the rating and certification of pilots and for the certification of airports, particularly those serving commercial air carriers. The FAA operates the nation's air traffic control system, including most air traffic control towers found at airports, and owns, installs, and maintains visual and electronic navigational aids found on and around airports. In addition, the FAA administers the majority of the rules that govern civil aviation and airport operations, as well as plays a large role in the funding of airports for improvement and expansion. The FAA is led by an administrator who is appointed by the secretary of transportation for a 5-year term.

The FAA is headquartered in Washington, D.C. Headquarter offices within the FAA include the offices of Air Traffic Services (ATS), Office of Security and Hazardous Materials (ASH), Commercial Space Transportation (AST), Regulation and Certification (AVR), Research and Acquisitions (ARA), and Airports (ARP).

Within the Office of Airports lies the Office of Airport Safety and Standards (AAS) and the Office of Planning and Programming (APP). It is in these offices where Federal Aviation Regulations and policies specific to airports are administered.

The FAA is also divided into nine geographic regions, as illustrated in Fig. 1-4. Within each region are two or more **Airport District Offices (ADOs).** ADOs keep in contact with airports within their respective regions to ensure compliance with federal regulations and to assist airport management in safe and efficient airport operations as well as in airport planning.

Many civil-use airports, including those that are not directly administered by the FAA, may be under the administrative control of their individual states, which in turn have their own departments of transportation and associated offices and regions. Airport management at individual airports should be familiar with all federal, state, and even local levels of administration that govern their facilities.

Airport management on an international level

Internationally, the recommended standards for the operation and management of civil-use airports are provided by the **International Civil Aviation**

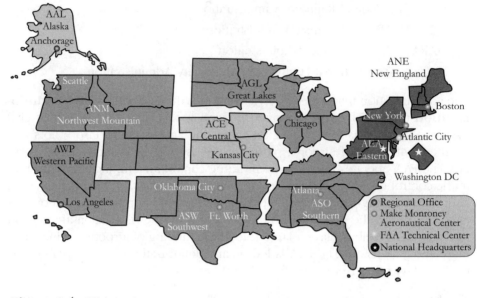

Figure 1-4. *FAA regions.*

Organization (ICAO). ICAO, headquartered in Montreal, Quebec, Canada, is a membership-based organization, comprised of 188 contracting states that span the world. ICAO came into existence as a part of the 1944 Chicago Convention on International Civil Aviation for the purpose of providing a source of communication and standardization among participating states with respect to civil aviation operations. ICAO publishes a series of recommended policies and regulations to be applied by individual states in the management of their airports and civil aviation systems.

In most individual countries, airports are managed directly by the federal government, most often under the ministry of transport. In some countries, including the United States, many airports are privately owned and operated, although, despite private ownership, they are still subject to much of the country's regulations regarding aviation operations.

The National Plan of Integrated Airport Systems

Since 1970, the FAA has recognized a subset of the 5,400 public-use airports in the United States as being vital to serving the public needs for air transportation, either directly or indirectly, and may be made eligible for federal funding to maintain their facilities. The **National Airport System Plan (NASP)** was the first such plan, which recognized approximately 3,200 such airports. In addition, the NASP categorized these airports on the basis of each airport's number of annual enplanements and the type of service provided. The NASP categorized airports as being "commercial service airports" if the airport enplaned at least 2,500 passengers annually on commercial air carriers or charter aircraft. Commercial service airports were subcategorized as "air carrier" airports and "commuter" airports, depending on the type of service dominant at a given airport. Airports that enplaned less than 2,500 passengers annually were classified as "general aviation airports." In 1983, the final year of the NASP, a total of 780 commercial service airports (635 air carrier airports and 145 commuter airports) and 2,423 GA airports were recognized under the NASP.

With the passage of the Airport and Airway Act of 1982, the FAA was charged with preparing a new version of the NASP, to be called the **National Plan of Integrated Airport Systems (NPIAS).** The NPIAS revised the method of classifying airports, primarily to reflect the extreme growth in annual enplanements that a relative few of the largest airports were experiencing at the time. As of 2008, a total of 3,411 airports in the United States were included in the NPIAS.

The categories of airports listed in the NPIAS are:

1. Primary airports
2. Commercial service airports
3. GA airports
4. Reliever airports

Figure 1-5 provides a geographic illustration of NPIAS airports throughout the United States (numbers include Alaska and Hawaii, although not illustrated).

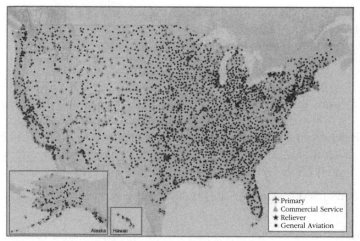

The NPIAS includes all commercial service, reliver (high-capacity general aviation airports in metropolitan areas), and select general aviation airports.

Figure 1-5. *NPIAS airports.* (Figure courtesy FAA)

Commercial service airports

Commercial service airports are those airports that accommodate scheduled air carrier service, provided by the world's certificated air carriers. Virtually all of the 770 million passengers who boarded domestic and international commercial aircraft in the United States in 2009 began, transferred through, and ended their trips at commercial service airports. Commercial service airports operate under very specific regulations enforced by the FAA and Transportation Security Administration (TSA), as well as state and local governments. In addition, other federal and local administrations, such as the Environmental Protection Agency, and local economic development organizations, indirectly affect how commercial service airports operate. The goal of commercial service airports, of course, is to provide for the safe and efficient movement of passengers and cargo between population centers through the nation's aviation system. In 2008, there were a total of 522 commercial service airports throughout the United States striving to fulfill this mission.

Primary airports are categorized in the NPIAS as those public-use airports enplaning at least 10,000 passengers annually in the United States. In 2008, there were 383 airports (less than 3 percent of the nation's total airports) categorized as primary airports.

Within this exclusive group of airports, the range of airport size and activity level is very wide, and the distribution of passenger enplanements is highly skewed. About half the primary airports handle relatively little traffic; the vast majority of passengers are enplaned through relatively few very large airports. This phenomenon is a direct result of the airline routing strategy, known as the "hub and spoke" system that was adopted by several of the nation's largest carriers. In fact, the top five airports in the United States, in terms of annual enplanements, boarded nearly 25 percent of all the passengers in the United States. The top two airports, the Hartsfield–Jackson Atlanta International Airport and Chicago's O'Hare Field, enplaned nearly 80 million (nearly 10 percent) of the nation's commercial air travelers in 2009 (Table 1-1).

Because of this wide range of size within the primary airport category, the NPIAS subcategorizes these airports into "hub" classifications. It should be noted that the term "hub" used by the FAA in the NPIAS is very different than the term used by the airline industry. Whereas the airline industry uses the term "hub" as an airport where the majority of an airline's passengers will transfer between flights to reach their ultimate destinations, the FAA defines hub strictly by the number of annual enplaned passengers to use the airport (Table 1-2).

The hub classifications used by the FAA in the NPIAS are:

1. Large hubs
2. Medium hubs
3. Small hubs
4. Nonhubs

Table 1-1. Passengers Boarded at the Top 50 U.S. Airports

Rank	Airport	2008 Total Enplaned Passengers
1	Atlanta, GA (Hartsfield-Jackson Atlanta International)	43,238,440
2	Chicago, IL (Chicago O'Hare International)	31,351,227
3	Dallas, TX (Dallas/Fort Worth International)	26,830,947
4	Denver, CO (Denver International)	23,919,713
5	Los Angeles, CA (Los Angeles International)	22,439,873
6	Las Vegas, NV (McCarran International)	19,887,290
7	Houston, TX (George Bush Intercontinental)	19,239,836
8	Phoenix, AZ (Phoenix Sky Harbor International	19,209,392
9	Charlotte, NC (Charlotte Douglas International)	17,185,243
10	New York, NY (John F. Kennedy International)	16,955,540
11	Detroit, MI (Detroit Metropolitan Wayne County)	16,794,472
12	Minneapolis, MN (Wold-Chamberlin International)	16,302,227
13	Orlando, FL (Orlando International)	16,122,383
14	Newark, NJ (Newark Liberty International)	16,105,083
15	San Francisco, CA (San Francisco International)	15,727,533
16	Philadelphia, PA (Philadelphia International)	15,257,081
17	Seattle, WA (Seattle-Tacoma International)	15,206,521
18	Miami, FL (Miami International)	13,577,782
19	Boston, MA (Logan International)	11,588,988
20	New York, NY (LaGuardia)	11,159,038
21	Fort Lauderdale, FL (Fort Lauderdale Hollywood International)	10,370,421
22	Baltimore, MD (Baltimore/Washington Thurgood Marshall)	10,078,747
23	Washington, DC (Dulles International)	9,917,944
24	Salt Lake City, UT (Salt Lake City International)	9,877,540
25	San Diego, CA (San Diego International)	8,931,211
26	Tampa, FL (Tampa International)	8,689,410
27	Washington, DC (Ronald Reagan Washington National)	8,599,934
28	Chicago, IL (Chicago Midway)	8,012,938

(continued)

Table 1-1. Passengers Boarded at the Top 50 U.S. Airports (*continued*)

Rank	Airport	2008 Total Enplaned Passengers
29	Honolulu, HI (Honolulu International)	7,785,515
30	Portland, OR (Portland International)	6,942,236
31	St. Louis, MO (Lambert-St. Louis International)	6,626,545
32	Cincinnati, OH (Cincinnati/Northern Kentucky International)	6,480,292
33	Oakland, CA (Oakland International)	5,482,324
34	Memphis, TN (Memphis International)	5,375,733
35	Kansas City, MO (Kansas City International)	5,346,702
36	Cleveland, OH (Cleveland-Hopkins International)	5,277,778
37	Sacramento, CA (Sacramento International)	4,891,967
38	Raleigh, NC (Raleigh-Durham International)	4,741,753
39	San Jose, CA (Norman Y. Mineta, San Jose International)	4,698,523
40	Nashville, TN (Nashville International)	4,615,999
41	San Juan, PR (Luis Munoz Marin International)	4,546,996
42	Santa Ana, CA (John Wayne - Orange County)	4,462,999
43	Pittsburgh, PA (Pittsburgh International)	4,264,809
44	Austin, TX (Austin Bergstrom International)	4,255,238
45	Houston, TX (William P. Hobby)	4,224,294
46	Dallas, TX (Love Field)	4,030,509
47	Indianapolis, IN (Indianapolis International)	4,025,647
48	New Orleans, LA (Louis Armstrong International)	3,976,840
49	San Antonio, TX (San Antonio International)	3,949,819
50	Milwaukee, WI (General Mitchell Field)	3,824,181

Courtesy RITA

Large hubs are those airports that account for at least 1 percent of the total annual passenger enplanements in the United States. In 2008, there were 30 large hub airports in the NPIAS. These 30 large hub airports accounted for nearly 70 percent of all passenger enplanements in the United States. **Medium hubs** are those airports that account for at least 0.25 percent but less than 1 percent of the total annual passenger enplanements. In 2008, there

Table 1-2. Airports by Level of Activity

Number of Airports	Airport Type	Percentage of 2008 Total Enplanements	Percentage of ALL Based Aircraft[1]	Percentage of NPIAS 2009–2013 Cost	Percentage of Population within 20 Miles of Airport
30	Large Hub Primary	68.7	0.9	36	26
37	Medium Hub Primary	20.0	2.6	14	18
72	Small Hub Primary	8.1	4.3	8	14
244	Nonhub Primary	3.0	10.9	10	20
139	Nonprimary Commercial Service	0.1	2.4	2	3
270	Relievers	0.0	28.2	7	56
2,564	General Aviation	0.0	40.8	19	69
3,356	Existing NPIAS Airports	99.9	89.8	100	98
16,459	Low Activity Landing Areas (Non-NPIAS)	0.1	10.2	N/A	N/C

[1] Based on active aircraft fleet of 221,942 aircraft in 2008.

N/A, not appropriate; N/C, not calculated.

Courtesy FFA

were 37 airports classified as medium hubs. **Small hubs** are defined as those airports accommodating greater than 0.05 percent but less than 0.25 percent of annual U.S. enplanements. Seventy-two NPIAS airports were categorized as small hubs. **Nonhub** primary airports are those airports that enplane at least 10,000 annual enplanements but less than 0.05 percent of the annual total U.S. enplanements. In 2008, 244 primary airports fell into the nonhub primary category.

Airports that handle at least 2,500 but less than 10,000 annual enplanements are categorized as nonprimary airports, or simply commercial service airports. In 2008, there were 139 nonprimary airports included in the NPIAS.

General aviation airports

Those airports with fewer than 2,500 annual enplaned passengers and those used exclusively by private business aircraft not providing commercial air carrier passenger service are categorized as **general aviation (GA) airports.** Although there are over 13,000 airports that fit this category, only a subset is included in the NPIAS. There is typically at least one GA airport in the NPIAS for every county in the United States. In addition, any GA airport that has at least 10 aircraft based at the airport and is located at least 20 miles away from the next nearest NPIAS airport is usually included in the NPIAS. In 2008, a total of 2,564 GA airports were included in the NPIAS.

Whereas commercial service airports accommodate virtually of the enplaned commercial passengers in the United States, GA airports account for the majority of aircraft operations. GA airports accommodate aviation operations of all kinds, from flight training, to aerial agricultural operations, to corporate passenger travel, to charter flights using the largest of civil aircraft. Pipeline patrol, search and rescue operations, medical transport, business and executive flying in fixed-wing aircraft and helicopters, charters, air taxis, flight training, personal transportation, and the many other industrial commercial and recreational uses of airplanes and helicopters take advantage of GA airports.

Similar to commercial service airports, GA airports vary widely in their characteristics. Many GA airports are small facilities, with typically a single runway long enough to accommodate only small aircraft, and are limited in their facilities. These small airports primarily serve as a base for a few aircraft.

Other GA airports have facilities and activity that rival their commercial service counterparts. These airports have multiple runways, at least one long enough to accommodate corporate and larger-size jet aircraft, and have a full spectrum of maintenance, fueling, and other service facilities. Many such GA airports even have rental car, restaurant, and hotel services to accommodate their customers.

An important aspect of GA airports is that they serve many functions for a wide variety of communities. Some GA airports provide isolated communities with valuable links to other population centers. This is particularly true in areas of Alaska where communities are often unreachable except by air, although many other parts of the United States, particularly in the west also depend heavily on GA as a mode of transportation. In such areas, the GA airport is sometimes the only means of supplying communities with necessities. In addition, the GA airport acts as the vital link to many emergency services.

The principal function of GA airports, however, is to provide facilities for privately owned aircraft to be used for business and personal activities. In most recent years, there has been a significant increase in the amount of small business jet aircraft using GA airports. Because of this growth, GA airports are continuously seeking to upgrade their facilities, from extending runways, to providing more services, to meeting the needs of the corporate jet traveler.

A GA airport is generally categorized as being either a **basic utility** or **general utility facility.** Basic utility airports are designed to accommodate most single-engine and small twin-engine propeller-driven aircraft. These types of aircraft accommodate approximately 95 percent of the GA aircraft fleet. General utility airports can accommodate larger aircraft, as well as the lighter, smaller aircraft handled by basic utility airports.

Table 1-3 identifies the busiest airports in the United States in terms of itinerant GA activity. Most of these airports fall in the general utility category. As illustrated in Table 1-3, several of these airports do also serve commercial service operations, and are thus identified in the NPIAS as primary or commercial service airports. Others, however, have no commercial service, and are thus considered GA or reliever airports in the NPIAS.

Reliever airports

Reliever airports comprise a special category of GA airports. Reliever airports, generally located within a relatively short distance (less than 50 miles) of primary airports, are specifically designated by the NPIAS as "general aviation-type airports that provide relief to congested major airports." To be classified as a reliever airport, the airport must have at least 100 aircraft based at the airport or handle at least 25,000 itinerant operations. As the name suggests, reliever airports are intended to encourage GA traffic to use the facility rather than the busier commercial service airport, which may be experiencing delays, by providing facilities of similar quality and convenience to those available at the commercial service airports.

In many major metropolitan areas, reliever airports account for a majority of airport operations. In the Atlanta, Georgia, **standard metropolitan**

Table 1-3. Top 50 Airports by # of Itinerant GA Operations, 2008

Rank	Airport	Name	# Ops
1	VNY	Van Nuys Airport, CA	258,155
2	DAB	Daytona Beach International Airport, FL	232,077
3	TMB	Kendall-Tamiami Executive Airport, FL	139,528
4	FXE	Fort Lauderdale Executive Airport, FL	137,403
5	RVS	Richard Lloyd Jones Airport, OK	136,382
6	FFZ	Falcon Field, AZ	135,382
7	LGB	Long Beach Airport, CA	133,576
8	DVT	Phoenix Deer Valley Airport, AZ	133,150
9	APA	Centennial Airport, CO	128,521
10	BFI	Boeing Field, Kind County Airport, WA	127,003
11	MYF	Montomery Field Airport, CA	124,079
12	PDK	Dekalb - Peachtree Airport, GA	121,055
13	CRQ	McClellan - Palomar Airport, CA	113,781
14	SNA	John Wayne - Orange County Airport, CA	113,763
15	TEB	Teterboro Airport, NJ	108,493
16	SDL	Scottsdale Airport, AZ	107,351
17	SEE	Gillespie Field Airport, CA	103,667
18	ADS	Addison Airport, TX	102,286
19	SFB	Orlando Sanford International Airport, FL	96,634
20	HPN	Westchecster County Airport, NY	96,631
21	VRB	Vero Beach Municipal Airport, FL	94,422
22	BED	Laurence G Hanscom Field Airport, CT	88,113
23	FRG	Republic Airport, NY	87,907
24	ISM	Kissimmee Gateway Airport, FL	84,531
25	DWH	David Wayne Hooks Memorial Airport, TX	83,487
26	SAT	San Antonio International Airport, TX	83,412
27	EVB	New Smyrna Beach Municipal Airport, FL	82,634
28	PRC	Ernest A. Love Field, AZ	82,536
29	MLB	Melbourne International Airport, FL	82,376
30	APF	Naples Municipal Airport, FL	81,794
31	FPR	St. Lucie County International Airport, FL	80,291
32	HOU	William P. Hobby Airport, TX	80,156
33	GYR	Phoenix Goodyear Airport, AZ	78,263

Table 1-3. Top 50 Airports by # of Itinerant GA Operations, 2008 (*continued*)

Rank	Airport	Name	# Ops
34	CMA	Camarillo Airport, TX	77,974
35	HIO	Portland-Hillsboro Airport, OR	76,256
36	TOA	Zamperini Field Airport, CA	75,896
37	CHD	Chandler Municipal Airport, AZ	75,280
38	PTK	Oakland County International Airport, MI	75,097
39	PBI	Palm Beach International Airport, FL	74,388
40	FAT	Fresno Yosemite International Airport, CA	73,707
41	SAC	Sacramento Municipal Airport, CA	73,525
42	OMN	Ormond Beach Municipal Airport, FL	73,328
43	MMU	Morristonwn Municipal Airport, NJ	73,058
44	DAL	Dallas Love Field, TX	72,731
45	FTW	Fort Worth Meacham International Airport, TX	72,334
46	GKY	Arlington Municipal Airport, TX	71,947
47	IWA	Phoenix-Mesa Gateway Airport, AZ	71,903
48	ORL	Orlando Executive Airport, FL	70,226
49	ANC	Ted Stevens Anchorage International Airport, AK	69,498
50	CRG	Craig Municipal Airport, FL	69,327

Source: AirportJournals.com

statistical area (SMSA), for example, the 11 designated reliever airports account for more operations than occur annually at the Hartsfield-Jackson Atlanta International Airport, the nation's busiest commercial service airport. Of the GA airports recognized in the NPIAS, 270 have been classified as reliever airports. These airports are home to over 28 percent of all GA aircraft.

Many of the more than 2,000 GA airports not formally included in the NPIAS are still recognized by the United States as public-use GA airports. However, those airports not formally in the NPIAS are not eligible for federal money for airport improvements. Of the nearly 1,900 airports open to the public but not in the NPIAS, most do not meet the minimum entry criteria of having at least 10 based aircraft or they may be located within 20 miles of an airport already in the NPIAS, are located at inadequate sites and cannot be expanded and improved to provide safe and efficient airport facilities, or cannot be adequately justified in terms of national interest. These airports are often included in state and local airport plans, and thus receive some levels of financial support. The more than 12,000 civil landing areas that are privately owned and not open to the general public are not included in the NPIAS, and are not funded by any public entity. They are considered part of the national airport system, however, because each facility is used to access the rest of the nation's system of air transportation, airways, and airports.

Many of the difficulties in planning a national airport system arise from its size and diversity. Each airport has unique issues, and each airport operator— although constrained by laws, regulations, and custom—is essentially an independent decision maker. Although airports collectively form a national system, the NPIAS system is not entirely centrally planned and managed. The FAA's role in planning the system has traditionally been one of gathering and reporting information on individual airport decisions and discouraging redundant development.

Since 1970, national airport plans have been prepared by FAA regional offices, working in conjunction with local airport management. The NPIAS presents an inventory of the projected capital needs of more than 3,200 airports "in which there is a potential federal interest and on which federal funds may be spent." Because funds available from federal and local sources are sufficient to complete only a fraction of the eligible projects, many of the airport improvements included in the NPIAS are never undertaken.

The criteria for the selection of the airports and projects to be included in the plan have come under criticism. Some have argued that most of the 3,300-plus airports in the NPIAS are not truly of national interest and that criteria should be made more stringent to reduce the number to a more manageable set. On the other hand, there are those who contend that the plan cannot be of national scope unless it contains all publicly owned airports. It is argued that because the NPIAS lists only development projects eligible for federal aid and not those that would be financed solely by state, local, and private sources, the total airport development needs are understated by the plan.

The rules that govern airport management

As with any system intended for use by the public, a complex system of federal, state, and often local regulations have been put in place by legislation to ensure the safe and efficient operation of public-use airports. All airports included in the NPIAS are subject to a variety of **Federal Aviation Regulations (FAR).** FARs are found in Title 14 of the United States **Code of Federal Regulations (CFR)** (14 CFR—Aeronautics and Space). The 14 CFR series is made up of over 100 chapters, known as parts, each of which provide regulatory mandates that govern various elements of the civil aviation system, including regulations for pilots, GA and commercial flight operations, and, of course, airport operations and management. Within airport management, regulations regarding airport operations, environmental policies, financial policies, administrative policies, airport planning, and other issues of direct concern to airports are covered.

Although all FARs are important to airport management, the following FARs are of specific importance to airport management, operations, and planning, and will be referenced in detail in this text:

14 CFR Part 1	Definitions and Abbreviations
14 CFR Part 11	General Rulemaking Procedures
14 CFR Part 36	Noise Standards: Aircraft Type and Airworthiness Certification
14 CFR Part 71	Designation of Class A, Class B, Class C, Class D, and Class E Airspace Areas; Airways, Routes, and Reporting Points
14 CFR Part 73	Special Use Airspace
14 CFR Part 77	Objects Affecting Navigable Airspace
14 CFR Part 91	General Operating and Flight Rules
14 CFR Part 93	Special Air Traffic Rules and Airport Traffic Patterns
14 CFR Part 97	Standard Instrument Approach Procedures
14 CFR Part 121	Operating Requirements: Domestic, Flag, and Supplemental Air Carrier Operations
14 CFR Part 129	Operations: Foreign Air Carriers and Foreign Operators of U.S. Registered Aircraft Engaged in Common Carriage
14 CFR Part 139	Certification of Airports
14 CFR Part 150	Airport Noise and Compatibility Planning
14 CFR Part 151	Federal Aid to Airports
14 CFR Part 152	Airport Aid Program
14 CFR Part 156	State Block Grant Pilot Program
14 CFR Part 157	Notice of Construction, Alteration, Activation, and Deactivation of Airports
14 CFR Part 158	Passenger Facility Charges
14 CFR Part 161	Notice and Approval of Airport Noise and Access Restrictions
14 CFR Part 169	Expenditure of Federal Funds for Nonmilitary Airports or Air Navigation Facilities Thereon (for airports not operated under FAA regulations)

In addition to the 14 CFR series, regulations regarding the security of airport and other civil aviation operations are published under Title 49 of the Code of Federal Regulations (49 CFR—Transportation) and are known as **Transportation Security Regulations (TSRs).** TSRs are enforced by the TSA. TSRs of specific importance to airport management include:

49 CFR Part 1500	Applicability, Terms, and Abbreviations
49 CFR Part 1502	Organization, Functions, and Procedures
49 CFR Part 1503	Investigative and Enforcement Procedures
49 CFR Part 1510	Passenger Civil Aviation Security Service Fees
49 CFR Part 1511	Aviation Security Infrastructure Fee
49 CFR Part 1520	Protection of Security Information (replaced FAR Part 191)
49 CFR Part 1540	Civil Aviation Security: General Rules
49 CFR Part 1542	Airport Security (replaced FAR Part 107)
49 CFR Part 1544	Aircraft Operator Security: Air Carriers and Commercial Operators (replaced FAR Part 108)
49 CFR Part 1546	Foreign Air Carrier Security (replaced parts of FAR Part 129)
49 CFR Part 1549	Indirect Air Carrier Security (replaced FAR Part 109)
49 CFR Part 1550	Aircraft Security Under General Operating and Flight Rules (replaced parts of FAR Part 91)

The volume of TSRs came into effect on November 19, 2001, with the signing of the Aviation and Transportation Security Act. Security regulations and policies under the TSA have been in a constant state of change, as the civil aviation industry adapts to increased threats of terrorism.

To assist airport management and other aviation operations in understanding and applying procedures dictated by federal regulations, the FAA makes available a series of **advisory circulars (ACs)** associated with each regulation and policies. The ACs specific to airports are compiled into the 150 Series of Advisory Circulars. There are over 100 current and historical ACs in the 150 series available to airport management. Those ACs of particular general interest to airport management are referenced throughout this text. Some of these include:

AC 150/5000-5C	Designated U.S. International Airports
AC 150/5020-1	Noise Control and Compatibility Planning for Airports
AC 150/5060-5	Airport Capacity and Delay
AC 150/5070-6A	Airport Master Plans
AC 150/5190-5	Exclusive Rights and Minimum Standards for Commercial Aeronautical Activities

AC 150/5200-28B	Notices to Airmen (NOTAMS) for Airport Operators
AC 150/5200-30A	Airport Winter Safety and Operations
AC 150/5200-31A	Airport Emergency Plan
AC 150/5300-13	Airport Design
AC 150/5325-4A	Runway Length Requirements for Airport Design
AC 150/5340-1H	Standards for Airport Markings
AC 150/5360-12C	Airport Signing and Graphics
AC 150/5360-13	Planning and Design Guidelines for Airport Terminal Facilities
AC 150/5360-14	Access to Airports by Individuals with Disabilities

ACs are constantly updated and often changed. The latest available ACs as well as FARs may be found by contacting the FAA. The latest information regarding TSRs may be found by contacting the TSA.

Airports are also subject to state and local civil regulations specific to the airport's metropolitan area. In addition, airport management itself may impose regulations and policies governing the operation and administration of the airport. Each airport is encouraged to have a published set of rules and regulations covering all the applicable federal, state, local, and individual airport policies to be made available for all employees and airport users on an as-needed basis. A complete list of current and historical Federal Aviation Regulations and Advisory Circulars may be found at the FAA website at http://www.faa.gov.

Organizations that influence airport regulatory policies

There are many national organizational and regional organizations that are deeply interested in the operation of airports. Most of these organizations are interested in developing and preserving airports because of their role in the national air transportation system and their value to the areas they serve. The primary goal of these groups is to provide political support for their causes with hopes to influence federal, state, and local laws concerning airports and aviation operations in their favor. In addition, these groups provide statistics and informational publications and provide guest speakers and information sessions to assist airport management and other members of the aviation community in order to provide support for civil aviation.

Each of these organizations is particularly concerned with the interests of their constituents; however there are numerous times when they close ranks and work together for mutual goals affecting the aviation community in general. The following is a brief listing of the most prominent associations. A complete listing can be found in the *World Aviation Directory* published by McGraw-Hill. These organizations, by virtue of the alphabetic acronyms they are most commonly referred by, make up the "alphabet soup" of aviation-related support organizations.

- *Aerospace Industries Association (AIA)—founded 1919.* Member companies represent the primary manufacturers of military and large commercial aircraft, engines, accessories, rockets, spacecraft, and related items.

- *Aircraft Owners & Pilots Association (AOPA)—founded 1939.* With more than 400,000 members, AOPA represents the interests of general aviation pilots. AOPA provides insurance plans, flight planning, and other services, as well as sponsors large fly-in meetings. In addition the AOPA's Airport Support Network plays a large role in the support and development of all airports, with particular support to smaller GA airports.

- *Air Line Pilots Association (ALPA)—founded 1931.* The Air Line Pilots Association is the oldest and largest airline pilots' union, supporting the interests of the commercial pilots and commercial air carrier airports.

- *Airports Council International–North America (ACI–NA)—founded 1991.* First established as the Airport Operators Council in 1947, the ACI–NA considers itself the "voice of airports" representing local, regional, and state governing bodies that own and operate commercial airports throughout the United States and Canada. As of 2003, 725 member airports throughout belong to ACI–NA. The mission of the ACI–NA is to identify, develop, and enhance common policies and programs for the enhancement and promotion of airports and their management that are effective, efficient, and responsive to consumer and community needs.

- *Air Transport Association of America (ATA)—founded 1936.* The ATA represents the nation's certificated air carriers in a broad spectrum of technical and economic issues. Promotes safety, industrywide programs, policies, and public understanding of airlines.

- *American Association of Airport Executives (AAAE)—founded 1928.* A division of the Aeronautical Chamber of Commerce at its inception, the AAAE became an independent entity in 1939. Membership includes individual representatives from airports of all sizes throughout the United States, as well as partners in the aviation industry and academia.

- *Aviation Distributors and Manufacturers Association (ADMA)—founded 1943.* Represents the interests of a wide variety of aviation firms including fixed-base operators (FBOs) who serve GA operations and aircraft component part manufacturers. The ADMA is a strong proponent of aviation education.

- *Experimental Aircraft Association (EAA)—founded 1953.* The EAA, with over 700 local chapters, promotes the interests of homebuilt and sport aircraft owners. EAA hosts two of the world's largest fly-in conventions each year, at Oshkosh, Wisconsin, and Lakeland, Florida.

- *Flight Safety Foundation (FSF)—founded 1947.*   The primary function of the FSF is to promote air transport safety. Its members include airport and airline executives and consultants.

- *General Aviation Manufacturers Association (GAMA)—founded 1970.*   GAMA's members include manufacturers of GA aircraft, engines, accessories, and avionics equipment. GAMA is a strong proponent of GA airports.

- *Helicopter Association International (HAI)—founded 1948.*   Members of HAI represent over 1,500 member organizations in 51 countries that operate, manufacture, and support civil helicopter operations.

- *International Air Transport Association (IATA)—founded 1945.*   IATA is an association of more than 220 international air carriers whose main functions include coordination of airline fares and operations. IATA annually assesses international airports for their service quality and publishes their findings industrywide.

- *National Agricultural Aviation Association (NAAA)—founded 1967.*   As the voice of the aerial application industry, NAAA represents the interests of agricultural aviation operators. The NAAA represents over 1,250 members including owners of aerial application businesses; pilots; manufacturers of aircraft, engines, and equipment; and those in related businesses.

- *National Air Transportation Association (NATA)—founded 1941.*   First known as the National Aviation Training Association and later Trades Association, NATA represents the interests of FBOs, air taxi services, and related suppliers and manufacturers.

- *National Association of State Aviation Officials (NASAO)—founded 1931.*   The NASAO represents departments of transportation and state aviation departments and commissions from 50 states, Puerto Rico, and Guam. NASAO encourages cooperation and mutual aid among local, state, and federal governments.

- *National Business Aviation Association (NBAA)—founded 1947.*   The NBAA represents the aviation interests of over 7,400 companies that own or operate GA aircraft as an aid to the conduct of their business, or are involved with some other aspect of business aviation.

- *Professional Aviation Maintenance Association (PAMA)—founded 1972.*   PAMA promotes the interest of airframe and power plant (A&P) technicians.

- *Regional Airline Association (RAA)—founded 1971.*   The RAA represents the interests of short- and medium-haul scheduled passenger air carriers, known as "regional airlines," and cargo carriers.

Concluding remarks

As described in these introductory remarks, the complex system of civil airports is made up of individual airport facilities of varying sizes, serving various purposes, all organized into plans of regional, national, and international levels. The range of rules, regulations, and policies, administered from varying levels of government, cover the full spectrum of airport and aviation system operations. Furthermore, a large number of professional and industry organizations play a large part in influencing the policies by which airport management must operate their facilities. By understanding where an airport manager's airport falls within the civil aviation system, what rules must be followed, and what sources of support and assistance exist, the task of efficiently managing the complex system that is an airport, becomes highly facilitated.

Key terms

joint-use civil-military airports

enplanements

deplanements

transfer passengers

aircraft operations

local operations

itinerant operations

based aircraft

Department of Transportation (DOT)

Federal Aviation Administration (FAA)

Airport District Office (ADO)

International Civil Aviation Organization (ICAO)

National Airport System Plan (NASP)

National Plan of Integrated Airport Systems (NPIAS)

commercial service airport

primary airport

standard metropolitan statistical area (SMSA)

large hub

medium hub

small hub

nonhub

general aviation (GA) airport

basic utility facility

general utility facility

reliever airport

Federal Aviation Regulations (FAR)

Code of Federal Regulations (CFR)

Transportation Security Regulations (TSR)

advisory circulars (AC)

Questions for review and discussion

1. How many airports exist in the United States?
2. Who owns airports in the United States?
3. What is the difference between a private airport and a public-use airport?
4. What are the different types of airports in the United States, as described in the NPIAS?
5. What are the leading airports in the United States in terms of enplaned passengers?
6. What are the leading airports in the United States in terms of aircraft operations?
7. What are the different hub classifications described in the NPIAS?
8. What are the requirements necessary for an airport to be classified as a reliever airport?
9. What purposes do general aviation airports serve?
10. What federal agencies exist in part to support and supervise airport operations?
11. What independent professional agencies exist to support airports?
12. What specific rules and regulations are used to operate airports?
13. What are advisory circulars? What purpose do they serve for airport management?

Suggested readings

de Neufville, Richard. *Airport System Planning*. London, England: Macmillan, 1976.

de Neufville, R., and Odoni, A. *Airport Systems: Planning, Design, and Management*. New York: McGraw-Hill, 2002.

Howard, George P., ed. *Airport Economic Planning*. Cambridge, Mass. MIT Press, 1974.

National Plan of Integrated Airport Systems (NPIAS), 2009–2013. Washington, D.C.: FAA, March 2008.

Sixteenth Annual Report of Accomplishments under the Airport Improvement Program. FY 1997. Washington, D.C.: FAA, April 1999.

Wiley, John R. *Airport Administration and Management.* Westport, Conn.: Eno Foundation for Transportation, 1986.

2

Airports and airport systems: Organization and administration

Outline

- Introduction
- Airport ownership and operation
  - Airport privatization
- The airport organization chart
  - Job descriptions
- Airport management as a career
  - Duties of an airport manager
  - Education and training
- The airport manager and public relations
  - The airport and its public
  - Public relations objectives

Objectives

The objectives of this section are to educate the reader with information to:

- Discuss the ownership structures of airports.
- Identify the various jobs that exist at airports.
- Understand an airport organization chart.
- Discuss airport management as a potential career.
- Understand the public relations issues that are associated with airport management.

Introduction

Whether privately owned or part of a public system, there are fundamental characteristics of the administrative and organizational structure of an airport. The number of people employed at a given airport can range from as few as one, at the smallest of general aviation facilities, to as many as 50,000 at the world's largest airport organizations.

Those airports that employ fewer numbers of people expect these people to accept a wider range of responsibilities. For example, an airport management employee at a small airport might be responsible for maintaining the airfield, managing finances, and maintaining good relations with the local public. At the larger airports, employees are typically given very specific responsibilities for a particular segment of airport management.

Airport ownership and operation

Public airports in the United States are owned and operated under a variety of organizational and jurisdictional arrangements. Usually, ownership and operation coincide: commercial airports might be owned and operated by a city, county, or state; or by more than one jurisdiction (a city and a county). In some cases, a commercial airport is owned by one or more of these governmental entities but operated by a separate public body, such as an airport authority specifically created for the purpose of managing the airport. Regardless of ownership, legal responsibility for day-to-day operation and administration can be vested in any of five kinds of governmental or public entities: a municipal or county government, a multipurpose port authority, an airport authority, a state government, or the federal government.

A typical **municipally operated airport** is city owned and run as a department of the city, with policy direction by the city council and, in some cases, by a separate airport commission or advisory board. County-run airports are similarly organized. Under this type of public operation, airport policy decisions are generally made in the broader context of city or county public investment needs, budgetary constraints, and development goals.

Some commercial airports in the United States are run by multipurpose port authorities. **Port authorities** are legally chartered institutions with the status of public corporations that operate a variety of publicly owned facilities, such as harbors, airports, toll roads, and bridges. In managing the properties under their jurisdiction, port authorities have extensive independence from the state and local governments. Their financial independence rests largely on the power to issue their own debt, in the form of revenue bonds, and on the breadth of their revenue bases, which might include fees and charges from marine terminals and airports as well as proceeds (such as bridge or tunnel tolls) from other port authority properties. In addition, some port authorities have the power to tax within the port district, although it is rarely exercised.

Another type of arrangement is the single-purpose **airport authority.** Similar in structure and in legal charter to port authorities, these single-purpose authorities also have considerable independence from the state or local governments, which often retain ownership of the airport or airports operated by the authority. Like multipurpose port authorities, airport authorities have the power to issue their own debt for financing capital development, and in a few cases, the power to tax. Compared to port authorities, however, they must rely on a much narrower base of revenues to run a financially self-sustaining enterprise.

Since the early 1950s, there has been a gradual transition from city- and county-controlled airports to the independent single or multipurpose authorities. The predominant form is still municipally owned and operated, particularly the smaller commercial and general aviation airports; however, there are reasons for this transition:

- Many airport market or service areas have outgrown the political jurisdiction whose responsibility the airport entails. In some cases there is considerable, actual or potential, tax liability to a rather limited area. In these cases the creation of an authority to "spread the potential or actual tax support" for the airport might be recommended. By spreading the tax base of support for the airport, more equitable treatment of the individual taxpayer can result and the taxpayers supporting the airport in most cases more nearly match the actual users of the facility.

- Another advantage of authority control of an airport is that such an organization allows the board to concentrate and specialize on airport matters.

- Aviation authorities can also provide efficiency of operation and economies of scale when several political jurisdictions, each with separate airport responsibilities, choose to combine these under one board. This has been done quite successfully in many areas of the country. Normally, the staff required by an airport authority will be quite small compared to the personnel requirements of a city or county government. This factor generally results in better coordination with the airport management team.

- Authorities can also provide on-scene decision makers, rates, and charges unclouded by off-airport costs, and with less political impact on the business of running the airport.

State-operated airports are typically managed by the state's department of transportation. Either general obligation or revenue bonding might be used to raise investment capital, and state taxes on aviation fuel might be applied to capital improvement projects.

Although several states run their own commercial airports, only a handful of large- and medium-size commercial airports are operated in this way, primarily in Alaska, Connecticut, Hawaii, Maryland, and Rhode Island. The federal government owns the airport at Pomona (Atlantic City), New Jersey, which is part of the Federal Aviation Administration (FAA) Technical Center. The South Jersey Transportation Authority operates this facility.

Airport privatization

Several airports in the United States are managed by private companies generally operating under a fixed-fee contract with a local government. By contrast, many U.S. airports are managed by the local government, but a significant number of airport functions are contracted out to private contractors, including janitorial, security, maintenance, and concession management. Neither of these situations is particularly controversial, nor are the economics of these airports unusual.

Privatization refers to shifting governmental functions and responsibilities, in whole or in part, to the private sector. The most extensive privatizations involve the sale or lease of public assets.

Airport privatization, in particular, typically involves the lease of airport property and/or facilities to a private company to build, operate, and/or manage commercial services offered at the airport. No commercial airport property in the United States has been completely sold to a private entity. Long-term operating leases are the standard privatization contract.

There have been attempts in recent years to completely privatize a few commercial service airports in the United States. In 1997, the FAA implemented the Pilot Program on Private Ownership of Airports, under which five public-use airports were to be operated under a private management group. The airports selected to participate in the program included Stewart International Airport in Newburgh, New York; Brown Field in San Diego, California; Rafael Hernández Airport in Aguadilla, Puerto Rico; New Orleans Lakefront Airport in New Orleans, Louisiana; and Niagara Falls International Airport in Niagara Falls, New York. The program was met with limited success, with only Stewart International Airport fully completing the privatization process. In 2008, Stewart International was acquired by the Port Authority of New York and New Jersey, an autonomous public agency that operates many of the region's ports, airports, bridges, and tunnels.

Coincidently, in 2008, an attempt was made to privatize Chicago's Midway Airport through a sale to a private consortium that included YVR Airport

Services, which owns and operates Vancouver International Airport and other airports worldwide. The attempt failed, however, when the consortium failed to receive the desired financing to complete the effort. However, as of early 2010, attempts to revive the privatization of Midway were initiated. Other privatization efforts under the Federal Airport Privatization Program that are in process as of 2010 include Gwinnett County – Briscoe Field airport in Lawrenceville, Georgia; Luis Muñoz Marin International Airport in San Juan, Puerto Rico; and Louis Armstrong New Orleans International Airport in New Orleans, Louisiana.

Although no U.S. commercial airport has been outright sold to a private entity, publicly owned airports, however, do have extensive private sector involvement. Most services now performed at large commercial airports, such as airline passenger processing, baggage handling, cleaning, retail concessions, and ground transportation, are provided by private firms. Some estimates indicate as many as 90 percent of the people working at the nation's largest airports are employed by private firms. The remaining 10 percent of the employees are local and state government personnel performing administrative or public safety duties; federal employees, such as FAA air traffic controllers and Transportation Security Administration (TSA) security screeners; or other public employees. Airports have been increasingly dependent on the private sector to provide services as a way to reduce costs and improve the quality and the range of services offered.

In the mid-1990s some public administrations contracted with private firms to manage their airports; most notably, in 1995, the Indianapolis Airport Authority contracted with a private firm, the British Airports Authority, to manage its system of airports, including the Indianapolis International Airport. Since 1995, several, but not many, airports have been contracted out for full private management. More commonly a portion of the airport, such as an airport terminal, concessions, parking, and so forth, has been subcontracted for management by private sector firms.

Airports are however relying more on private financing for capital development. Airports have sought to diversify their sources of capital development funding, including the amount of private sector financing. Traditionally, airports have relied on the airlines and federal grants to finance their operations and development. However, in recent years, airports, especially the larger ones, have sought to decrease their reliance on airlines while increasing revenue from other sources. Nonairline revenue, such as concession receipts, now account for more than 50 percent of the total revenue larger airports receive.

In most other countries, the national government owns and operates airports. However, a growing number of countries, including Canada, Australia, and India, have been implementing strategies to more extensively involve the private sector as a way to provide capital for development and improve efficiency.

These privatization activities range from contracting out services and infrastructure development, in a role similar to private sector activities at U.S. airports, to the sale or lease of nationally owned airports.

For example, Mexico passed legislation in 1995 to lease 58 major airports on a long-term basis. Most countries' privatization efforts do not transfer ownership of airports to the private sector, but involve long-term leases, management contracts, the sale of minority shares in individual airports, or the development of runways or terminals by the private sector. Only the United Kingdom has sold major airports to the private sector. To privatize, the United Kingdom sold the government corporation British Airports Authority (BAA) and the seven major airports it operated (including London's Heathrow and Gatwick Airports) in a $2.5 billion public share offering. Proceeds from this sale were used to reduce the national debt. Even after privatization, the airports have remained subject to government regulation of airline access, airport charges to airlines, safety, security, and environmental protection. The government also maintains a right to veto new investments in, or divestitures of, airports. BAA has generated profits every year since it assumed ownership of the United Kingdom's major airports in 1987.

Several factors have motivated interest in expanding the role of the private sector at commercial airports in the United States. First, privatization advocates believe that private firms would provide additional capital for development. Second, proponents believe that privatized airports would be more profitable because the private sector would operate them more efficiently. Last, advocates believe that privatization would financially benefit all levels of government by reducing demand on public funds and increasing the tax base.

The enthusiasm toward full airport privatization has appeared to wane since the late 1990s, as the overall economy of the United States has declined. However, the concepts that drive private enterprises toward competitive and efficient operations are becoming embraced by publicly owned and managed airports. As a result, more efficient organizational structures and management responsibilities have resulted in more streamlined and efficient airport management organizational structures.

The airport organization chart

An **organization chart** shows the formal authority relationships between superiors and subordinates at various levels, as well as the formal channels of communication within the organization. It provides a framework within which the management functions can be carried out. The chart aids employees to perceive more clearly their positions in the organization in relation to others and how and where managers and workers fit into the overall organizational structure.

Airport management organization charts range from the very simple to the very complex, depending primarily on the size, ownership, and management structure of the airport.

The organization chart is a static model of an airport's management structure; that is, it shows how the airport is organized at a given point in time. This is a major limitation of the chart, because airports operate in a dynamic environment and thus must continually adapt to changing conditions. Some old positions might no longer be required, or new positions might have to be created in order that new objectives can be reached; therefore, it is necessary that the chart be revised and updated periodically to reflect these changing conditions.

The duties, policies, and theories that govern the job of airport management vary widely over time. In addition, many such policies vary from airport to airport on the basis of individual airport operating characteristics. As a result, it is difficult to say that any organization chart is typical or that the chart of one airport at any particular time is the one still in effect even a few months later; however, all airports do have certain common functional areas into which airport activities are divided. Understandably, the larger the airport, the greater the specialization of tasks and the greater the departmentalization. Figure 2-1 shows the major functional areas and typical managerial job titles for a commercial airport.

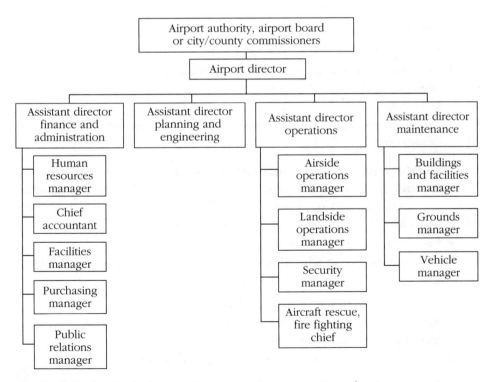

Figure 2-1. *Typical airport management organization chart.*

Job descriptions

The following is a brief job description for each position shown in Fig. 2-1.

Airport director   The **airport director** is responsible for the overall day-to-day operation of the airport. He or she reports directly to the airport authority, the airport board, or governmental commission charged with the development and administration of the airport. This individual directs, coordinates, and reviews through subordinate supervisors, all aircraft operations, building and field maintenance, construction plans, community relations, and financial and personnel matters at the airport. The airport director also:

- Supervises and coordinates with airline, general aviation, and military tenants use of airport facilities.
- Reviews airport tenant activities for compliance with terms of leases and other agreements.
- Supervises enforcement of aircraft air and ground traffic and other applicable regulations.
- Confers with airlines, tenants, the FAA, and others regarding airport regulations, facilities, and related matters.
- Participates in planning for increased aircraft and passenger volume and facilities expansion.
- Determines and recommends airport staffing requirements.
- Compiles and submits for review an annual airport budget.
- Coordinates airport activities with construction, maintenance, and other work done by departmental staff, tenants, public utilities, and contractors.
- Promotes acceptance of airport-oriented activities in surrounding communities.

Assistant director—finance and administration   The **assistant director—finance and administration** is charged with the responsibility for overall matters concerning finance, personnel, purchasing, facilities management, and office management. Specifically, this individual's duties include:

- Fiscal planning and budget administration.
- Accomplishment of basic finance functions such as accounts receivable and payable, auditing, and payroll.
- Administration of the purchasing function.
- Administration and use of real property including negotiation of tenant leases and inventory control.
- Personnel functions including compensation, employee relations, and training.

- Adequate telephone and mail service.
- Public relations.

Human resources manager The **human resources manager** is responsible for administering the airport personnel program. In such capacity, this individual's duties include:

- Dealing with personnel problems involving position classification, compensation, recruitment, placement, transfers, layoffs, promotions, leaves of absence, supervisor-subordinate relationships, and working conditions.
- Serving as equal rights and equal opportunity officer for the airport.
- Handling worker's compensation cases.
- Evaluating the organization pattern, reviewing and recommending proposed departmental organizational changes, and preparing position descriptions.
- Conferring with employees and their supervisors on personnel problems.
- Preparing personnel documents and maintaining personnel records.
- Interviewing or supervising the interviewing of applicants for airport positions.

Chief accountant The **chief accountant** is responsible for financial planning, budgeting, accounting, payroll, and auditing. The principal duties include:

- Coordinating, consolidating, and presentation of financial plans.
- Administering basic accounts such as general accounts, cost accounting, and accounts receivable and payable.
- Administering budget; reviewing and analyzing actual performance at budget review sessions.
- Supervising all receipts and disbursements.
- Administering payroll.
- Conducting periodic internal audit of all airport functions.

Facilities manager The **facilities manager** establishes criteria and procedures for the administration of all airport property. In this capacity, he or she is responsible for inventory control of all equipment and facilities. Principal duties and responsibilities of this individual include:

- Identification and control of all property and equipment including periodic audits.
- Evaluating and making recommendations concerning the most efficient use of airport real property utilization.
- Soliciting tenants and concessionaires.

- Developing policy and rate structure applicable to use of property by tenants and concessionaires.
- Coordinating with purchasing and legal staff concerning tenant and concessionaire leases.

Purchasing manager   The **purchasing manager** directs the procurement of materials and services to support the airport; he or she prepares, negotiates, interprets, and administers contracts with vendors. This individual's principal duties include:

- Coordinating requirements for materials and services to be purchased.
- Purchasing all materials and services.
- Establishing bidding policies and procedures.
- Working closely with the facilities' chief and legal staff regarding contracts associated with purchasing equipment.

Manager of public relations   The **manager of public relations** is the chief liaison officer between the airport and the surrounding community. In this capacity, he or she is responsible for all public relations activities including the development of advertising and publicity concerning the airport. This individual is also responsible for handling all noise and other environmental matters. Principal duties include:

- Consulting with and advising airport management regarding public relations policies and practices.
- Coordinating all publicity releases to the various media.
- Supervising all airport guides and information booths.
- Coordinating VIP visits to the airport.
- Receiving and analyzing all public complaints regarding such things as noise and other environmental concerns.
- Preparing answers to complaints and advising management as appropriate.
- Sponsoring activities and special events to generate goodwill and public acceptance.

Assistant director—planning and engineering   The **assistant director—planning and engineering** provides technical assistance to all airport organizations, and ensures the engineering integrity of construction, alteration, and installation programs. This individual also establishes industrial safety standards. Principal duties and responsibilities include:

- Developing standards and specifications for construction, alteration, and installation programs; monitors such programs to ensure compliance therewith.

- Reviewing all construction plans to determine technical integrity and conformance to aesthetic design standards.

- Developing and publishing standards and procedures for industrial safety.

- Participating in the negotiation of construction contracts.

Assistant director—operations   The **assistant director—operations** is responsible for all airside and landside operations including security, and crash, fire, and rescue operations. Principal duties include:

- Directing the operations and security programs for the airport.

- Coordinating and supervising security activities with field maintenance personnel, police and fire departments, federal agencies, and airport tenants.

- Recommending and assisting in promulgating operational rules and procedures.

- Supervising investigations of violations of airport regulations.

- Preparing annual operations budget.

- Directing monitoring of noise levels and coordinating noise level studies.

- Participating in special programs relating to or affecting airport operations, such as studies of height limits around airport property and studies of noise control.

Airside operations manager   The **airside operations manager** is responsible for all airfield operations. In this capacity, principal duties include:

- Enforcing operating and security rules, regulations, and procedures concerning landing, taxiing, parking, servicing loading and unloading of aircraft, operation of vehicular traffic on the airfield, airline activities, and emergency situations.

- Inspecting conditions of airfield lighting, runways, taxiways, and ramp areas.

- Correcting hazardous conditions.

- Coordinating airfield activities with maintenance and security personnel.

- Assisting in all airfield emergency calls and disasters by notifying control tower to close runways, directing maintenance personnel, directing security officers in crowd control, and overseeing other safety considerations and activities necessary to resume normal airport operations.

- Investigating and reporting on complaints and disrupted airport operations, including unscheduled plane arrivals, aircraft accidents, rule and procedure violations, airline activities, and other operations of the airport.

- Assigning gate and parking spaces to all aircraft.

- Coordinating special arrangements for arrivals and departures of important persons.
- Completing all report forms pertaining to operations activities on assigned shifts.
- Assisting in directing noise level studies with other departmental personnel.

Landside operations manager   The **landside operations manager** is responsible for all landside operations. In this capacity, principal duties include:

- Enforcing operating and security rules, regulations, and procedures concerning buildings, access roads, and parking facilities.
- Exercising authority to halt hazardous or unauthorized activities by tenants, employees, or the public in violation of safety regulations and procedures.
- Answering inquiries and explaining terminal use procedures and safety regulations to tenants.
- Coordinating terminal building and other facility activities with maintenance and security personnel.
- Coordinating all parking facility activities with tenants and transit companies.
- Preparing personal injury and property damage reports and general incident reports.
- Completing all report forms pertaining to operations activities on assigned shifts.

Security manager   The **security manager** enforces interior security, traffic, and safety rules and regulations and participates in law enforcement activities at the airport. This individual also works closely with federal security officials assigned to the airport. Principal duties include:

- Enforcing ordinances and regulations pertaining to parking, traffic control, safety, and property protection.
- Patrolling facilities to prevent trespass and unauthorized or hazardous use.
- Preventing public entry into dangerous or restricted areas.
- Issuing citations and warnings for violations of specific provisions of airport rules and regulations.
- Securing gates and locks and watching buildings and facilities for indications of fire, dangerous conditions, unauthorized entry, and vandalism.
- Responding to emergencies and taking immediate action to control crowds, direct traffic, assist the injured, and turn in alarms.

- Responding to calls for police service; participating in arrests; apprehending, or assisting members of the police department in apprehending, law violators.
- Providing information to the public regarding locations and operations of the airport.
- Assigning uniformed and armed personnel to patrol and stand watch, on a 24-hour basis, to protect and safeguard all persons in the airport and property on the airport.

Aircraft rescue/fire fighting chief The **aircraft rescue/fire fighting chief** develops procedures and implements accident, fire, and disaster plans. Principal duties include:

- Conducting a training (continuing) program for all aircraft rescue, fire fighting personnel.
- Developing and implementing all aircraft rescue and fire fighting programs.
- Staffing and operating all aircraft rescue and fire fighting equipment on the airport.
- Inspecting and testing all types of fixed fire prevention and extinguishing equipment on the airport.
- Inspecting all facilities for fire and/or safety hazards.

Assistant director—maintenance   The **assistant director—maintenance** is responsible for planning, coordinating, directing, and reviewing the maintenance of buildings, facilities, vehicles, and utilities. Principal duties include:

- Developing, directing, and coordinating policies, programs, procedures, standards, and schedules for buildings, utilities, vehicle maintenance, and field facilities.
- Coordinating work done by tenants and contractors.
- Inspecting maintenance work for compliance with plans, specifications, and applicable laws.
- Making recommendations as to adequacy, sufficiency, and condition of buildings, facilities, and vehicles.
- Overseeing maintenance contracts.

Buildings and facilities manager   The **buildings and facilities manager** is responsible for ensuring that buildings are adequately maintained with a minimum of cost. Types of maintenance required are primarily electrical, mechanical, plumbing, painting, carpentry, masonry, and cement work. Principal duties include:

- Developing an approved maintenance schedule for all building maintenance requirements.

- Assigning qualified personnel to perform maintenance.
- Inspecting work for adequacy and compliance with requirements.
- Developing special maintenance methods where necessary.

Grounds manager   The **grounds manager** is responsible for ensuring that the grounds are maintained in good repair and that the landscape is adequately maintained. Principal duties include:

- Developing approved schedules for maintaining all airport surface areas including paving, landscaping, and drainage systems.
- Assigning qualified personnel to accomplish ground maintenance.
- Inspecting work for adequacy and compliance with maintenance standards.

Vehicle manager   The **vehicle manager** is responsible for the maintenance of all vehicles utilized by the airport. Vehicle maintenance includes tune-up, minor maintenance, washing and polishing, tires and batteries, lubrication, and fueling. Principal duties include:

- Developing an approved vehicle maintenance schedule.
- Coordinating schedule with users of airport vehicles.
- Assigning qualified personnel to perform maintenance.
- Inspecting all work to determine compliance with established maintenance standards.
- Coordinating with purchasing to obtain vendor services as required.
- Maintaining vehicle usage and maintenance records.
- Coordinating with purchasing in developing a vehicle disposal and replacement program.

Although the aforementioned positions represent a typical managerial structure at a commercial airport, there are numerous employees with a wide variety of job skills reporting to them. Some of the typical job titles found at major airports include the following:

Accountant	Buyer
Administrative assistant	Carpenter
Air-conditioning mechanic	Cement finisher
Airport guide	Civil engineer
Airport noise abatement officer	Construction inspector
Architect	Contract analyst
Auditor	Custodian
Auto mechanic	Drafter

Bus driver	Electrician
Elevator mechanic	Plumber
Equipment mechanic	Public relations representative
Equipment operator	Secretary
Facility planner	Security officer
Financial analyst	Sheet metal mechanic
Firefighter	Software engineer
Groundskeeper	Student intern
Heavy-duty equipment operator	Supervisor of operations
Industrial engineer	Tilesetter
Maintenance and construction laborer	Toolroom keeper
Maintenance foreman	Traffic painter and sign poster
Operations assistant	Tree surgeon
Painter	Truck operator
Personnel representative	Welder
Plasterer	Window cleaner

Airport management as a career

There are many career paths within the field of airport management as evidenced by the wide variety of job descriptions under the previous section. Even the job of airport manager varies greatly. At one extreme is the manager of a large metropolitan airport, an appointee or civil service employee of the city government or airport authority, who heads a large staff of assistants and specialists through which he or she manages a highly complex organization. At the other extreme is the owner-manager of a small private field near a rural community. The latter might combine activities as airport manager with work in some other business.

Between these two extremes is the manager of a municipally owned or privately owned airport where there are a limited number of scheduled airline flights each day. Based at the airport are several fixed-base operators (FBOs) and a number of aircraft owned by individuals and corporations. The typical manager of a medium-size airport deals with all segments of the aviation community including the airlines, general aviation, and federal and state agencies.

In the early days of aviation, an individual could become an airport manager if he or she was a pilot and had several years of experience in some segment of the industry. Although the individual had to be able to manage the operation for the owner, his or her experience was likely to be in some area of flying rather than in business management.

Today an airport manager must be primarily a skilled and experienced executive with a broad background in all facets of aviation and management in general. It is no longer necessary that the manager be a pilot. Almost every airport manager's job situation is unique in some major respects because of the wide variety of size of airport and type of ownership and operation. There are also wide variations in government procedures in different communities. This sometimes causes the responsibilities, salaries, and authority of airport managers to be completely different from one city to the next. Even the job title varies. Director of aviation, airport superintendent, executive director, airport director, general manager, and other titles are often used instead of airport manager.

Duties of an airport manager

An airport manager is often part landlord and part business executive. As a landlord, the safe condition and operation of the airport is the manager's greatest responsibility. The maintenance of the airport buildings and land is also important. As a business executive, the manager is in charge of public relations; financial planning; profitable and efficient day-to-day operation; and coordination of airline, concession, and airport facilities to best serve the tenants and flying public.

The airport manager's primary duty is the safe and efficient operation of the airport and all its facilities regardless of its size. However, at least in the larger commercial airports, the manager does not have direct control over most aeronautical activities. He or she must deal with all groups and individuals who use the airport facilities. These include representatives of the airlines who schedule flights, maintain and service their aircraft, and process passengers; all segments of the general aviation community, including FBOs and individual and corporate owners and operators of aircraft; federally operated and contracted air traffic management personnel; and federal customs, security, and border patrol employees.

All of these groups may be regarded as tenants of the airport, carrying on their independent activities. Besides dealing with the companies and individuals directly concerned with air travel, the manager is in contact with concessionaires who operate restaurants, shops, and parking facilities, and with the traveling public.

The size of the airport and the services it offers its tenants and the public play an important part in determining the airport manager's specific duties. A manager must formulate fiscal policy, secure new business, recommend and enforce field rules and regulations, make provisions for handling spectators and passengers, oversee construction projects, and see that the airport is adequately policed and that airplane and automobile traffic is regulated.

The manager interprets the functions and activities of the airport to the city or other local government and to the public; he or she is a public relations expert as well as a business manager. This public relations function is extremely important and will be taken up at the end of this chapter in a separate section.

Not all of these duties are required of all managers of airports. Many airports are too small to support air traffic management facilities, such as air traffic control towers. Others have no scheduled airline flights. In these airports the job may seem less complex from an organizational perspective, but the manager must usually do all of the work personally. In large airports, the manager has a larger staff to whom much of the work is delegated.

The job of airport manager is obviously not limited to the typical "9:00 to 5:00" business hours. The hours are often irregular and most managers have some weekend and holiday work. They will often have to work at night. In emergency situations they will usually work additional hours. Difficult weather conditions, labor problems, personnel irregularities, and flight schedule changes are only some of the things that will affect job hours. Even when not actually working, most airport managers are on call 24 hours a day.

Education and training

The major requirement for the job of airport manager is business and administrative ability; this means the ability to make decisions, to coordinate details, to direct the work of others, and to work smoothly with many kinds of people. Perhaps the best college program to follow is one that leads to a degree in aviation management. College courses in engineering; management; accounting; finance and economics; business and aviation law; and airline, general aviation, and airport management are good preparation for a career in airport management. Many schools that are members of the University Aviation Association (UAA) offer programs and courses that can be applied to the problems of managing an airport.

A number of the primary airports in the United States have 1- or 2-year internship programs that train college graduates for various aspects of airport management. Other individuals have started at a small general aviation airport where they become involved in all aspects of airport management—from maintenance and repair to attending city commission meetings. Some college graduates have taken jobs with aviation or airport consulting firms and after several years have moved into airport management. Many others have acquired experience in some other area of aviation before entering the field.

Career advancement in airport management is often described as a "diagonal" path. Typically, an airport professional will begin his/her career at a smaller airport, and then take a more senior position at a larger airport, until he/she eventually achieves a senior management position at a large airport. Others proceed vertically, advancing their careers at the same airport. At smaller airports, however, advancement is limited simply because of the size of the staff. At larger airports, similar limitations are found, simply because there are so many employees who may be seeking promotions to relatively few positions. A typical airport manager's career may span several airports around the country, and increasingly, around the world.

The important public service an airport provides along with its economic advantages to a community has caused city governments to recognize the need for professional management of airports. To meet the need, the American Association of Airport Executives (AAAE) initiated a program of accreditation for those currently employed in airport management. A minimum of 1 year of work experience at a public-use airport, a 4-year college degree or 8 years of civil airport management experience, an original paper on an airport problem, and the completion of a comprehensive examination are the major requirements of the accreditation program. The applicant must also be at least 21 years of age and of good moral character. Once an airport manager has completed these requirements the person may use the initials A.A.E. (Accredited Airport Executive) after his or her name and is eligible to vote at the business meetings of the AAAE. The airport managed by the person may be designated as an AAAE airport.

For those not currently employed at an airport, the AAAE offers a Certified Member (C.M.) certification. The C.M. certification may be achieved by completing the comprehensive written examination with a passing score.

Many career opportunities in airport management should become available in the years ahead because of expansion of facilities and attrition. As the number of new airports increases and the facilities of many existing airports expand, new managerial positions will be created. Many of these will not be top jobs, but the airport of the future will require assistant managers specializing in one part of the huge operation. Also, qualified people will be needed to replace those who retire. In addition, there is an increasing need for experienced and educated airport managers internationally, particularly in the Middle East and Asia.

Because the job of managing a medium-to-large commercial airport is a fascinating one that requires high qualifications, there will be tough competition for jobs; however, the motivated individual with a solid educational background and varied experience in the fields of aviation and management will find openings in a field of work that is and will remain comparatively small—but one that provides an interesting and challenging profession.

The airport manager and public relations

Unquestionably, one of the most important and challenging aspects of an airport manager's job is that of public relations. **Public relations** is the management function that attempts to create goodwill for an organization and its products, services, or ideals with groups of people who can affect its present and future welfare. The most advanced type of public relations not only attempts to create goodwill for the organization as it exists, but also helps formulate policies, if needed, that of themselves result in a favorable reaction.

Aviation and airports have such great impact on our lives, and on the life of our nation, that it is difficult to find a person who has no knowledge or opinion of airports. Despite the tremendous growth in all segments of aviation over the past 50 years, and the resulting challenges, problems, and opportunities, aviation has not been exempted from the controversies that inevitably are part of any endeavor affecting or touching the lives of a large number of people. This controversy is the reason why every opinion, whether positive or negative, will be a strong one. The net result is that every airport has an image—either good or bad.

The great problems of airports are always related to the original and elemental images resulting from the collective opinions of the public. These images are really the balancing or compensating factors that correspond with the problems the public encounters with airports. The images are deposits representing the accumulated experience of jet noise, hours of struggle to reach the airport on clogged highways under construction, the frustration of trying to find a close-in parking place, the lines to obtain tickets, the time waiting for luggage, and other inconveniences, not to mention the ever-increasing impacts of enhanced airport security procedures and airline capacity consolidation, and even the impacts of natural disasters. Very often, operational issues that are the responsibility of others are perceived as the airports' responsibilities, and thus public relations issues for the airports themselves.

In this respect, some of the public will have an image of the airport as a very exciting place that makes major contributions to our society through commercial channels, and even more valuable contributions of a personal nature, by offering a means to efficient travel, and thus greater personal development and greater enjoyment of life. Despite the hundreds of positive impacts of aviation, negative images do arise. Perhaps such negative images result from the fact that the industry has been so intent on the technological aspects of resolving problems that it has overlooked the less tangible components. The industry has the technology and resources to resolve many of the problems of the airport-airway system; however, the important link or catalyst in bringing together technology and community opinion is the airport public relations effort.

Both the airport and the community have a responsibility to work together to solve their mutual problems, attain desired goals, and ultimately achieve a better community. It takes continuing contributions—and sometimes sacrifices as well—to the general welfare on the part of individual citizens and the aviation industry to earn the opportunities and rewards of a good community for the public. This two-way relationship has its problems too. Many are spawned by misunderstanding that can arise and grow to disproportionate size, and in our context, result in a negative image for the airport and a loss of public confidence in the aviation industry. Ensuring that problems are met head-on, with full and explicit information made continuously available to the public to prevent misunderstanding, is the point at which airport public relations enters the picture.

Regardless of the size of an airport, there are several basic principles underlying the public relations process:

- Every airport and every company and interest on the airport has public relations, whether or not it does anything about them.
- Public goodwill is the greatest asset that can be enjoyed by any airport, and public opinion is the most powerful force. Public opinion that is informed and supplied with facts and fair interpretation might be sympathetic. Public opinion that is misinformed or uninformed will probably be hostile and damaging to an airport.
- The basic ingredient of good relations for any airport is integrity. Without it, there can be no successful public relations.
- Airport policies and programs that are not in the public interest have no chance of final success.
- Airport public relations can never be some kind of program that is used only to respond to a negative situation. Good public relations have to be earned through continuing effort.
- Airport public relations go far beyond press relations and publicity. Public relations must interpret the airport interests to the public, and should be a two-way flow with input and interpretation of public opinion to airport management and community leadership. Public relations must use many means of reaching the various segments of the public interested in airport operations, and must try to instill the public relations spirit into all facets of the airport's operation.

The airport and its public

Basically, every airport has four types of constituents with which it deals, and despite the wide variance in size and scope of activities of airports, these publics are basically the same for all airports:

- *The external business public.*   These are the past, present, and future airport customers for all the services offered on an airport. It includes all segments of the business, government, educational, and general flying public.

- *The external general public.* These are the local citizens and taxpayers, many of whom have never been to the airport but who vote on airport issues or who represent citizen groups with particular concerns.
- *Airport users and tenants.* These are the businesses and enterprises whose interests are tied directly to the airport—the airlines, FBOs, other members of the general aviation community, government officials, and other aviation and travel-oriented local businesses and trade organizations, and the employees of all of these enterprises.
- *Airport employees.* This group includes everyone who works for the airport and its parent organization.

These are the most important airport constituents. These are the sources of vital information that management must have in order to know what and how it is doing, and they are the ones who must be informed and persuaded if any airport objective is to be achieved.

Public relations objectives

The primary objectives of an airport's public relations activities are as follows:

- Establishing the airport in the minds of the external public as a facility that is dedicated to serving the public interest: Many airports work closely with the local chamber of commerce in developing a brochure or pamphlet citing various accomplishments and activities at the airport that would be of interest to the local business community and the community in general.
- Communicating with the external public with the goal of establishing and building goodwill: The airport manager and other members of his or her staff often serve as guest speakers at various civic and social organizations. They also become active members of local or civic organizations in order to informally promote the airport and determine the pulse of the community. Public announcements of new developments at the airport are made through all media. This is a continuing part of the communications process.
- Answering general and environmental complaints on an individual basis: It is important that the airport develop a good rapport with its neighbors and concerned citizen groups. Working closely with the airlines and other internal business publics, airport management attempts to work out such problems as noise by changing traffic patterns and adjusting hours of flight operation. Tours of the airport are given to various community groups in order for them to get a better understanding of operations. Civic-oriented activities are also

conducted at the airport to improve relations with airport neighbors and address their concerns. Citizen participation in airport planning and public hearings is another means by which airport management is continually apprised of community feelings about airport-related activities.

- Establishing good working relationships with internal business publics whose interests are similar to those of airport management.

- Promoting programs designed to enhance and improve employee morale.

Like any other facility that serves the total community, an airport requires total community understanding. A well-executed public relations program can make the community aware of the airport and its benefits and create an atmosphere of acceptance. Attitudes are not changed overnight, so the public relations effort must be a continuous campaign to build understanding and develop attitudes of acceptance.

Concluding remarks

Each airport in the United States is unique in its organizational and administrative structure. In addition, each airport is uniquely subject to rules, regulations, and policies applicable to the airport's operational characteristics, the ownership structure, and the laws of the local municipality, region, and state in which it's located.

Conversely, each airport is subject to fundamental regulations mandated by the FAA, the TSA, and state Departments of Transportation, and functions under basic organizational structures that allow for the safe and efficient movement of aircraft, passengers, and cargo in and around the airport.

The great challenge in airport management is to establish an ownership and organizational structure that meets the needs of each of the airport's "publics," from direct users of the airport, to airport employees, to the local community. In addition, the ownership and organizational structure of an airport must be flexible to adapt to the changing needs of the airport's publics. This is not an easy task, although it is one that maintains the excitement of airport management.

Key terms

municipally operated airport
port authorities
airport authority
state-operated airports
organization chart

airport director

assistant director—finance and administration

human resources manager

chief accountant

facilities manager

purchasing manager

manager of public relations

assistant director—planning and engineering

assistant director—operations

airside operations manager

landside operations manager

security manager

aircraft rescue/fire fighting chief

assistant director—maintenance

building and facilities manager

grounds manager

vehicle manager

public relations

Questions for review and discussion

1. Who typically owns airports in the United States?
2. Who typically owns airports in countries other than the United States?
3. What is privatization?
4. What is the difference between a port authority and an airport authority form of airport ownership and operation?
5. What is the purpose of an organization chart?
6. What are the principal duties of a typical airport manager at a medium-size commercial airport?
7. Why is public relations such an important function of airport management?
8. What are some of the basic principles underlying the public relations process at an airport?
9. What are the primary objectives of an airport's public relations process?
10. What types of formal training and education programs exist for current and future airport managers?

Suggested readings

Ashford, Norman, H. P. Martin Stanton, and Clifton A. Moore. *Airport Operations*. London: Pitman, 1993.

Doganis, Rigas. *The Airport Business*. New York: Routledge, Chapman and Hall, 1992.

Eckrose, Roy A., and William H. Green. *How to Assure the Future of Your Airport*. Madison, Wis.: Eckrose/Green Associates, 1988.

Gesell, Laurence E. *The Administration of Public Airports*. San Luis Obispo, Calif.: Coast Aire, 1981.

Odegard, John D., Donald I. Smith, and William Shea. *Airport Planning and Management*. Belmont, Calif.: Wadsworth, 1984.

Wiley, John R. *Airport Administration and Management*. Westport, Conn.: Eno Foundation for Transportation, 1986.

Young, *Airport Workforce Development Practices*, ACRP Synthesis 18, Airport Cooperative Research Program, FAA, 2010.

Web references

FAA Airport Privatization Program:
http://www.faa.gov/airports/airport_compliance/privatization/

AAAE Accredited Airport Executive Program:
http://www.aaae.org/training_professional_development/professional_development/accredited_airport_executive_program/

3

Airports and airport systems: A historical and legislative perspective

Outline

- Introduction
- The formative period of aviation and airports: 1903–1938
  - The birth of civil aviation: 1903–1913
  - World War I: 1914–1918
  - Early airmail service: 1919–1925
  - The Air Commerce Act: 1926–1938
  - The Civil Aeronautics Act: 1938–1939
- Airport growth: World War II and the postwar period
  - The Federal Airport Act: 1946
- Airport modernization: The early jet age
  - The Airways Modernization Act of 1957
  - The Federal Aviation Act of 1958
  - The Department of Transportation: 1967
  - The Airport and Airway Development Act of 1970
  - The National Airport System Plan
  - The Airport and Airway Development Act Amendments of 1976
- Airport legislation after airline deregulation
  - The Deregulation Acts of 1976 and 1978
  - The Airport and Airway Improvement Act of 1982
  - The Aviation Safety and Capacity Expansion Act of 1990
  - Military Airport Program (MAP)

- The Aviation Security Improvement Act of 1990
- The Airport and Airway Safety, Capacity, Noise Improvement, and Intermodal Transportation Act of 1992
- The AIP Temporary Extension Act of 1994
- The Federal Aviation Administration Act of 1994
- The Federal Aviation Reauthorization Act of 1996
- Airports in the twenty-first century: From peacetime prosperity to terror insecurity
  - AIR-21: The Wendell H. Ford Aviation Investment and Reform Act for the Twenty-First Century
  - The Aviation and Transportation Security Act of 2001
  - Homeland Security Act of 2002
  - Vision 100—Century of Aviation Reauthorization Act of 2003

Objectives

The objectives of this section are to educate the reader with information to:

- Discuss the various acts of legislation that have influenced the development and operation of airports since the early days of civil aviation.
- Highlight several important political events that have influenced civil aviation.
- Describe the development of national administrations that have regulated civil aviation throughout its history.
- Describe the various funding programs that have existed to support airports over the course of history.
- Discuss some of the current and future issues concerning airports and how the U.S. government might address these issues.

Introduction

The relatively short but very rich history of civil aviation has made dramatic impacts on society. The growth of civil aviation in general, and airports in particular, has paralleled industrial, technical, economic, and sociopolitical events and has been associated with legislation to adapt to an ever-changing world. This section highlights the growth of aviation and airports through a legislative perspective including decisions and other acts of Congress that have financially, technically, economically, and politically regulated the industry through its first 100 years.

The formative period of aviation and airports: 1903–1938

The birth of civil aviation: 1903–1913

December 17, 1903, the day Orville and Wilbur Wright succeeded in achieving flight with a fixed-wing, heavier-than-air vehicle at Kitty Hawk, North Carolina, has gone down in history as being the "birth of aviation." Their first airplane flight occurred on a large field, with sufficient room for the aircraft to take off and land. There were no paved runways, gates, fuel facilities, lights, or air traffic control. There was no terminal building and there was no automobile-parking garage. There were no rules and regulations governing the flight. That field in Kitty Hawk was, however, the first airport.

In the 10 years following the Wright brothers' first flight, the aviation world evolved in a very slow and hesitant manner, with most of the advances focusing on improving aircraft technology, and much of the efforts trying to promote the technology. Little, if any, consideration was focused on creating facilities for aircraft to take off and land.

As a result, by 1912, there were only 20 recognized landing facilities in the country, all of which were privately owned and operated. The earliest operational airfields date as far back as 1909, although they were generally indistinguishable from, and often also functioned as, local athletic fields, parks, and golf courses. Construction and maintenance of early airfields were, in general, considered local responsibility, and with limited municipal funds, and the very low level of aviation activity, priorities to build "airports" were understandably low.

World War I: 1914–1918

The outbreak of World War I in 1914 opened up initial opportunities for fixed-wing aircraft to serve in a military capacity. The effort to use aviation as a military force in World War I resulted in the production of thousands of aircraft (most of which were produced and served in France, Germany, and England), and hundreds of military pilots, to first fly reconnaissance and later fighting missions. As a result, the U.S. military built 67 airports for the war effort. These predominantly grass fields provided facilities to base, fuel, and maintain aircraft, as well as provide sufficient room for takeoff and landing—but required little other infrastructure. After the war, 25 of these military airfields remained operational, and the rest were decommissioned.

Early airmail service: 1919–1925

After the end of World War I in 1918, many of the aircraft and airmen that had served in the military turned their talents towards civil uses. One of the first civil applications of aviation was that of providing air transportation for the

U.S. mail. The first regular airmail route in the United States was established on May 15, 1918, between New York City and Washington, D.C. The service was conducted jointly by the United States War Department and the Post Office Department. The War Department operated and maintained the aircraft and provided trained airmen, and the Post Office Department attended to the sorting of the mail and its transportation to and from the airfield, and the loading and discharge of the aircraft at the airfields. This joint arrangement lasted until August 12, 1918, when the Post Office Department took exclusive responsibility for the development of the mail service on a larger scale (Fig. 3-1).

Communities suddenly became aware of the importance of having an aerial connection to the rest of the country, and as a result, municipalities began constructing and operating local airports. By 1920, there were 145 municipally owned airports. A nationwide airport system was beginning to form. Domestic airmail service grew considerably between 1918 and 1925. Facilities for air transportation had been established, and the desirability of continued direct government operation or private operation under contract with the government was widely discussed.

The policy of the U.S. government in the intercity transportation of mail had traditionally been to arrange with the other, more popular and widely accepted means of intercity transportation at the time, railroads and steamships. In order to facilitate the use of aviation to transport mail, similar formal arrangements needed to be made between the government and the airmail carriers (Fig. 3-2).

Figure 3-1. *A Curtis JN4-H prepares to carry mail on a northward flight from Washington's Polo Field.* (Photo courtesy Smithsonian Institution, National Air and Space Museum)

Figure 3-2. *Pilots of National Air Transport, one of the companies that would later become United Airlines, preparing for departure from Chicago's Midway Airport in 1927.* (Photo courtesy Landrum & Brown)

The first formal airmail arrangement was ushered in with the Contract Air Mail Act of 1925. This act, also known as the **Kelly Act,** authorized the postmaster general to enter into formal contracts with private persons or companies for the transportation of the mail by air. Contracts were let for a number of feeder and auxiliary main lines during 1925 and 1926, and for the portions of the transcontinental airmail route in 1927 (Fig. 3-2).

The Air Commerce Act: 1926–1938

The potential for growth of the airmail industry in particular, and in aviation activity in general, resulted in the need to have aviation managed, controlled, and regulated as a comprehensive system so that its potential for widespread growth would be met. On May 20, 1926, President Calvin Coolidge signed the **Air Commerce Act of 1926** into law. The object of the Air Commerce Act was to promote the development and stability of commercial aviation in order to attract adequate capital into the business and to provide the fledgling industry with the assistance and legal basis necessary for its growth. The act made it the duty of the secretary of commerce to encourage air commerce by establishing civil airways and other navigational facilities to aid aerial navigation and air commerce. Under the act, the Department of Commerce was charged with encouraging local and municipal development of airfields, for the purpose of economic growth, and to contribute to the infrastructure that would allow the growth of airmail service, as well as to provide safe landing facilities for the newly formed U.S. Army Air Service.

The regulation of aviation provided for in the act included the licensing, inspection, and operation of aircraft; the marking of licensed and unlicensed aircraft; the licensing of pilots and of mechanics engaged in aircraft work; and the regulation of the use of public airways (Fig. 3-3).

In July 1927, a director of aeronautics was appointed who, under the general direction of the assistant secretary of commerce for aeronautics, was in charge

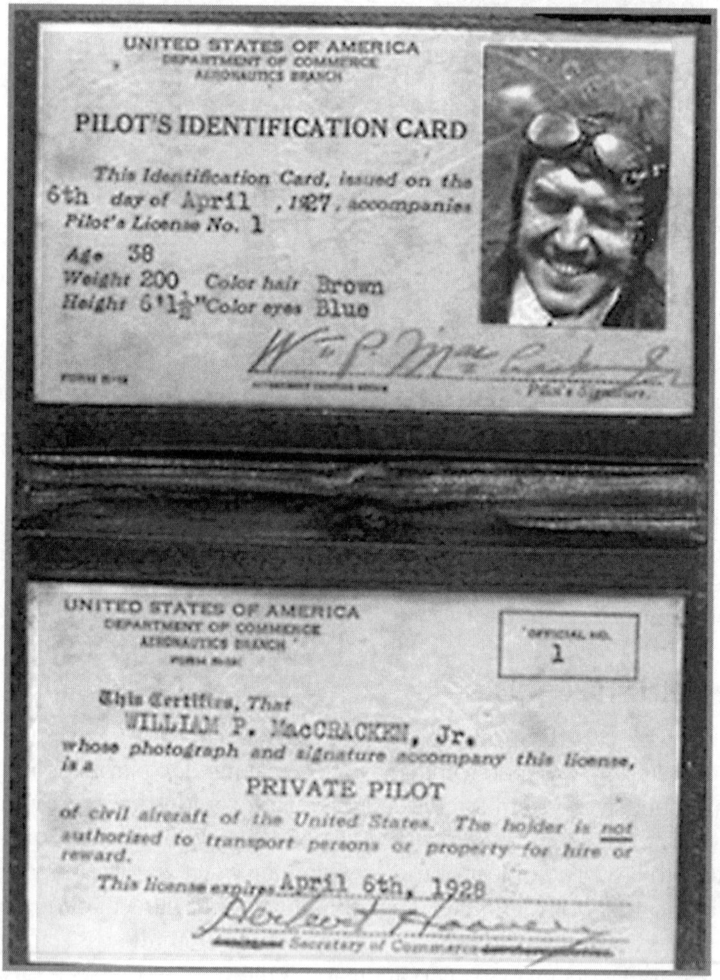

UNITED STATES OF AMERICA
DEPARTMENT OF COMMERCE
AERONAUTICS BRANCH

PILOT'S IDENTIFICATION CARD

This Identification Card, issued on the 6th day of April, 1927, accompanies Pilot's License No. 1

Age 38
Weight 200 Color hair Brown
Height 6'1½" Color eyes Blue

Pilot's Signature.

UNITED STATES OF AMERICA
DEPARTMENT OF COMMERCE
AERONAUTICS BRANCH

OFFICIAL NO. 1

This Certifies, That
WILLIAM P. MacCRACKEN, Jr.
whose photograph and signature accompany this license, is a
PRIVATE PILOT
of civil aircraft of the United States. The holder is not authorized to transport persons or property for hire or reward.

This license expires April 6th, 1928

Secretary of Commerce

Figure 3-3. *The Aeronautics Branch of the Department of Commerce began pilot certification with this license, issued on April 6, 1927. The recipient was the chief of the branch, William P. MacCracken, Jr.* (Photo courtesy FAA)

of the work of the Department of Commerce in the administration of the Air Commerce Act. By November 1929 it was necessary to decentralize the organization from the commerce department, primarily because of the increasing volume of work incident to the rapid development of aviation. Three assistants and the staffs of employees of the divisions under their respective jurisdictions were assigned to the assistant secretary of commerce for aeronautics. These included a director of air regulation, a chief engineer of the airways division, and a director of aeronautics development, to assist in aeronautical regulation and promotion. The organization became known as the Aeronautics Branch of the Department of Commerce.

Authority over civil aviation was further reassigned by executive order of the president in 1933 to place the promotion and regulation of aeronautics in a separately constituted Bureau of the Department of Commerce. An administrative order of the secretary of commerce provided for the establishment of the **Bureau of Air Commerce** in 1934. The bureau consisted of two divisions, the division of air navigation and the division of air regulation.

A revised plan of organization for the Bureau of Air Commerce, adopted in April 1937, placed all the activities of the bureau under a director, aided by an assistant director, with supervision over six principal divisions: airway engineering, airway operation, safety and planning, administration and statistics, certification and inspection, and regulation.

From 1926 until 1938, the federal government was prohibited by the Air Commerce Act of 1926 from participating directly in the establishment, operation, and maintenance of airports. One exception to this law occurred, however, in response to the great depression. The Civil Works Administration—from the autumn of 1933 up until it was superseded by the Federal Emergency Relief Administration (FERA) in April of 1934—spent approximately $11.5 million establishing 585 new airports, mostly in smaller communities. The FERA spent its appropriation on 943 airport projects, mostly in smaller cities, with 55 new airports receiving aid.

In July 1935, the **Works Progress Administration (WPA)** took over the federal airport development work. Under the WPA, there was an emphasis on spending for larger airports and projects of a more permanent nature. Under the WPA, about half the expense for materials and equipment was borne by the sponsors. The remainder of the expenses, including labor, was supplied by the federal government.

The Civil Aeronautics Act: 1938–1939

On June 23, 1938, the **Civil Aeronautics Act of 1938** was approved by President Franklin Delano Roosevelt. This act substituted a single federal statute for the several general statutes that had up to this time provided for the regulation of civil aviation. The act placed all regulation of aviation and air transportation into one authority. The act created an administrative agency consisting of three partly autonomous bodies. The five-man **Civil Aeronautics Authority** was principally concerned with the economic regulation of the new passenger air carrier industry. The Air Safety Board was an independent body for the investigation of accidents. The administrator of civil aviation was concerned with construction, operation, and maintenance of the airway system as a whole.

The transfer of responsibilities, personnel, property, and unexpended balances of appropriations of the Bureau of Air Commerce to the Civil Aeronautics Authority, effected in August 1938, under the provisions of the act, brought to a close a 12-year period during which the development and regulation of civil aviation was under the jurisdiction of the Department of Commerce.

During the first year and half of its existence, a number of organizational difficulties arose within the Civil Aeronautics Authority. As a result, President Roosevelt, acting within the authority granted to him in the Reorganization Act of 1939, reorganized the Civil Aeronautics Authority by creating two separate entities. The five-man authority initially known as the Civil Aeronautics Authority remained an independent operation and became known as the **Civil Aeronautics Board (CAB).** The Air Safety Board was abolished and its functions given to the CAB. The administrator of the old Civil Aviation Authority became the head of an office within the Department of Commerce known as the **Civil Aeronautics Administration (CAA).** The duties of the original five-man authority were unchanged, except that certain responsibilities, such as accident investigation, previously handled by the Air Safety Board, were added. The administrator, in addition to retaining the function of supervising construction, maintenance, and operation of the airways, was required to undertake the administration and enforcement of safety regulations, and the administration of the laws with regard to aircraft operation. Subsequently, the administrator became directly responsible to the secretary of commerce. The term CAA, which originally identified the Civil Aeronautics Authority, became the common moniker of the Civil Aeronautics Administration.

Airport growth: World War II and the postwar period

Section 303 of the Civil Aeronautics Act of 1938 authorized the expenditure of federal funds for construction of landing areas provided the administrator certified "that such a landing area was reasonably necessary for use in air commerce or in the interests of national defense." When war broke out in Europe in September 1939, the administrator certified the necessity of federal aid because of issues of national defense. As a result, Congress in 1940 authorized the appropriation of $40 million for the **Development of Landing Areas for National Defense (DLAND).** Under DLAND, with the approval of the secretaries of war, commerce, and the Navy, the CAA was authorized to construct no more than 250 airports. In actuality, in 1941, work on the construction of 200 airports began, with an additional 149 added to the program later in the year. Under this program, government subdivisions furnished the land and agreed to operate and maintain the improved field, and the essential landing facilities were developed by federal funds. The DLAND program was coordinated with the work and funds of other programs and other government sources, including the U.S. Army, throughout the war years.

In 1940, the Army Air Corps started an aggressive expansion program. This expansion quickly resulted in the need for a far greater number of airports than was appropriated under the original DLAND. As a result, reauthorizations allowed DLAND to expand to the construction of 504, then 608, and finally a total of 986 airports receiving aid.

During the war years the federal government, through the CAA, spent over $353 million for the repair and construction of military landing areas in the continental United States, not including funds spent by the military. During the same period, the CAA spent $9.5 million for the development of landing areas in the United States solely for the use of civil aviation.

Many of the new airports constructed for the military during the war were planned so as to be useful for civil aviation purposes after the war. As a result, more than 500 airports constructed for the military by the CAA were declared military surplus after the war ended and were subsequently handed over to the authorities of cities, counties, and states for civil aviation use. An understanding was reached between the federal government and the sponsor that the facilities would be available to the public without discrimination and to the government in the event of a national emergency.

Section 302(c) of the Civil Aeronautics Act directed the civil aeronautics administrator to make a field survey of the existing system of airports and to report to Congress with recommendations as to future federal participation in airport construction, improvement, development, operation, or maintenance. To perform this survey, an advisory committee composed of representatives of interested civil and military federal agencies, state aviation officials, airport managers, airline representatives, and others, was appointed. The first survey and report, made in 1939, did not result in congressional action, but a revised plan and recommendations submitted in November 1944 were influential in calling attention to the private airport deficiencies of inadequate distribution and inadequate facilities. This 1944 plan became known as the first **National Airport Plan (NAP).** This first NAP formed the basis for airport system planning and federal funding programs for the construction and improvement of airports in the United States.

The Federal Airport Act: 1946

After the war, Congress formalized legislation considering the NAP and established the first formal continuous federal airport funding programs with the signing of the **Federal Airports Act of 1946** on May 13, 1946. It was the purpose of the Federal Airport Act to formally recognize the civil-use airports in the United States as a comprehensive system of airports, administered by the Civil Aeronautics Administration. Small communities that had inherited surplus military airfields, or that needed new airports to help develop their social and economic structure, were theoretically supposed to benefit from this program.

Congress appropriated $500 million for airport aid over a 7-year period beginning July 1, 1946, with no more than $100 million to be appropriated in any one year. Of the total appropriations, 25 percent was placed in a discretionary fund to be used as the civil aeronautics administrator saw fit for airport construction. Half the remaining 75 percent was apportioned to the states, on the basis of population, and the other half, on the basis of land area. The discretionary fund allowed the administrator to choose the projects regardless of their location. The funds appropriated were restricted to construction of operational facilities such as runways and taxiways.

This federal-aid program, known as the **Federal-Aid Airport Program (FAAP),** provided that the federal government would pay as much as 50 percent of the cost of moderate to major airport construction projects, with the balance of the costs paid by the airport sponsor, typically the local municipality, county, or state. This policy worked well for larger cities, because they could issue and sell bonds to pay for their share of major expenses. For smaller cities, the burden of having to share 50 percent of the costs of a major project often made airport construction prohibitive.

For an airport or governmental unit to be available for such aid, it was necessary that the airport be considered part of the National Airport Plan. Under the Federal Airport Act, the administrator had to take into account the needs of both air commerce and private aviation, as well as technological development, probable growth, and any other considerations found appropriate, when considering airports to be part of the National Airport Plan.

As a condition to funding approval of a project, the administrator had to receive in writing a guarantee that the following provisions would be adhered to:

1. The airport would be available for public use without unjust discrimination.

2. The airport would be suitably operated and maintained.

3. The aerial approach would be cleared and protected and future hazards would be prevented.

4. Proper zoning would be provided to restrict the use of land adjacent to the airport.

5. All facilities developed from federal aid would be made available to the military.

6. All project accounts would be kept in accordance with a standard system.

7. All airport records would be available for inspection by an agent of the administrator upon reasonable request.

The Federal-Aid Airport Program progressed from the paper to the construction stage during 1947, and by February of the following year the CAA had made 133 grant offers to local sponsors totaling $13.3 million. This marked the beginning of the federal government's continuous and current participation in construction of airport facilities.

On August 3, 1955, President Dwight Eisenhower signed Public Law 211, making minor changes in the Federal-Aid Airport Program and removed the 1958 expiration date prescribed by the Federal Airport Act, as amended in 1950. These changes established a 4-year program with authorizations amounting to $63 million for each fiscal year 1957–1959, made all types and sizes of airports eligible for aid, included airport buildings as eligible items of development, and provided that funds apportioned yearly to states continue on the area-population formula.

During its 24-year lifetime (1946–1969), the Federal-Aid Airport Program generated $1.2 billion in federal aid to airports, all of it drawn from the general treasury. Most of the money, nearly $1 billion, was used to build runways, taxiways, and roadways, while the rest was spent on land, terminal buildings, and lighting systems. For all of its success, however, the program failed to anticipate the travel boom beginning in the late 1950s, which overloaded the country's commercial air routes and prompted carriers to expand their fleets.

Airport modernization: The early jet age

The Airways Modernization Act of 1957

Recognizing that the demands on the federal government in the years ahead would be substantial, the director of the Bureau of the Budget requested a review of aviation facilities problems in 1955, William B. Harding was appointed as a consultant to the director. Harding, in turn, solicited the help of a number of prominent individuals in aviation to form his committee. In late December 1955, Harding submitted his report. Reporting that the need to improve air traffic management had already reached critical proportions, the group recommended that an individual of national reputation, responsible directly to the president, be appointed to provide full-time, high-level leadership in developing a program for solving the complex technical and organizational problems facing the government and the aviation industry.

Following approval of the Harding Committee recommendations, President Eisenhower appointed Edward P. Curtis as his special assistant for aviation facilities planning. His assignment was to direct and coordinate a "long-range study of the nation's requirements," to develop "a comprehensive plan for meeting in the most effective and economical manner the needs disclosed by the study," and "to formulate legislative, organizational, administrative, and budgetary recommendations to implement the comprehensive plan."

In 1956, a Trans World Airlines Super Constellation and a United Airlines Douglas DC-7 collided in midair over the Grand Canyon, killing 128 people (Fig. 3-4). As a result of this high-profile accident, the public outcry for reform leading to a safer air traffic management system became louder and clearer. Furthermore, the threat made visible by the collision of two relatively slow piston aircraft was projected to be far greater with the introduction of jet aircraft into the civil aviation system.

On May 10, 1957, Curtis submitted his report entitled "Aviation Facilities Planning" to the president. The report warned of "a crisis in the making" as a result of the inability of the current airspace management system to cope with the complex patterns of civil and military traffic that filled the sky. The growing congestion of airspace was inhibiting defense and retarding the process of air commerce. Concluding that many excellent plans for improving the nation's aviation facilities had failed in the past to mature because of the inability of our governmental organization to keep pace with aviation's dynamic growth, Curtis recommended the establishment of an independent Federal Aviation Agency "into which are consolidated all the essential management functions necessary to support the common needs of the military and civil aviation of the United States." Until such a permanent organization could be created, Curtis recommended the creation of an **Airways Modernization Board** as an independent office responsible for developing and consolidating the requirements for future systems of communications, navigation, and traffic control needed to accommodate U.S. air traffic.

Congress was receptive to this recommendation and passed the **Airways Modernization Act of 1957** on August 14, 1957. The purpose of the act was "to provide for the development and modernization of the national system of

Figure 3-4. *The wreckage of a United Airlines DC-7 after it collided with a TWA Constellation over the Grand Canyon in Arizona on June 30, 1956. All 128 people on both planes were killed.* (Photo courtesy www.planecrashinfo.com)

navigation and traffic control facilities (many of which exist on the property of civil airports) to serve present and future needs of civil and military aviation." The act further provided for its own expiration on June 30, 1960. Appointment of Elwood R. Quesada as chairman of the Airways Modernization Board was confirmed by the Senate on August 16, 1957.

The Federal Aviation Act of 1958

On May 20, 1958, a military jet trainer and a civilian transport plane collided over Brunswick, Maryland, killing 12, the third major air disaster within a period of three and 1/2 months. This tragedy spurred governmental action to establish a comprehensive Federal Aviation Agency. Instead of taking a predicted 2 or 3 years to create a single aviation agency, Congress immediately took action to enact legislation. As a result, the **Federal Aviation Act of 1958** was signed by the president on August 23, 1958. Treating comprehensively the federal government's role in fostering and regulating civil aeronautics and air commerce, the new statute repealed the Air Commerce Act of 1926, the Civil Aeronautics Act of 1938, the Airways Modernization Act of 1957, and those portions of the various presidential reorganization plans dealing with civil aviation.

The law provided for the retention of the CAB as an independent office including all its functions except the safety rule-making powers, which were transferred to the new **Federal Aviation Agency.** The Federal Aviation Agency was created with an administrator responsible to the president. The agency incorporated the function of the Civil Aeronautics Administration and the Airways Modernization Board.

Section 103 of the act concisely stated the administrator's major powers and responsibilities:

1. The regulation of air commerce in such manner as to best promote its development and safety and fulfill the requirements of national defense.
2. The promotion, encouragement, and development of civil aeronautics.
3. The control of the use of the navigable airspace of the United States and the regulation of both civil and military operations in such airspace in the interest of the safety and efficiency of both.
4. The consolidation of research and development with respect to air navigation facilities, as well as the installation and operation thereof.
5. The development and operation of a common system of air traffic control and navigation for both military and civil aircraft.

On November 1, 1958, Elwood R. Quesada, special assistant to the president for aviation matters and chairman of the Airways Modernization Board, became the first administrator of the Federal Aviation Agency (Fig. 3-5).

The Department of Transportation: 1967

For many years it was argued that there had been unrestrained growth and considerable duplication of federal activities regarding transportation. In 1966, President Lyndon Johnson chose to deliver a special transportation message to Congress. He focused in on the need for coordination of the national transportation system, reorganization of transportation planning activities, and active promotion of safety. In his address, President Johnson contended that the U.S. transportation system lacked true coordination and that this resulted in system

Figure 3-5. *On November 1, 1958, Elwood R. Quesada took the oath as FAA's first administrator.* (Photo courtesy FAA)

inefficiencies. He advocated the creation of a federal department of transportation to promote coordination of existing federal programs and to act as a focal point for future research and development efforts in transportation.

Congressional hearings were held on several bills involving most of President Johnson's recommendations. Although some opposition was expressed to specific proposals, there was general support for creation of the **Department of Transportation (DOT).** The legislation creating the DOT was approved on October 15, 1966. DOT commenced operations on April 1, 1967, and Alan S. Boyd was appointed the first secretary of transportation.

The agencies and functions transferred to the DOT that related to air transportation included the Federal Aviation Agency in its entirety and the safety functions of the CAB, including the responsibility for investigating and determining the probable cause of aircraft accidents and safety functions involving review on appeal of the suspension, modification, or denial of certificates or licenses. The Federal Aviation Agency was placed under the DOT and renamed the Federal Aviation Administration (FAA). The administrator of the FAA was still appointed by the president but from then on reported directly to the secretary of transportation.

The act also created within the new department a five-member **National Transportation Safety Board (NTSB).** The act charged the NTSB with (1) determining the cause or probable cause of transportation accidents and reporting the fact, conditions, and circumstances relating to such accidents; and (2) reviewing on appeal the suspension, amendment, modification, revocation, or denial of any certificate or license issued by the secretary or by an administrator.

The Airport and Airway Development Act of 1970

The tremendous growth in all segments of aviation during the late 1960s put a strain on the existing airway system. Air delays getting into and out of major airports began to develop rapidly. Along with the delays in the air, congestion was taking place in parking areas and terminal buildings. Public indignation at the failure of the system to keep pace with the demand for air transportation reached a peak in 1969. It was undoubtedly hastened by the widely publicized touchdown of the first of the new family of wide-bodied jets, including the Boeing 747. President Richard Nixon told Congress in 1969 that stacks of airplanes over the nation's airports were ample evidence that something needed to be done.

It was evident that to reduce congestion, substantial amounts of money would have to be invested in airway and airport improvements. For airports alone it was estimated that $11 billion in new capital improvements would be required for public airports in the 10-year period 1970–1980. The amount of money authorized by the Federal Airport Act of 1946 was insufficient to assist in financing such a vast program. The normal and anticipated sources of revenue available to public airports were also not sufficient to acquire the required funds for capital expenditures.

Congress responded with an idea it borrowed from the interstate highway program: a trust fund supported by taxes on people who used the national aviation system. Such a mechanism, according to its proponents, would shift the cost of increasing the system's capacity from taxpayers to those groups that benefited most directly: passengers, shippers, and aircraft owners.

On May 21, 1970, President Nixon signed a two-title law that was to run for 10 years. Title I was the **Airport and Airway Development Act of 1970,** and Title II was the **Airport and Airway Revenue Act of 1970.** The new legislation assured a fund estimated at the time to generate more than $11 billion in funds for airport and airway modernization during the decade. By establishing an **Airport and Airway Trust Fund,** modeled on the existing successful Highway Trust Fund, it freed airport and airway development from having to compete for general treasury funds, the basic reason for the funding uncertainties and inadequacies of the past. Into the trust would go new revenues from aviation user taxes levied by the Airport and Airway Revenue Act, and other funds that Congress might choose to appropriate to meet authorized expenditures.

The Airport and Airway Trust Fund was funded by levies on aviation users, including:

1. An 8 percent tax on domestic passenger fares
2. A $3 surcharge on passenger tickets for international flights originating in the United States
3. A tax of 7 cents per gallon on both gasoline (Avgas) and jet fuel (Jet-A) used by aircraft in noncommercial "general" aviation
4. A 5 percent tax on airfreight waybills
5. An annual registration fee of $25 on all civil aircraft, plus (1) in the case of piston-powered aircraft weighing more than 2,500 pounds, 2 cents per pound for each pound of maximum certificated takeoff weight; or (2) in the case of turbine-powered aircraft, 3.5 cents per pound of maximum takeoff weight

The principal advantages to this user-fee approach to funding were that it (1) provided a predictable and increasing source of income, more commensurate with need; (2) permitted more effective and longer-range planning; and (3) assured that the tax revenues generated by aviation would not be diverted to nonaviation interests.

Two grant-in-aid programs were provided for under the Airport and Airway Development Act: The **Planning Grant Program (PGP)** and the **Airport Development Aid Program (ADAP).** These grant programs were fund-matching assistance programs in which the federal government paid a predetermined share of approved airport planning and development project costs, and the airport owners at the various state and local levels, who were eligible to participate in the program, paid the remainder of the expenses. The act also provided that the funding authority of the grant-in-aid programs would expire on June 30, 1975. The object of this was to see what, if any, changes needed to be made before further funds were authorized for the remainder of the program.

The major weaknesses of the 1946 Federal Airport Act, which was repealed by the 1970 Airport and Airway Development Act, were the overall inadequacy of the resources provided, as well as the nature of the formula used for distributing those scarce resources. The annual authorization under the 1946 Act totaled only $75 million, and of this total, less than $8 million were appropriated in a truly discretionary manor, a funding level far too small to make any significant impact on critical needs.

By contrast, the Airport and Airway Development Act increased the total annual authorizations by nearly four times for each of the first 5 years, to $280 million, and provided a distribution formula improved in the light of the experience of the Federal Airport Act. Of the $280 million, $250 million would be available annually for modernization and improvement programs at air carrier and reliever airports, and $30 million annually for general aviation airports (see Table 3-1).

The National Airport System Plan

In its provisions concerning planning, the new legislation reflected not only certain lessons of experience, but also the emergence of certain new planning factors. For example, experience under the Federal Airport Act with the National Airport Plan, which covered a period of 5 years and was revised annually, led to the requirement in the new law for a **National Airport System Plan (NASP).** The NASP called for a 10-year program to be revised only as necessary. Notable among factors explicitly mentioned for the secretary's consideration in relation to the NASP were, among others, (1) the relationship of each airport to the local transportation system, to forecasted technological developments in aeronautics, and to developments forecasted in other modes of intercity transport; and (2) factors affecting the quality of the natural environment. The NASP effectively defined the United States' **National Airspace System (NAS).**

Airway modernization also benefited from the increased funding authorized by the Airport and Airway Development Act. Throughout the 1960s, appropriations for airway facilities and equipment averaged $93 million per year. The new legislation authorized "not less than" $250 million per year for the first 5 fiscal years for acquiring, establishing, and improving air navigation facilities. A principal beneficiary of this more-generous authorization would be the FAA's automation of the air traffic control system of the NAS.

Table 3-1. ADAP Spending (millions), 1971–1975

FY	Amount Permitted under 1970 Act (authorizations)	Amount Approved by Congress Each Year (appropriations)	Amount Actually Spent by FAA (obligations)
1971	$280	$170	$170
1972	280	280	280
1973	280	280	207
1974	310	300	300
1975	310	335	335
Total	$1,460	$1,365	$1,292

Source: FAA.

ADAP funding under the act, for which the federal share for large and medium air carrier hubs had been 50 percent, and for the smaller air carrier, general aviation, and reliever airports, 75 percent, had initially been $280 million per year. In 1973, under the amendments to the act of that year, the funding level rose to $310 million.

Total ADAP funds obligated under the act over the 5-year period totaled nearly $1.3 billion, a figure that exceeded by $100 million the $1.2 billion airport development aid funds disbursed by the federal government in the entire 24-year history of the earlier Federal-Aid Airport Program. The $1.3 billion had made it possible for the FAA to provide and fund a total of 2,434 ADAP projects during the 5-year period. Of this number, 1,528 projects were completed at 520 air carrier airports, 757 projects at 624 general aviation airports, and 149 projects at 81 reliever airports. At air carrier airports, over $1.09 billion was expended, at general aviation airports, $212.8 million, and at reliever airports, $61.6 million.

With this infusion of additional federal money, 85 new airports were built and more than 1,000 others significantly improved. The improvement included the construction of 178 new runways, 520 new taxiways, 201 runway extensions, hundreds of miles of security fencing, and fleets of aircraft rescue and fire fighting (ARFF) equipment. They also comprised some of the most advanced navigational aid equipment available, including 28 Instrument Landing Systems (ILS), 141 runway end identifier lighting systems (REIL), and 471 visual approach slope indicators (VASI).

The act had served its purpose well during its first 5 years. Nevertheless, as its authority began drawing to a close, it was clear that the required review was fortunate in its timing. With a sharp increase in air carrier and general aviation operations, mounting environmental and terminal access problems, along with increasing inflation, there was no time to be lost in getting a legislative review underway.

The Airport and Airway Development Act Amendments of 1976

On July 12, 1976, President Gerald Ford signed into law the **Airport and Airway Development Act Amendments of 1976.** ADAP funding levels for the remaining 5 years under the 1970 act were sharply increased.

Some of the important amendments included under the 1976 act were as follows:

1. Expanded the types of airport development projects eligible for ADAP funding. These now included (1) snow removal equipment, (2) noise suppressing equipment, (3) physical barriers and landscaping to diminish the effects of aircraft noise, and (4) the acquisition of land to ensure environmental

compatibility. Non-revenue-producing public-use terminal area facilities for the movement of passengers and baggage at airports CAB certificated air carriers also became eligible for ADAP funding, except that in such cases the federal share would be greater than 50 percent.

2. Established the "commuter service airport," a new carrier airport category comprised of approximately 130 airports that served noncertificated air carriers, and enplaned at least 2,500 passengers annually. This new airport category was created in recognition of the substantial growth of commercial commuter services during the previous 5-year period, their potential for future growth, and the resulting need to assure airports serving them proper development funding.

3. Directed that the reliever airports in the National Airport System, previously grouped for funding purposes with the air carrier airports, be included instead with the general aviation airport category because, aside from their usefulness as relievers, they functioned primarily as general aviation airports.

4. Increased the federal share for ADAP grants. For smaller commercial service and general aviation airports, the federal share of funding was 90 percent for fiscal years 1976 through 1978, and 80 percent for fiscal years 1979 through 1980. For the 67 largest commercial service airports in the National Airport System, sharing of funding rose from 50 percent to 75 percent for the entire 5-year period.

5. Increased the federal share for PGP grants from 66.7 to 75 percent.

6. Ordered the preparation and publication of a major revision of the National Airport System Plan by January 1, 1978. Last submitted to Congress in 1973, the NASP comprised more than 4,000 locations, including 649 facilities served by the certificated air carriers.

7. Directed the initiation of a series of studies having to do with (1) the feasibility of "landbank-banking" as an expedient in airport development; (2) the case for sound proofing public institutions located near airports; (3) the identification of places in the United states where major new airports would be needed, and alternative approaches to their financing; and (4) the identification of needed airports across the nation, which for economic reasons were threatened with closure, with an analysis in the individual case of what could be done to keep them open.

8. Authorized a 5-year appropriation (1976–1980) from the Airport and Airway Trust Fund for disbursement in annual increments of the following sums: (1) up to $1.1.5 billion to cover the costs of flight checking and maintaining the air navigation facilities of the federal airway system; (2) $1.275 million to assist the states in developing their own general aviation airport standards; and (3) $1.3 billion for the purpose of acquiring, establishing, and improving federal air navigation facilities (see Table 3-2).

In signing the new legislation, President Ford stated, "The Airport and Airway Development Act of 1976 will make possible the continuing modernization of

Table 3-2. ADAP Spending (millions), 1976–1980

FY	Air Carrier Airports	General Aviation Airports	Total
1976	$435	$65	$500
1977	440	70	510
1978	465	75	540
1979	495	80	575
1980	525	85	610
Total	$2,360	$375	$2,735

Source: FAA.

our airways, airports, and related facilities in communities throughout our fifty states."

Airport legislation after airline deregulation

The Deregulation Acts of 1976 and 1978

The passage of the **Air Cargo Deregulation Act of 1976** and, more important, the **Airline Deregulation Act of 1978,** signaled an end to the 40-year history of economic regulation of the airline industry. Deregulation of airlines was part of a general trend gaining momentum in the 1970s to reduce government regulation of private industry. By this time, many observers in Congress and elsewhere had begun to doubt that federal regulation was encouraging orderly competition and had come to suspect that the regulatory process was imposing unnecessary costs and creating distortions in the marketplace. Even before Congress passed the deregulation acts, the CAB itself had conducted a number of experimental reductions of certain types of regulation in order to encourage competition. With the 1978 act, the market was opened to new entrants, and carriers gained much greater freedom to enter or leave markets, to change routes, and to compete on the basis of price. The 1978 act also called for the "sunset" of the CAB by the end of 1984, with transfer of its few remaining essential functions to the DOT and other departments.

Airline deregulation has had a profound effect on the nation's airports. Once air carriers were permitted to change route strategies without CAB approval, many less-profitable markets were dropped, confirming the fears of deregulation that air service to smaller communities would suffer. Service to some smaller cities continued under the **Essential Air Service (EAS)** provisions of the Deregulation Act, provisions which provided subsidies to the last remaining carrier in a market so as to prevent selected cities from losing air service altogether. In many cases, small commuter carriers entered the markets abandoned by larger carriers. In addition, the airlines' new freedoms have greatly changed their relationships with airport operators, who can no longer depend on the stability of service by the air carriers that serve their airports, must accommodate new entrants, and must handle the unfortunate situations when existing air carriers decide to significantly reduce or even completely remove operations from the market.

Perhaps the most profound impact of airline deregulation was the proliferation of a *hub and spoke* routing strategy by most of the largest air carriers. Under a hub and spoke routing strategy, air carriers arranged flight schedules and routes so that a large number of aircraft would arrive from outlying *spoke* airports, over a short period of time into a hub facility, where passengers would deplane to transfer to aircraft bound for their final destinations. This routing strategy afforded air carriers the ability to serve more markets with a given fleet of aircraft and crew. Figure 3-6 illustrates the point-to-point route structures reminiscent of preregulation and Fig. 3-7 illustrates a route structure common to the hub and spoke route structure.

The hub and spoke route network resulted in significant increases in aircraft operations and total passenger movements at those airports selected as hub facilities by the air carriers. Smaller spoke airports, on the other hand, often suffered from reduced service, particularly nonstop service, to destinations served prior to airline deregulation.

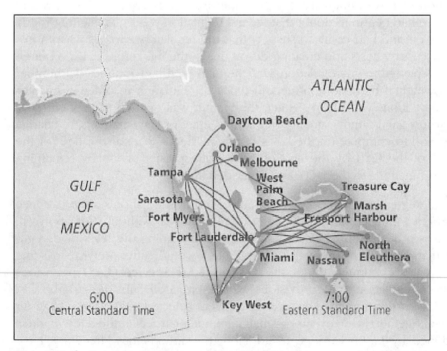

Figure 3-6. *Point-to-point route network.* (Courtesy Gulfstream Airlines)

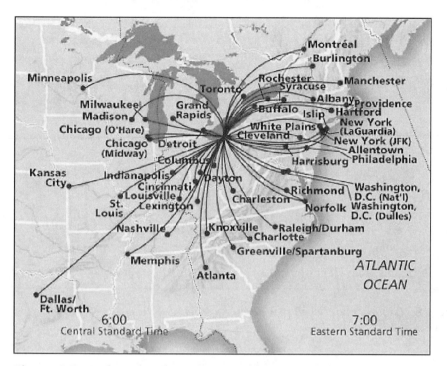

Figure 3-7. *Hub and spoke route network.* (Courtesy Continental Express Airlines)

Figure 3-8 illustrates the increase in passenger activity as a result of postderegulation hub and spoke strategies. Such increases in passenger and aircraft activity resulted in the need for significant airport expansion in a very short period of time.

The Airport and Airway Improvement Act of 1982

Between 1971 and 1980, the Airport and Airway Trust Fund received approximately $13.8 billion, of which $4.1 billion was invested in the airport system through ADAP grants. The Airport and Airway Development Act expired in 1980. During fiscal years 1981 and 1982, the taxing provisions of the trust fund were reduced and the funds raised were deposited in the General Fund and Highway Trust Fund. Congress appropriated approximately $900 million over these 2 years.

The expiration of ADAP funding compounded with the deregulated operations of the major air carriers created new funding needs for airport improvements. In particular, airports that had recently become operating "hubs" for major air carriers required funding for major expansions of airfields, terminals, and ground access facilities. Interestingly, debate in Congress ensued regarding whether or

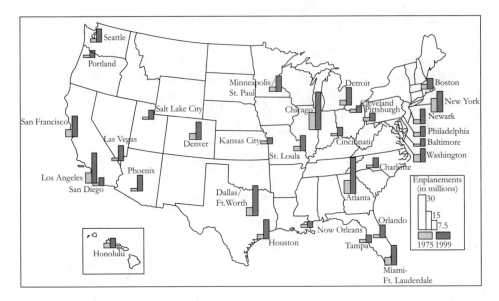

Figure 3-8. *Enplanement levels pre- versus postderegulation at selected airports.* (Courtesy U.S. Bureau of Transportation Statistics)

not these larger airports should receive federal aid at all on the grounds that increased revenues from the extraordinary growth in airline service should be sufficient to cover necessary improvements. This idea of "defederalizing" the largest airports was eventually discarded, with the finding that capital improvements required to accommodate the airlines' hub and spoke route networks far exceeded any projected increases in airport revenues. In fact, the final version of the signed **Airport and Airway Improvement Act of 1982** favored increased federal funding for these largest of airports.

The Airport and Airway Improvement Act of 1982 reestablished the operation of the Airport and Airway Trust Fund, although with a revised schedule of user fees. Operators of piston aircraft, for example, were required to pay 12 cents per gallon for "avgas," an increase of 5 cents over the 1970 tax rate. Turbine aircraft paid 14 cents per gallon of jet fuel, an increase of 7 cents.

The act authorized a new capital grant program, called the **Airport Improvement Program (AIP).** In basic philosophy, AIP was similar to ADAP. It was intended to support a national system of integrated airports that recognizes the role of large and small airports together in a national air transportation system. Maximized joint use of underutilized, nonstrategic U.S. military airfields was also encouraged.

As part of the act, the NASP was reorganized as the **National Plan of Integrated Airport Systems (NPIAS).** This reorganization added the categories of large hub, medium hub, small hub, and nonhub to commercial service airports. Airports were assigned a hub category on the basis of the percentage of the total annual enplanements in the United States to occur at the airport. AIP funding was appropriated on the basis of where airports fit within the NPIAS.

The 1982 act also contained a provision to make funds available for noise compatibility planning and to carry out noise compatibility programs as authorized by the **Aviation Safety and Noise Abatement Act of 1979.**

The Airport and Airway Improvement Act has been amended several times. In October 1982, the **Continuing Appropriations Act** added a section providing authority to issue discretionary grants in lieu of unused apportioned funds under certain circumstances, and the **Surface Transportation Assistance Act,** passed in January 1983, increased the annual authorizations for the AIP for fiscal years 1983 to 1985. Overall, the Airport and Airway Improvement Act of 1982 authorized a total of $4.8 billion in airport aid for fiscal years 1983 through 1987 (see Table 3-3).

The **Airport and Airway Safety and Capacity Expansion Act of 1987** extended the authority for the AIP for 5 years. The act authorized $1.7 billion each fiscal year through 1990 and $1.8 billion each year for fiscal years 1991 and 1992. This act also authorized a new procedure in which an airport sponsor is advised of federal intentions to fund long-term, high-priority capacity projects as appropriations allow and to reimburse sponsors for certain specified work performed before a grant is received. This procedure is implemented through a letter of intent issued to sponsors.

Another provision of the Airport and Airway Safety and Capacity Expansion Act established a requirement that 10 percent of the funds made available under the AIP be given to small business concerns owned and controlled by socially and economically disadvantaged individuals, known as the **Disadvantaged Business Enterprise (DBE) Program.**

The Aviation Safety and Capacity Expansion Act of 1990

Whereas AIP funding under the Airport and Airway Improvement Act had benefited the largest of airports, smaller commercial service airports, especially those that had experienced a decline in air service since airline deregulation, suffered from stagnant and often decreasing levels of funding associated with their stagnant and often declining enplanement levels. The argument was posed that the largest hub airports could generate funding revenue on an individual airport basis, perhaps through some sort of head tax.

Airports had historically been permitted to impose head taxes on passengers utilizing their facilities, until 1973 when Congress imposed blanket "antihead

Table 3-3. AIP Funding by Airport Type (millions), 1982–1998

| FY | Congressional Authorization | Appropriations by Type of Airport | | | | Total Appropriations |
		Primary	Nonprimary Commercial	General Aviation	Reliever	
1982	$460.0	$312.3	$31.5	$62.4	$48.2	$454.4
1983	800.0	465.0	69.2	155.1	98.7	788.0
1984	993.5	502.8	62.0	146.5	103.6	814.9
1985	987.0	623.4	52.4	154.1	110.1	940.0
1986	1,017.0	542.0	58.9	146.5	100.8	848.2
1987	1,017.2	525.6	72.2	155.8	129.7	883.3
1988	1,700.0	1,082.9	47.7	190.9	135.1	1,456.6
1989	1,700.0	1,013.5	43.9	178.0	171.2	1,406.6
1990	1,700.0	1,010.6	43.7	168.5	138.0	1,360.8
1991	1,800.0	1,210.1	45.5	248.7	211.1	1,715.4
1992	1,900.0	1,203.4	56.4	249.2	166.5	1,675.5
1993	2,025.0	1,296.4	41.2	199.1	180.6	1,717.3
1994	2,070.3	1,316.1	41.4	181.1	133.2	1,671.8
1995	2,161.0	1,166.3	32.5	157.6	85.7	1,442.1
1996	2,214.0	1,025.3	27.8	145.6	105.6	1,304.3
1997	2,280.0	1,209.3	57.7	140.1	114.6	1,521.7
1998	2,347.0	956.7	39.1	185.5	127.8	1,309.1
Subtotal	$27,172.0	$14,145.6	$823.1	$2,864.7	$2,160.5	$21,310.0
		72.6%	3.9%	13.4%	10.1%	100.0%
Planning funds and state block grants, 1982–1998						$754.1
Total						$22,064.1

tax" legislation. As early as 1978, efforts were made to lift the prohibition of head tax legislation; however, this was partnered with the termination of any other federal funding for larger airports.

In 1990, Secretary of Transportation Samuel R. Skinner embarked on creating legislation to effectively remove the antihead tax without eliminating other current levels of federal funding. In the spring of 1990, Secretary Skinner asked Congress to enact legislation that would allow airports to impose **passenger facility charges (PFCs)** of up to $3 per passenger as part of the extension of the Airport and Airway Improvement Act programs. Having identified airport capital improvement needs of $50 billion over the next 5 years, the Airport Operators Council International (AOCI) and the American Association of Airport Executives (AAAE) mounted a major legislative campaign for PFCs.

In November 1990 Congress passed legislation authorizing PFCs as part of the **Aviation Safety and Capacity Expansion Act of 1990.** Some of the important provisions under the act were as follows:

1. The airport operator may propose collecting $1, $2, or $3 per enplaned passenger, domestic or foreign. No intermediate amounts (e.g., $2.50) are permitted.

2. PFCs will be collected by the air carriers.

3. PFCs are limited to no more than two charges on each leg of a round trip at airports at which passengers enplaned an aircraft.

4. PFC revenue must be spent at the designated airport controlled by the same body that imposes the fee.

5. The revenue from PFCs may be used to finance only the allowable costs of any approved projects.

6. Revenues from PFCs can be used for airport planning and development projects eligible for AIP funding. In addition, PFC revenue may be used for the preparation of noise compatibility plans and measures.

7. The legislation requires that AIP funds apportioned to a large or medium hub airport be reduced if a PFC is imposed at that airport.

As of December 2002, 309 airports collected PFCs totaling over $10.9 billion in actual charges collected. Over 1,000 airport projects have been accepted for participation in the program, with approved collection levels totaling over $30 billion.

As a result, the PFC program has been successful in allowing the largest hub airports to achieve necessary levels of improvement funding by drawing less from AIP, thereby providing more AIP funds to be allocated to smaller airports with fewer levels of enplaned passengers.

Military Airport Program (MAP)

Also authorized under the Aviation Safety and Capacity Expansion Act was the **Military Airport Program (MAP).** The MAP is a special set-aside (currently 4 percent) of the discretionary portion of the AIP to be used for capacity- and/or conversion-related projects at current and former military airports. The MAP allows the secretary of transportation to fund capital development at current or former military airports that have been designated as civil commercial or reliever airports in the NPIAS. Specifically, the criterion requires that approved projects at any MAP location must be able to reduce delays at an existing commercial service airport that has more than 20,000 hours of annual delays in commercial passenger aircraft takeoffs and landings. The designated airports remain eligible to participate in the program for 5 fiscal years after their initial designation as participants. A maximum of 12 airports (now 15) was initially allowed for participation in the MAP during any given year. Airports participating in the MAP support approved projects such as land acquisition, airfield construction and improvements, lighting and terminal developments, and other projects that facilitate the conversion of military air bases to civil aviation facilities.

The FAA is continuing to pursue a series of initiatives with the Department of Defense (DOD), individual states, and local governments for joint civil and military use of existing military airfields, as well as the conversion of military airfields being closed by the DOD. Over 50 military airfields have closed or are slated for closure since the MAP was authored. It is anticipated that approximately 40 of these airfields will be converted to civil airports. It has been estimated that to replicate the infrastructure at these airfields would require a total investment of nearly $50 billion. The small fraction of AIP funds that has been allocated for the MAP will facilitate the conversion of these airfields at a much lower level of investment.

The Aviation Security Improvement Act of 1990

In 1988, two independent deadly incidents involving sabotage of commercial aircraft introduced two new threats to the security of civil aviation. On December 7, 1988, a disgruntled former employee of Pacific Southwest Airlines boarded a PSA BAE 146-200 aircraft bound for San Francisco from Los Angeles, killed the flight crew and crashed the aircraft, killing a total of 43 people. The success of the killer's intentions was attributed in part by his ability to access the aircraft with a lethal weapon, despite existing security screening measures at Los Angeles International Airport. On December 21, Pan American Airlines Flight 103, a Boeing 747, exploded over Lockerbie, Scotland, killing all on board. The ensuing investigation revealed the cause of the explosion was a bomb disguised in a radio/cassette player that was stowed in checked luggage on the aircraft. The bomb was originally loaded onto an Air Malta Aircraft in Malta and eventually transferred onto flight 103 from another Pan American aircraft in London. The "passenger" associated with the explosive did not board Pan Am 103. His only intention was to perform this act of terrorism.

As a result of these incidents, President George H. Bush established the President's Commission on Aviation Security and Terrorism to assess the overall effectiveness of the U.S. Civil Aviation Security System. The outcome from the commission's report was the basis for establishing the Aviation Security Improvement Act of 1990.

The 1990 Aviation Security Improvement Act directed the FAA to accelerate explosives detection research and development. As a result, the FAA created a security research and development program intended to defeat the threat of terrorism and criminal acts targeting aviation.

The Airport and Airway Safety, Capacity, Noise Improvement, and Intermodal Transportation Act of 1992

The **Airport and Airway Safety, Capacity, Noise Improvement, and Intermodal Transportation Act of 1992** authorized the extension of AIP at a funding level of $2.1 billion through 1993. The act also included a number of changes in AIP funding. The primary changes included the expanded eligibility of development under the MAP; eligibility for relocation of air traffic control towers and navigational aids (including radar) if they impede other projects funded under the AIP; the eligibility of land, paving, drainage, aircraft deicing equipment, and structures for centralized aircraft deicing areas; and to comply with the Americans with Disabilities Act of 1990, the Clean Air Act, and the Federal Water Pollution Control Act. The act also increased the number of states that may participate in the State Block Grant Program from three to seven and extended that program through 1996.

The AIP Temporary Extension Act of 1994

The **AIP Temporary Extension Act of 1994** extended the authorization for AIP funding through June 1994. It provided that the minimum amount to be apportioned to a primary airport on the basis of passenger enplanements would be $500,000. The act also made modifications to the percentage of AIP funds that must be set aside for reliever airports (reduced from 10 percent to 5 percent); for commercial service, nonprimary airports (reduced from 2.5 percent to 1.5 percent); and for system planning projects (increased from 0.5 percent to 7.5 percent). Eligibility for terminal development was expanded to allow the use of discretionary funds at reliever airports and primary airports enplaning less that 0.05 percent of annual national enplanements.

The Federal Aviation Administration Authorization Act of 1994

The **Federal Aviation Administration Authorization Act of 1994** extended AIP funding through September 1996. Significant changes to AIP included increasing the number of airports that can be designated in the MAP from 12 to 15, but required that FAA find that projects at newly designated airports will reduce delays at airports with 20,000 hours of delay or more, expanded eligibility to include universal access control and explosives detection security devices, and required a number of actions by FAA and airport sponsors regarding airport rates and charges and airport revenue diversion.

The Federal Aviation Reauthorization Act of 1996

The **Federal Aviation Reauthorization Act of 1996** extended AIP through September 1998. Various changes were made to the formula computation of primary and cargo entitlements, state apportionment, and discretionary set-asides. Specifically, under primary airport entitlements, the formula was adjusted by changing the credit for the number of enplaning passengers over 500,000 from $0.65 to $0.65 for the passengers from 500,000 up to 1 million and $0.50 for each passenger over 1 million. Cargo entitlements were decreased from 3.5 percent of AIP to 2.5 percent of AIP. The previous cap of 44 percent of AIP for primary and cargo entitlements was removed.

State apportionments were increased from 12 percent of AIP to 18.5 percent, with the previous set-asides for reliever and nonprimary commercial service airports removed. The eligibility for use of state apportionments was expanded to include nonprimary commercial service airports. The system planning set-aside was also eliminated.

The noise and MAP set-aside computations were also changed from 12.5 percent and 2.5 percent of total AIP, respectively, to 31 percent and 4 percent of the discretionary fund. In addition, previously there was a minimum level of $325 million for the discretionary fund after subtraction of the various apportioned funds and set-asides. The new act changed the minimum level to $148 million over the payments necessary for letters of intent payments (for letters of intent issued prior to January 1, 1996) from the discretionary fund.

Three new pilot programs for innovative financing techniques, pavement maintenance, and privatization of airports were added to the program. Other changes included changes to the MAP in the number of airports under the program, criteria for selection, project eligibility, and permission to extend MAP participants for an additional 5-year period. The state block grant program was formally adopted by removing the designation of "pilot" and the number of participant states was increased from seven to eight states in 1997 and to nine states in 1998.

The act also aligned PFC and AIP to permit both to be used for funding projects to comply with federal mandates and to relocate navigational aids and air traffic control towers. These relocations are eligible only when needed in conjunction with approved airport development using AIP or PFC funding. Finally, new provisions for revenue diversion enforcement were added to FAA's authority.

Airports in the twenty-first century: From peacetime prosperity to terror insecurity

The years following the 1996 reauthorization of AIP were a time of unparalleled economic prosperity in the United States. The information technology revolution of the late 1990s coincided with record levels of commercial aviation activity. Major carriers experienced exponential growth in revenues and bottom line profits. Much of these revenues was reinvested by the airlines in the form of increased numbers of aircraft, increased markets, and increased service frequencies. Much of the new aircraft fleet being put into service was newly designed regional jets, which were utilized by airlines to provide higher-frequency service between major markets.

The rapid expansion of airline activity exceeded the pace of expansion and modernization of air traffic control systems and many of the airports that served the airlines. As a result, the commercial aviation industry experienced record levels of congestion and delay. Much of the congestion was concentrated at the largest "hub" airports and airports in major metropolitan areas (see Table 3-4).

Ironically, while many of the larger airports in the United States were experiencing record levels of congestion, many of the commercial service airports serving smaller communities were having difficulty maintaining current levels of service, particularly in the form of nonstop service to more than one or two destinations, because the strategy of many major air carriers was to consolidate operations toward feeding their individual hubs. As a result, the disparity between the needs of airports became extreme, with the larger airports struggling for funding to accommodate increased demands and smaller airports struggling for funding to attract much needed air service for their communities.

AIR-21: The Wendell H. Ford Aviation Investment and Reform Act for the Twenty-First Century

A lack of sufficient funding from current AIP provisions was cited as one of the root causes of the slow pace of airport construction and development and air traffic control modernization. In response, Congress began developing legislation to increase funding levels and encourage infrastructure enhancements.

Table 3-4. Most Congested Airports, 2000

Airport	Total Flights	Percentage Delayed	No. of Delays
Newark	463,000	7.89	36,553
LaGuardia	368,311	7.73	28,474
O'Hare	897,290	5.48	49,202
San Francisco	441,606	4.79	21,187
JFK	355,677	3.80	13,547
Atlanta	909,840	3.59	32,737
Philadelphia	480,279	3.02	14,516
Boston	502,822	2.98	14,989
Phoenix	570,788	2.08	11,919
Detroit	559,509	2.05	11,522

Source: Federal Aviation Administration.

The resulting legislation, the **Wendell H. Ford Aviation Investment and Reform Act for the Twenty-First Century,** known as **AIR-21,** was signed into law on April 5, 2000. AIR-21 increased annual levels of funding for aviation investments by $10 billion, with most of the funding to be appropriated toward air traffic control modernizations and much needed airport construction and improvement projects. In addition, AIR-21 provided guaranteed funding for aviation projects through legislative point-of-order provisions, effectively protecting the full investment of aviation taxes and user fees into aviation improvements. The total authorized funding for federal aviation programs for the 3-year appropriation of the act totaled nearly $40 billion, with $33 billion of the funding to be guaranteed from the aviation trust fund.

Funding provided through AIR-21 was designed to assist both the larger congested and smaller underutilized airports. For larger airports, minimum levels of annual AIP funding were doubled from $500,000 to $1 million per airport, and the maximum amount of annual funding for a large airport was increased from $22 million to $26 million. For smaller airports, minimum AIP funding levels were also doubled. In addition, funds were guaranteed for improvements at general aviation and reliever airports.

In addition, AIR-21 increased the cap on passenger facility charges, which were increased from $3.00 to $4.50. This increase was intended to benefit both larger and smaller airports. For larger airports, funds generated from increased PFC charges would provide significant resources for airport improvements without drawing from AIP funds. As a result, more AIP resources could be allocated to the smaller airports that draw less funding from PFCs because of their relatively low enplanement levels.

Upon the passage of AIR-21, many airports embarked on plans for major capital improvement projects, totaling several billions of dollars in planned expenditures (see Table 3-5).

The Wendell Ford Aviation Reform Act was authorized for the fiscal years 2001 through 2003, a period projected to enjoy continued increased growth in demand for air travel and economic activity in the United States.

A sharp downturn in the U.S. economy beginning in late 2000, spurred in part by the "burst of the Internet industry bubble," resulted in the commencement of financially troubled times for many of the nation's largest air carriers. This, in turn, invoked concerns about the expansion plans of airports. Initially, however, the focus of the aviation industry in general, and airports in particular, was to improve and modernize the system to reduce congestion and improve system efficiency.

The Aviation and Transportation Security Act of 2001

Even though issues concerning the security of the civil aviation system in general, and airports in particular, have been recognized and addressed with various levels of intensity since the early days of civil aviation, no single event in history did more to affect how the civil aviation system operates with respect to ensuring a secure travel environment than the terrorists attacks on the United States on September 11, 2001 (Fig. 3-9).

Between the hours of 8:00 and 9:00 on the Tuesday morning of September 11, 2001, four commercial airliners, departing from three major U.S. airports, were hijacked and subsequently used in suicide attack missions to destroy major

Table 3-5. Major Airport Expansion Projects, 1998–2003

Project	Project Budget, $billions
Chicago O'Hare runway reconfiguration/expansion	6.6
Hartsfield Atlanta runway/terminal expansion	5.4
Newark International Airport expansion	3.8
Washington Dulles International Airport	3.4
Seattle-Tacoma International Airport runway expansion	3.3
Minneapolis-St. Paul International Airport	3.1
Dallas/Fort Worth Airfield terminal expansion	2.6
San Francisco International Airport expansion	2.4
Las Vegas-McCarran International Airport	2.0
Baltimore-Washington International Airport	1.8
Cleveland Hopkins International Airport	1.1

Figure 3-9. *United Airlines Flight 175 crashes into the south tower of the World Trade Center, September 11, 2001.* (Photo courtesy www.cnn.com.)

landmarks in New York City and Washington D.C. The hijacking of American Airlines Flight 11 and United Airlines Flight 175, both Boeing 767 aircraft that departed Boston's Logan International Airport, were flown by suicide hijackers into the two 110-story towers of New York's World Trade Center, causing the eventual collapse of the two towers and surrounding buildings, resulting in the deaths of nearly 3,000 people and causing billions of dollars of structural damage to New York's financial district. Hijacked American Airlines Flight 77, a Boeing 757 that departed Washington D.C.'s Dulles International Airport, was flown into the side of the Pentagon, headquarters of the U.S. Department of Defense, killing nearly 300 people. The final aircraft to be hijacked, United Airlines Flight 93, a Boeing 757 that departed Newark International Airport, apparently targeted to attack a landmark in Washington, D.C., perhaps the

White House or the U.S. Capitol Building, crashed in an open field in Shankes-ville, Pennsylvania, after passengers on board the aircraft, receiving news of the attacks on the World Trade Center while talking on their cellular phones, attempted to combat the hijackers and recover the aircraft. The September 11, 2001, suicide hijackings marked the single largest attack and resulting number of fatalities involving commercial airlines in the history of aviation, and in fact marked one of the deadliest days on United States soil in history.

As governmental administrations became aware of the events that were unfold-ing on September 11, the FAA ordered a complete shutdown of the civil aviation system, including both commercial and general aviation activity, directing all aircraft currently in flight to land at the nearest available airport, and all aircraft on the ground to cancel all activity until further notice. All aircraft outside U.S. airspace were prohibited from entering the United States, forcing hundreds of aircraft inbound for U.S. cities from overseas to land in Canada or Mexico, or return to their originating locations (Fig. 3-10). By noon on September 11, there were zero civilian aircraft in the air over the United States, marking the first time in history that the FAA had completely shut down civil aviation.

Initial investigations attempting to identify the methods that were employed by the suicide hijackers to carry out their mission identified the following:

1. Nineteen hijackers, later found to be associated with the Al-Qaida terrorist organization, boarded aircraft as ticketed passengers at Boston Logan International Airport, Newark International Airport, and Washington Dulles International Airport. It was also determined that at least two of the hijackers initially boarded a flight to Boston Logan Airport as ticketed passengers at the Portland, Maine, International Airport, to transfer onto American Airlines Flight 11.

2. Hijackers used knives and box cutters to attack passengers and flight crew, with the intention of overtaking control of the aircraft.

3. Several of the hijackers received flight training in preparation for their attack mission. In addition, geographic identification of landmarks was performed prior to the attack to aid in direct navigation to their intended targets.

4. An automobile owned by one of the hijackers was found in the parking lot of Boston Logan International Airport. Inside the automobile was a pass allowing access to the aircraft apron at the airport.

5. A search of other commercial aircraft immediately after the attack revealed knives and box cutters found in the seat backs of at least two other aircraft at Boston Logan Airport as well as at the Hartsfield Atlanta International Airport.

6. Suspects thought to be accessories to the September 11 attacks were detained in New York's LaGuardia and John F. Kennedy Airports with uni-forms and credentials belonging to American Airlines crew members.

Figure 3-10. *Aircraft inbound to the United States grounded at Halifax International Airport, September 12, 2001.* (Courtesy Halifax International Airport.)

The initial investigations revealed suspicion of:

- Hostile sabotage of aircraft in flight via unlawful entrance to the cockpit using nonfirearm weapons.
- Planting of weapons on aircraft prior to hijacker boarding.
- Significant/worldwide plans of attacks.
- Further attacks using knowledge of commercial and general aviation operations.

From a security standpoint, the attacks of September 11, 2001, were the largest infiltration of the United States Civil Aviation Systems through multiple breaches of aviation security.

Immediately following the initial investigations a series of emergency security directives were imposed by the federal government, some affecting aircraft operations, and others specifically targeting airport operations.

Mandatory aircraft operations directives included modifications to aircraft, including the fortification of cockpit doors to deny access from the cabin during flight, mandatory pre- and postflight security inspection procedures, and absolute strict adherence to identification verification of all crew and other employees boarding the aircraft. In addition, the federal air marshal program, a program which was initiated in the 1970s to protect against hijackings but had over time been significantly reduced, was expanded in total force to include use of federal air marshals on domestic flights.

At airports the following emergency directives were implemented:

- Passengers were banned from carrying knives, box cutters, and any other potential nonfirearm weapons onto aircraft.
- Only ticketed passengers were allowed to proceed through airport security screening checkpoints within airport terminals.
- All curbside check-in facilities were closed.
- All automobile parking facilities located within 300 feet of the airport terminal were ordered closed.
- National Guard troops were deployed at each of the airports serving commercial carriers upon reopening of civil aviation activity to provide a presence of enhanced security for passengers.

While these emergency directives were implemented, the U.S. Congress directed itself to develop formal legislation to address the issue of aviation security. Drawing upon the knowledge and experiences of previous security threats and incidents, and the resulting legislation, recommendations, and policies that had been implemented with varying levels of effectiveness, and the new threats of suicide hijackings, Congress drafted the **Aviation and Transportation Security Act (ATSA) of 2001.** Stating that the legislation offers "permanent and aggressive steps to improve the security of our airways," President George W. Bush signed the ATSA into law on November 19, 2001 (Fig. 3-11).

The fundamental tenet of the law was the establishment of a federal agency tasked with the goal of ensuring the security of the nation's transportation systems. As such, the **Transportation Security Administration (TSA)** was established upon the signing of the ATSA. In addition, the ATSA prescribed a

Figure 3-11. *President George W. Bush at the signing of the Aviation and Transportation Security Act at Ronald Reagan Washington International Airport, November 19, 2001.* (Source:TSA.)

series of deadlines for security enhancements to be met by the newly established agency. These deadlines included:

November 19, 2001   All airport and airline employees with access to security-sensitive areas must undergo new federal background checks before receiving access clearance.

January 18, 2002   All checked baggage in U.S. airports must be screened by either explosive detection systems, passenger bag matching, manual searches, canine units, or other approved means.

February 17, 2002   The TSA is to officially assume all civil aviation functions from FAA.

November 21, 2002   All passengers and carry-on baggage must be screened by TSA-employed screening staff at the nation's 429 largest commercial air carrier airports (in terms of passenger enplanements).

December 31, 2002   All checked-in baggage must be screened by use of certified explosive detection equipment by TSA-employed screening staff at the nation's 429 largest commercial air carrier airports (in terms of passenger enplanements).

To meet these deadlines, the TSA invested over $5 billion toward the hiring of more than 50,000 federally employed airport passenger- and baggage-screening staff, administrative staff, and equipment necessary to accomplish the required goals of the ATSA while maintaining a system that can still provide the efficient travel of passengers through the national aviation system.

Over time, several of the emergency directives implemented since September 11, 2001, were lifted. The National Guard ceased their airport presence in May 2002. Curbside check-in facilities were reopened, and prohibition against automobile parking in designated spaces within 300 feet of airline terminals were lifted. As of 2010, only ticketed passengers were allowed through airport terminal passenger security screening checkpoints, and although the specific list of prohibited items continued to change, many sharp and heavy items such as knives, box cutters, baseball bats, and bricks remained prohibited from being transported in passenger carry-on baggage.

To fund the Transportation Security Administration, the ATSA authorized a surcharge on air carrier passenger tickets of $2.50 per flight segment, with a maximum charge of $10 per round-trip itinerary.

Homeland Security Act of 2002

On November 25, 2002, the Homeland Security Act of 2002 was signed into law. This act established the Department of Homeland Security (DHS) in an effort to coordinate the work of several agencies responsible for protecting the nation's homeland under one cabinet-level department. These agencies included Customs and Border Protection, Immigration and Customs Enforcement, the Federal Emergency Management Agency (FEMA), the Secret Service, the Coast Guard, and the TSA.

Legislation and operations with respect to airport security continues to evolve, in part to adapt to the ever-changing threats to the traveling public. More detailed information regarding the historical, current, and future airport security environment may be found in Chap. 8 of this text.

Vision 100—Century of Aviation Reauthorization Act

The history of civil aviation in general, and on airport development in particular, is one that has been dynamic since its beginnings in the early twentieth century. The history of civil aviation in the early part of the twenty-first century, however, may be characterized as being more dynamic, and certainly more volatile than in the entire twentieth century. The beginning of the year 2000 experienced record levels of enplanements, aircraft operations, airline profits, and airport expansion plans. By the beginning of 2003, the aviation industry was suffering from the aftereffects of extreme acts of terrorism, a down economy, record financial losses, and bankruptcies by a number of commercial air carriers, resulting in a new set of challenges for airports.

As the AIR-21 legislation, authorized during a period of relative prosperity for civil aviation, was due to expire on September 30, 2003, one of the initial tasks of the 108th Congress was to provide reauthorization legislation to reassess the needs of the aviation industry, and allocate funding to meet those needs. In response to these needs, the Bush administration signed into law the "Vision 100—Century of Aviation Reauthorization Act" on December 12, 2003. The Vision 100 Act is best known for reauthorizing an AIP funding of $3.4 billion in fiscal year 2004, increasing the funding to $3.7 billion in fiscal year 2007.

Vision 100 also established the Next Generation Air Transportation System Joint Planning and Development Office (JPDO), an organization within the FAA but that is closely integrated with several other federal organizations, including the DOD, the DOT, the DHS, the Department of Commerce, the White House Office of Science and Technology, and NASA. The mission of the JPDO was to create and carry out an integrated plan for a "Next-Generation Air Transportation System," known as "NextGen," which would meet the potential demand for air travel in the year 2025. As will be discussed in further detail in Chap. 5 of this text, NextGen focuses on the implementation of twenty-first-century digital and satellite-based communications and navigation technologies to provide for more efficient management of air traffic within the national airspace system. Because of the enormity and complexity of the task of completely reforming the nation's air transportation system, the JPDO was formed to coordinate the efforts of all government organizations with interests and influence over the air transportation system in the United States. As of 2010, the JPDO continues to coordinate efforts in the development, testing, and incremental implementation of new air traffic management technologies. More information regarding the JPDO may be found at http://www.jpdo.gov.

The years following the Vision 100 Act continued to be tumultuous for the aviation industry. Competition from a growing sector of commercial service carriers, known as "low-cost carriers," continued to drive down airfares, creating hardships not only for older "legacy" carriers, who had higher operating costs, but also for airports, as funding through the AIP program became scarce. Under the current model, funding into the Aviation Trust fund began to decrease just as the needs for AIP funding to improve airport infrastructure began to increase again. Because of this, significant debate occurred over how to reauthorize the AIP program upon the expiration of the Vision 100 Act in 2007.

As of 2010, the debate over AIP reauthorization continues, without resolution. As such, as of 2010, the AIP program continues to operate under a "continuing resolution," which has limited the increase in AIP funding, any discussed increases in PFC authorization, and other policies to further invest in the aviation system.

In March 2010, the U.S. Senate unanimously passed its version of the bill known as the "FAA Reauthorization Act of 2010"; a similar version of the act was debated in the U.S. House of Representatives. The act focuses on enhancing aviation safety in the wake of a number of fatal accidents involving regional commercial service carriers, and on increasing airport capacity, in part by improving funding by increasing the allowable PFC that airports may charge. As of the end of 2010, however, the act had not been passed into law, leaving future aviation legislation still in question.

Concluding remarks

As has been witnessed over the history of civil aviation, current events and economic climates, as well as social issues, tend to have significant impacts on the continuous development of civil aviation, particularly when it comes to its legislative history. In the later years of the twenty-first century's first decade, four fundamental issues affecting civil aviation were in the forefront: economic recession in the United States and other industrialized nations, attention paid to environmental issues, and the issues of aviation security, and those of safety. Economic downturns have resulted in the downsizing and consolidation of commercial air carriers and a sharp decline in general aviation activity. These issues have affected the financial health of the nation's airports and aviation system in general. Despite the economic challenges that exist, there has been a push to improve the environmental sustainability of aviation. The initiative includes making airports more energy efficient. It is expected that airport managers and planners will be developing ways to further design their facilities to minimize energy usage and seek operating methods to reduce the impact of airport activity on the environment, all while needing to address the ever-changing threats to security, and seeking to constantly improving the safety of airport operations.

Finally, in the distant future, there will always be the need to further increase system capacity and efficiency, as it is clear that upon the revival of the world's economies, the demand for air travel will increase, and the world's aviation and airport systems must be prepared to accommodate the demand. It is thought that the development of new aviation system technologies will significantly contribute to enhancing system capacity and efficiency, including technologies that will be present at every airport.

Whatever the future of civil aviation legislation brings, it will no doubt be developed on the basis of the rich history that is the first 100+ years of civil aviation. As such, it is the historical formation of legislation based on technological, economic, and political events, as well as concerns for system efficiency, capacity, safety, and security that must be understood to best manage and create legislation for the future of civil aviation.

Key acts of legislation

1925	Kelly Act
1926	Air Commerce Act
1938	Civil Aeronautics Act
1946	Federal Airport Act
1957	Airways Modernization Act
1958	Federal Aviation Act
1966	Department of Transportation Act
1970	Airport and Airway Development Act/Airport and Airway Revenue Act
1976	Airport and Airway Development Act Amendments
1976	Air Cargo Deregulation Act
1978	Airline Deregulation Act
1979	Aviation Safety and Noise Abatement Act
1982	Airport and Airway Improvement Act
1982	Continuing Appropriations Act
1983	Surface Transportation Assistance Act
1987	Airport and Airway Safety and Capacity Expansion Act
1990	Aviation Safety and Capacity Expansion Act
1992	Airport and Airway Safety, Capacity, Noise Improvement, and Intermodal Transportation Act
1994	AIP Temporary Expansion Act
1994	Federal Aviation Administration Authorization Act
1996	Federal Aviation Administration Reauthorization Act
2000	AIR-21: Wendell H. Ford Aviation Investment and Reform Act for the Twenty-First Century
2001	Aviation and Transportation Security Act
2002	Homeland Security Act
2003	Vision 100—Century of Aviation Reauthorization Act

Key organizations and administrations

Bureau of Air Commerce

Works Progress Administration (WPA)

Civil Aeronautics Authority

Civil Aeronautics Board (CAB)

Civil Aeronautics Administration (CAA)

Airways Modernization Board

Federal Aviation Agency (FAA)

Department of Transportation (DOT)

National Transportation Safety Board (NTSB)

Transportation Security Administration (TSA)

Department of Homeland Security (DHS)

Joint Planning and Development Office (JPDO)

Key plans, programs, and policies

DLAND (Development of Landing Areas for National Defense)

NAP (National Airport Plan)

FAAP (Federal-Aid Airport Program)

Airport and Airway Trust Fund

PGP (Planning Grant Program)

ADAP (Airport Development Aid Program)

NASP (National Airport System Plan)

NAS (National Airspace System)

EAS (Essential Air Service Program)

AIP (Airport Improvement Program)

NPIAS (National Plan of Integrated Airport Systems)

DBE (Disadvantaged Business Enterprise) Program

PFC (Passenger Facility Charge)

MAP (Military Airport Program)

NextGen (Next-Generation Air Transportation System)

Questions for review and discussion

1. Who established the first airmail service in the United States? How long did it last?

2. What was the primary purpose of the Kelly Act?

3. What was the primary purpose of the Air Commerce Act of 1926?

4. How did the Bureau of Air Commerce become established?

5. When did the federal government first give financial support for the development of airports?

6. What was the overriding purpose of the Civil Aeronautics Act of 1938?

7. What was the difference between the Civil Aeronautics Board and the Civil Aeronautics Administration?

8. What was the function of DLAND?

9. What was the purpose of the Federal Airport Act of 1946?

10. What were some of the provisions that had to be adhered to before federal aid was granted under the Federal-Aid Airport Program?

11. What was the main concern of the aviation industry that led to the Federal Aviation Act of 1958?

12. What were the federal aviation administrator's major responsibilities under the 1958 act?

13. What was the purpose of the Department of Transportation Act of 1966?

14. How were revenues raised under the Airport and Airway Revenue Act of 1970?

15. What are some advantages of the user charge/trust fund approach?

16. What is the PGP? How does it work?

17. What were some of the important amendments under the Airport and Airway Development Act Amendments of 1976?

18. How did the Airline Deregulation Act of 1978 affect the airport system in the United States?

19. What is the "essential air service" program?

20. What is the purpose of the Aviation Safety and Noise Abatement Act of 1979?

21. What was the primary purpose of the Airport and Airway Safety and Capacity Expansion Act of 1987?

22. What were some of the features of the Aviation Safety and Capacity Act of 1990?

23. How did funding to airports change under AIR-21?

24. Why were Passenger Facility Charges (PFCs) finally approved by Congress? List some of the provisions under the Aviation and Capacity and Expansion Act of 1990 pertaining to PFCs.

25. What is the State Block Grant Pilot Program?

26. The Military Airport Program?

27. How did the tremendous growth in air travel during the 1980s and 1990s affect the airport system?

28. What are some of the biggest problems faced by the airport system during the early part of the twenty-first century?

29. What was the purpose of the ATSA?

30. How has airport security legislation changed since the events of September 11, 2001?

31. What was the purpose of forming the Department of Homeland Security?

32. What was the purpose of the JPDO?

33. What is NextGen?

34. Why has there been so much debate over reauthorization since the expiration of the Vision 100 Act in 2007?

Suggested readings

Air Commerce Act of 1926, Public Law 254, 69th Congress, May 20, 1926.

Airport and Airway Development Act of 1970 (Title 1) and the Airport and Airway Revenue Act of 1970 (Title II), Public Law 258, 91st Congress, May 21, 1970.

Airport and Airway Development Act Amendments of 1976, Public Law 353, 94th Congress, July 21, 1976.

Airport and Airway Improvement of 1982, Public Law 248, 97th Congress, September 15, 1982.

Airport and Airway Safety and Capacity Expansion Act of 1987, Public Law 223, 100th Congress, December 30, 1987.

Arey, Charles K. *The Airport,* New York: Macmillan, 1943.

Aviation Safety and Capacity Expansion Act of 1990, Public Law 508, 101st Congress, November 8, 1990.

Briddon, Arnold E., Ellmore A. Champie, and Peter A. Marraine. *FAA Historical Fact Book: A Chronology 1926–1971.* DOT/FAA Office of Information Services. Washington, D.C.: U.S. Government Printing Office, 1974.

Civil Aeronautics Act of 1938, Public Law 706, 76th Congress, June 23, 1938.

Department of Transportation Act of 1966, Public Law 670, 89th Congress, October 15, 1966.

Department of Transportation, Thirteenth Annual Report of Accomplishments under the Airport Improvement Program—Fiscal Year 1994, October 1995.

Federal Airport Act of 1946, Public Law 377, 79th Congress, May 13, 1946.

Federal Aviation Act of 1958, Public Law 726, 85th Congress, August 23, 1958.

Frederick, John H. *Airport Management.* Chicago: Richard D. Irwin, 1949.

Kelly Air Mail Act of 1925, Public Law 359, 68th Congress, February 2, 1925.

National Plan of Integrated Airport Systems (NPIAS) 2009–2013, Washington, D.C.: FAA, 2009.

Richmond, S. *Regulation and Competition in Air Transportation.* New York; Columbia University Press, 1962.

Sixteenth Annual Report of Accomplishments under the Airport Improvement Program. FY 1997, Washington D.C.: FAA, May 1999.

Smith, Donald I., John D. Odegard, and William Shea. *Airport Planning and Management.* Belmont, CA.: Wadsworth, 1984.

VISION 100—Century of Aviation Reauthorization Act of 2003 Public Law 108-176, 108th Congress, December 12, 2003.

Part II

Airport operations management

4

The airfield

Outline

- The components of an airport
- The airfield
  - Runways
    - Runway configuration
    - Runway designation
    - Runway length and width
    - Runway pavements
    - Runway markings
    - Runway safety areas, protection zones, and "imaginary" obstruction surfaces
      - A runway's imaginary surfaces
  - Taxiways
    - Taxiway markings
  - Other airfield markings
  - Other airfield areas
  - Airfield signage
- Airfield lighting
  - Runway lighting
  - Taxiway lighting
  - Other airfield lighting
- Navigational aids (NAVAIDS) located on airfields
  - Nondirectional radio beacons (NDB)
  - Very-high-frequency omnidirectional range radio beacons (VOR)
  - Instrument Landing Systems (ILS)

- Air traffic control and surveillance facilities located on the airfield
  - Air traffic control towers
  - Airport surveillance radar
  - Airport surface detection equipment
- Weather reporting facilities located on airfields
  - Wind indicators
- Security infrastructure on airfields

Objectives

The objectives of this section are to educate the reader with information to:

- Identify the various facilities located on an airport's airfield.
- Discuss the specifications and types of airport runways.
- Understand the importance of runway orientation.
- Identify an airport's reference code.
- Be familiar with airfield lighting, signage, and markings.
- Describe the various navigational aids that exist on airfields.
- Describe the infrastructure existing to increase the security of the airfield.

The components of an airport

An airport is a complex transportation facility, designed to serve aircraft, passengers, cargo, and surface vehicles. Each of these users is served by different components of an airport. The components of an airport are typically placed into two categories.

The **airside** of an airport is planned and managed to accommodate the movement of aircraft around the airport as well as to and from the air. The airside components of an airport are further categorized as being part of the local airspace or the airfield. The airport's **airfield** component includes all the facilities located on the physical property of the airport to facilitate aircraft operations. The **airspace** surrounding an airport is simply the area, off the ground, surrounding the airport, where aircraft maneuver, after takeoff, prior to landing, or even merely to pass through on the way to another airport.

The **landside** components of an airport are planned and managed to accommodate the movement of ground-based vehicles, passengers, and cargo. These components are further categorized to reflect the specific users being served. The airport **terminal** component is primarily designed to facilitate the movement of passengers and luggage from the landside to aircraft on the airside. The airport's **ground access** component accommodates the movement of ground-based vehicles to and from the surrounding metropolitan area, as well as between the various buildings found on the airport property.

No matter what the size or category of an airport, each of the above components is necessary to properly move people from one metropolitan area to another using air transportation. The components of an airport are planned in a manner that allows for the proper "flow" from one component to another. An example of a typical "flow" between components is illustrated in Fig. 4-1. Figure 4-1 further identifies some of the facilities located on the airfield and ground access components of the airport.

The airfield

The area and facilities on the property of an airport that facilitate the movement of aircraft are said to be part of the airport's airfield. The airfield of any given airport is planned, designed, and managed to specifically accommodate

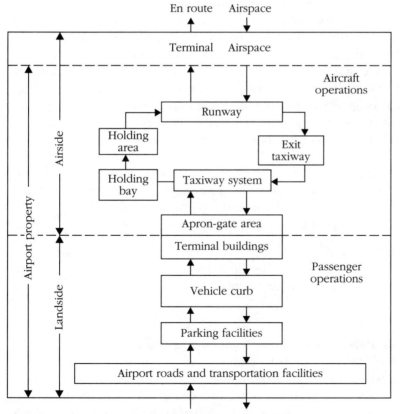

Figure 4-1. *The components of an airport.*

the volume and type of aircraft that utilize the airport. As one would expect, the planning and management of airfields at small general aviation airports is very different from that of large commercial service airports, although many of the fundamental principles that govern the planning and management of each type of airfield are very similar.

The most prominent facilities that are located on an airport's airfield are runways, taxiways, aircraft parking areas, navigational aids, lighting systems, signage, and markings. In addition, facilities to aid in the safe operation of the airport, such as **aircraft rescue and fire fighting (ARFF) facilities,** snow plowing and aircraft de-icing stations, and fuel facilities may be located on or closely near the airfield. The smallest of airports may have very simple airfield infrastructures, such as a single unlit runway with very minimal markings, no taxiways, and little in the way of signage or aircraft parking areas, whereas larger airport airfields may have complex systems of multiple runways and taxiways, various airfield lighting systems and navigational aids, and the highest levels of ARFF and other facilities. Particularly at airports with an operating control tower, the airfield is typically divided into the **movement area** and the **nonmovement area**. The movement area is the part of the airfield that is under the direct authority of the air traffic control tower for the movement of aircraft and ground vehicles. In fact, air traffic control's permission is required for any aircraft or vehicle to enter the movement area. Facilities in the movement area include the airfield's runways, most taxiways, and safety areas surrounding these primary aircraft movement facilities. The nonmovement areas include the ramps and taxi lanes located closest to aircraft parking areas. In these areas, aircraft and ground vehicles often move without direct routing instructions from air traffic control. While these areas are not so explicitly defined at "nontowered" or "uncontrolled" airports (i.e., those areas without an active air traffic control tower managing the movement of aircraft about the airfield), airport management and pilots alike often consider the runway and taxi environment of the airfield as a more sensitive area when it comes to the safe movement of vehicles.

Much of the information regarding the infrastructure and facilities located on the airfield of public-use airports in the United States may be found on the airports' **FAA Form 5010 - Master Record**.

Runways

Perhaps the single most important facility on the airfield's movement area is the **runway.** After all, without a properly planned and managed runway, desired aircraft would be unable to use the airport. Regulations regarding the management and planning of runway systems are some of the most comprehensive and strict in airport management. For example, strict design guidelines must be followed when planning runways, with particular criteria for the length, width, orientation (direction), configuration (of multiple runways), slope, and even pavement thickness of runways, as well as the immediate airfield area surrounding the runways to assure that there are no dangerous obstructions preventing the safe operation of aircraft. Runway operations are facilitated by systems of markings, lighting systems, and associated airfield signage that identify runways and provide directional guidance for aircraft taxiing, takeoff, approach, and landing. Strict regulations regarding the use of runways, including when and how the aircraft may use a runway for takeoff and landing, are imposed on airfield operations.

The design and operation of runways are determined in part by the type of aircraft using the runway. Runways designed to handle operations of propeller-driven aircraft weighing 12,500 pounds or less are known as *utility runways*. Runways that are not utility runways are designed to handle operations of aircraft greater than 12,500 pounds.

Runway configuration

When the Wright brothers made their first flight at Kitty Hawk in 1903, there were no runways to facilitate the flight. However, certain conditions existed during the flight that led directly to the orientation of today's runways. The Wright brothers knew that, since fixed-wing aircraft rely on airflow over the aircraft's wings to achieve flight, the appropriate direction to take off an aircraft was into whichever way the wind was blowing. This allows aircraft to achieve the desired amount of airflow over the wings with the least amount of ground speed and takeoff distance. Similarly, the safest direction in which to land an aircraft is also into the wind. As a result of this physical property of aircraft, airport runways are typically oriented into the prevailing winds of the area. While many airports have runways that are oriented in different directions, the runway(s) that is oriented into the prevailing winds is known as the **primary runway(s).**

Just as it is most appropriate for aircraft to take off and land into the wind, that is, with a *headwind,* it is least appropriate, and in fact sometimes highly unsafe, to land or take off with a wind blowing directly perpendicular to the direction of travel, that is, with a direct *crosswind*. Smaller, lighter, slower-moving aircraft tend to be much more sensitive to crosswinds than larger aircraft. As a result, airports that are located in areas with winds that blow from various directions at sufficient wind speeds and/or accommodate primarily smaller aircraft are also planned with runways oriented toward the most common crosswind directions. These runways are known as **crosswind runways.** The planning of primary and crosswind runways with respect to runway orientation is discussed in more detail in Chap. 11 of this text.

Although many airports have only one runway, airports that typically serve smaller aircraft tend to have additional runways in the form of crosswind runways (Fig. 4-2). Airports serving higher volumes of primarily larger aircraft tend to have additional runways in the form of **parallel primary runways** or simply "parallel runways" (Fig. 4-3). Airports which serve a high volume of both

Figure 4-2. *Flagler County Airport in Bunnell, Florida, has multiple crosswind runways to accommodate smaller aircraft in variable wind conditions.* (Photo courtesy Seth Young)

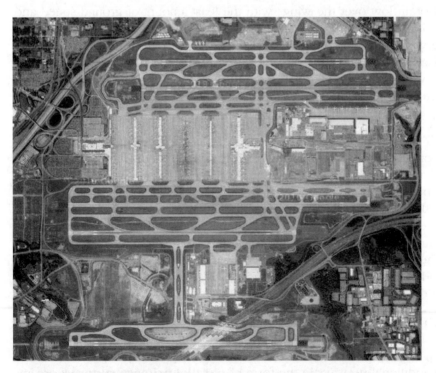

Figure 4-3. *Atlanta's Hartsfield International Airport has five primary parallel runways to accommodate high-volume large-aircraft operations.* (Courtesy Google maps)

larger and smaller aircraft operations might have both parallel and crosswind runways.

Runway designation

Runways are in fact designated by their orientation with respect to magnetic north. Runways are identified by their degrees from magnetic north, divided by 10, rounded to the nearest integer. For example, a runway oriented to the east, that is, 90 degrees from magnetic north, would be identified as runway 9. A northerly oriented runway is identified as runway 36. Often, the planning of runways is considered so that aircraft may also operate with headwinds when the winds at an airport blow from the opposite direction to that of the prevailing winds. When runways are planned in such a manner, the runway is identified by both of its possible operating directions. For example, a runway whose primary orientation is easterly but also may be used in a westerly direction (i.e., 270 degrees from magnetic north) would be identified as runway 9-27. The lower number is always identified first, regardless of which direction is actually the primary operating orientation. When an airfield has parallel runways, each runway designator is allocated a letter to identify whether it is the Left (L) or Right (R) runway when viewed from an approaching aircraft. For example, for two runways with an easterly runway orientation, the northern parallel runway would be designated 9L and the southern runway, 9R. If the two runways were operated in both east and west directions, the northern and southern runways would be designated 9L-27R and 9R-27L, respectively.

In the case of multiple-parallel runways, the standard procedure is to designate two runways in accordance with the standard described in the previous paragraph, and for additional pairs the designator itself is changed by using an adjacent designator. So, for example, if a new runway was built parallel to 9L-27R and 9R-27L, it would be designated 8-26 (if located to the north of the original runways) or 10-28 (if located to the south of the original runways). Less often, a set of three parallel runways will be designated L, R, and C (for center).

An example of multiple-parallel runways is found in Fig. 4-3. The northernmost parallel runways are designated 8L-26R and 8R-26L, the south parallel runways are designated 9L-27R and 9R-27L, and the new fifth parallel runway is designated 10-28.

It is interesting to note that in the United States, single-digit runway designators (1 through 9) are placed on runways without a leading zero, while internationally, ICAO standards dictate that a leading zero, for example, "runway 09-27", is used.

For planning purposes, runways are identified by the one or two allowable operating directions. For operating purposes, however, runways are identified only by the current direction of operations.

Runway length and width

Because aircraft require given minimum distances to accelerate for takeoff and to decelerate after landing, runways are planned with specific lengths to accommodate aircraft operations. Characteristics that determine the required length of a runway include the performance specifications of the runway's design aircraft and the prevailing atmospheric conditions. Specifically, the *maximum gross takeoff weight,* acceleration rate, and safe lift off velocity of aircraft are considered. In addition, the elevation above sea level (known as mean sea level or **MSL**) of the airport, along with the outside air temperature significantly affect required runway lengths. This is due to the fact that air at higher elevations and at higher temperatures is less dense that cooler air and air closer to sea level. The density of air is a significant determinant in the takeoff performance of aircraft.

Most air carrier jet aircraft require between 6,000 and 10,000 feet of runway length for takeoff at a typical airport located at sea level. Many smaller general aviation aircraft have the ability to utilize runways as short as 2,500 feet (or in some cases even shorter).

As with runway length, the width of a runway is determined by the design aircraft. Specifically, the wingspan of the largest aircraft performing 500 annual itinerant operations determines the width of a runway. Runway widths at public-use airports vary from 50 to 200 feet, whereas the most common runway width planned to accommodate commercial service air carrier operations is 150 feet.

Runway pavements

In 1903, the relatively light weight of the Wright brothers' first flyer allowed the aircraft, and all other aircraft of the time, the ability to operate on grass. Even today, many of the lighter aircraft in use have the ability to take off and land on any of the hundreds of grass runways that exist. However, with the creation of heavier aircraft, it became necessary to stabilize and strengthen the runway environment. Today, virtually all commercial service airports have at least one paved runway to accommodate the full fleet of commercial and general aviation aircraft.

The first paved runway was constructed in 1928 at the Ford Terminal in Dearborn, Michigan. During the next 5 years paved runways were constructed in Cheyenne, Wyoming; Glendale, California; Louisville, Kentucky; and Cincinnati, Ohio. By the middle of the 1930s, paved runways and airfields became popular at civilian as well as military airports. With the introduction of larger aircraft in the years following World War II, runway pavements became a necessity rather than a luxury. Today, the thickness of runway pavements ranges from 6 inches for runways serving lighter aircraft to over 3 feet for runways serving large commercial service aircraft.

Runways may be constructed of **flexible (asphalt)** or **rigid (concrete)** materials. Concrete, a rigid pavement that can remain useful for 20 to 40 years, is typically found at large commercial service airports and former military base airfields. Runways made of rigid pavements are typically constructed by aligning a series of concrete slabs connected by joints that allow for pavement contraction and expansion as a result of the loading of aircraft on the pavement surface, and as a result of changes in air temperature. Runways constructed from flexible pavement mixtures are typically found at most smaller airports. Flexible pavement runways are typically much less expensive to construct than rigid pavement runways. The life of asphalt runways typically lasts between 15 and 20 years, given proper design, construction, and maintenance.

The planning and management of runway pavements is a vital operation in itself. Careful pavement maintenance management of properly planned runways will result in many years of healthy use. Without proper management, however, runway pavements can prematurely fail, resulting in the inability to safely accommodate aircraft operations. Further details regarding pavement management are discussed in Chap. 6 of this text.

Runway markings

There are three types of markings for runways: visual (also known as "basic"), nonprecision instrument, and precision instrument. These marking types reflect the types of navigational aids associated with assisting aircraft on approach to land on the runway. A visual runway is intended solely for aircraft operations using visual approach procedures. A nonprecision instrument runway is one having an instrument approach procedure using air navigation facilities with only horizontal guidance or for which a *straight-in* nonprecision instrument approach procedure has been approved by the FAA. A precision instrument runway is one having an instrument approach procedure using a precision instrument landing system (e.g., ILS) or Precision Approach Radar (PAR) that provide both horizontal and vertical guidance to the runway.

Visual, nonprecision, and precision instrument runway markings include runway designators and centerlines. Nonprecision instrument runways also include runway threshold markings and aiming points (used be called fixed-distance markers) (Fig. 4-4). Threshold markings are also found on visual runways intended to accommodate international commercial operations. Aiming points are also found on visual runways of at least 4,000 feet in length and are used by jet aircraft. Precision instrument runways also include touchdown zone markers and side stripes (Fig. 4-5). All runway markings are painted in white.

Runway designators identify the name of the runway by the runway's orientation. The runway number is the whole number nearest one-tenth the magnetic azimuth of the centerline of the runway, measured clockwise from magnetic north. The letters differentiate among left (L), right (R), or center (C) parallel runways, as applicable.

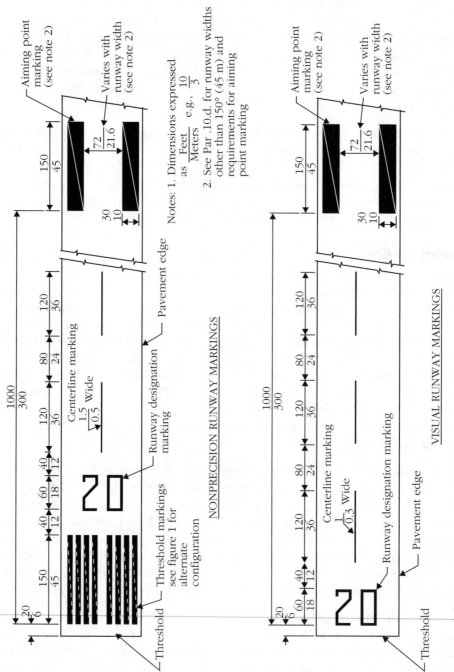

Figure 4-4. *Visual and nonprecision runway markings.* (Source: Horonjeff et al.)

110

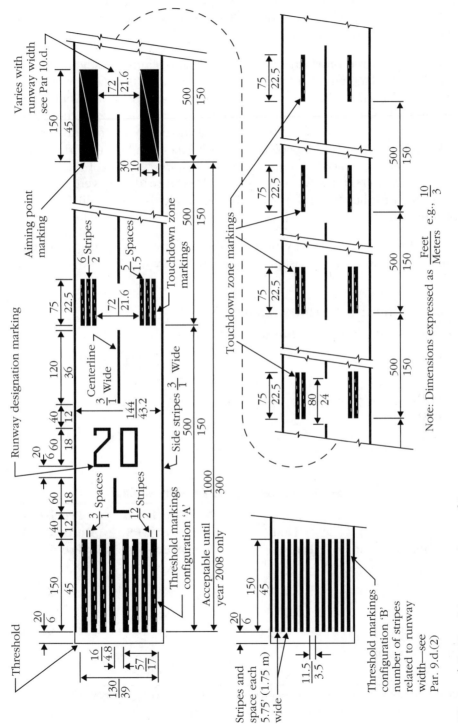

Figure 4-5. *Precision runway markings.* (Source: FAA AIM)

111

Runway centerlines identify the center of the runway and provide alignment guidance during takeoff and landings. The centerline consists of a line of uniformly spaced stripes and gaps.

Runway threshold markings help identify the beginning of the runway that is available for landing. In some instances, the landing threshold may be *relocated* or *displaced* up the runway from the actual beginning of pavement. Runway threshold markings consist of a number of stripes related to the width of the runway. Table 4-1 relates runway width to the number of runway threshold marking stripes. It should be noted that, as illustrated in Fig. 4-5, prior to 2008, runway threshold markings were allowed to simply consist of eight uniform stripes. Since 2008, runways with these traditional markings were required to be modified to the standard as described in Table 4-1. As of 2010, the traditional uniform threshold markings could still be found at several U.S. airports.

Table 4-2 provides a summary of the types of runway markings found on visual, nonprecision, and precision instrument runways.

Table 4-1. Number of Runway Threshold Stripes

Runway Width, ft (m)	Number of Stripes
60 (18)	4
75 (23)	6
100 (30)	8
150 (45)	12
200 (60)	16

Table 4-2. ILS Ceiling and Visibility Minima

Marking Element	Visual Runway	Nonprecision Runway/GPS Nonprecision	Precision Runway/GPS Precision
Designation	X	X	X
Centerline	X	X	X
Threshold marking	X[1]	X	X
Aiming point	X[2]	X[2]	X
Touchdown zone			X
Side stripes	X[3]	X[3]	X

[1] Only required on runways used, or intended to be used, by international commercial transport.

[2] On runways 4,000 feet (1,200 m) or longer used by jet aircraft.

[3] Used when the full pavement width may not be available as a runway.

Sometimes construction, maintenance, or other activities require the threshold to be relocated up the runway from the original threshold. This **relocated threshold** is marked by a *runway threshold bar.* The runway threshold bar is a 10-foot-wide white-painted stripe that extends across the width of the runway. The distance between the beginning of the runway pavement and the relocated threshold is marked by yellow-painted chevrons, which denote that the pavement is unusable for landing, takeoff, or taxiing of aircraft. This chevroned area is also known as a **blast pad**, as illustrated in Fig. 4-6.

A **displaced threshold** is also a threshold located at a point on the runway other than the designated beginning of the runway. Unlike a relocated threshold, a displaced threshold only reduces the available runway length for landing. The portion of the runway behind a displaced threshold is available for taxiing and takeoffs in either direction and landings from the opposite direction. A 10-foot-wide white threshold bar is located across the width of the runway at the displaced threshold. White arrows are located along the centerline in the area between the beginning of the runway and the displaced threshold.

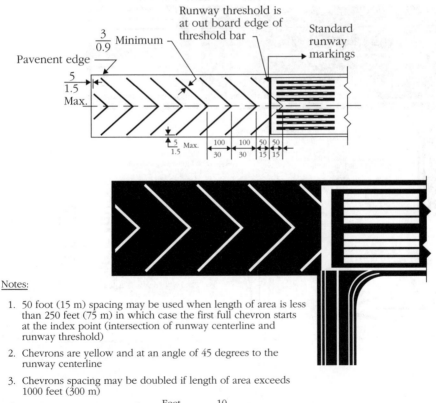

Notes:

1. 50 foot (15 m) spacing may be used when length of area is less than 250 feet (75 m) in which case the first full chevron starts at the index point (intersection of runway centerline and runway threshold)

2. Chevrons are yellow and at an angle of 45 degrees to the runway centerline

3. Chevrons spacing may be doubled if length of area exceeds 1000 feet (300 m)

4. Dimensions are expressed as $\frac{\text{Feet}}{\text{Meters}}$ e.g., $\frac{10}{3}$

Figure 4-6. *Relocated threshold markings.* (Source: FAA AIM)

White arrowheads are located across the width of the runway just prior to the threshold bar, as illustrated in Fig. 4-7.

Runway aiming points serve as visual aiming points for a landing aircraft. These two rectangular markings consist of a broad white stripe located on each side of the runway centerline and approximately 1,000 feet (300 m) from the landing threshold, that is, the beginning of the runway allowable for landing.

Runway touchdown zone markings identify the touchdown zone for landing operations. They are coded to provide distance information in 500-foot (150 m) increments for a distance of 2,500 feet from the threshold. These markings consist of groups of one, two, and three rectangular bars, symmetrically arranged in pairs about the runway centerline. For runways having touchdown zone markings at both ends, those pairs of markings that extend to within 900 feet (270 m) of the midpoint between the thresholds are eliminated.

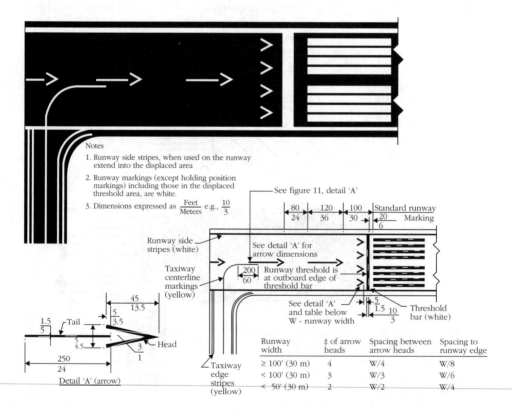

Figure 4-7. *Displaced threshold markings.* (Source: FAA AIM)

Runway side stripes delineate the edges of the runway. They provide a visual contrast between the runway and the abutting terrain or shoulders. Side stripes consist of continuous white stripes located on each side of the runway. Runway shoulder stripes may also be used to supplement runway side stripes to identify pavement areas contiguous to the runway sides that are not intended for use by aircraft. Runway shoulder stripes are yellow stripes marked at a 45-degree angle to the direction of the runway, upward in the direction of operation, from the threshold to the midpoint of the runway, as illustrated in Fig. 4-8.

Runway safety areas, protection zones, and "imaginary" obstruction surfaces

While the physical pavement is clearly the most visible and most utilized element of the runway environment, essential elements of the runway environment that are less visible, and in fact, quite invisible, exist for the safety and protection of all users of the airfield. These areas include runway shoulders and safety areas that provide clear areas for aircraft that overrun the side or end of a runway upon a landing or aborted takeoff, object-free zones and runway protection zones to protect both the aircraft and those on the ground from

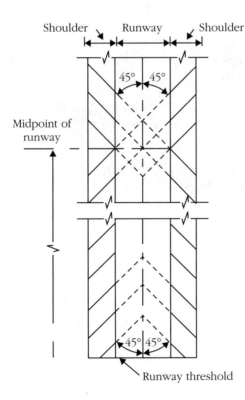

Figure 4-8. *Runway shoulder markings.* (Source: FAA AIM)

aircraft that may be departing from or approaching a runway at too shallow a descent or departure angle, and imaginary surfaces to protect runways from the encroachment of taller structures in the vicinity.

Runway shoulders   Similar to roads, **runway shoulders** provide for space between the runway itself and associated signage and lighting systems, as well as provide space between the pavement and non-treated ground to reduce the dispersion of rocks, dirt, and dust from jet blast, and accommodate the passage of maintenance and emergency vehicles near the runway environment. Runway shoulders range from 10 feet to 25 feet off the end of the runway itself.

Runway safety areas   **Runway Safety Areas (RSA)** are areas surrounding the runway defined by the FAA as the surface surrounding the runway suitable for reducing the risk of damage to airplanes in the event of an undershoot, overshoot, or excursion from the runway. The RSA is typically twice the width of the runway pavement and extends for up to 1,000 feet beyond each end of the runway. The actual required RSA dimensions depend on the size of the aircraft that typically uses the runway. The RSA is an area clear of all objects and typically defines where structures may be located near a runway on the airfield. The dimensions of an RSA are found in FAA-published airport design "advisory circulars" (ref. FAA Advisory Circular: 150/5300–13 "Airport Design"). At airports serving commercial service operations under FAR Part 139, RSAs of specific dimensions are required (Fig. 4-9).

ICAO defines the RSA as the safety area off the side of runways, and defines the term **Runway End Safety Area (RESA)** as the safety area off the ends of the runway pavement.

Some airports supplement the safety areas off the end of runways with an **Engineered Material Arresting System (EMAS).** EMAS is a soft, easily crushable concrete material that acts as an emergency braking agent for aircraft that

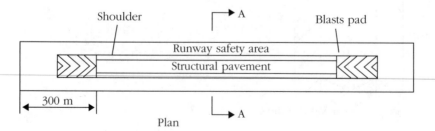

Figure 4-9. *RSA Illustration.*

overrun the runway on landing or aborted takeoff. EMAS is required at commercial airports operating under FAR Part 139 when a "nonstandard" RSA exists. Non-standard RSAs exist primarily at airports that have had urban development, natural terrain, or bodies of water closely adjacent to runway ends. As of 2010, more than 30 airports have installed EMAS on more than 45 runway ends and have safely arrested six aircraft in overrun incidents since the first EMAS installation in 1996. **An example of an EMAS is illustrated in Fig. 4-10.**

Runway protection zones   Off the end of each runway pavement exists a **runway protection zone (RPZ).** The RPZ is a trapezoidal area beginning 200 feet beyond the end of a runway extending for up to 2,500 feet beyond the runway. Its function is to "enhance the protection of people and property on the ground." While not necessarily part of the airfield itself, the FAA strongly recommends that the land encompassing the RPZ be owned, or at least controlled, by the airport. The FAA prohibits residential development, as well as any facilities that would encourage "public assembly" from being located within the RPZ. **Figure 4-11 illustrates the configuration of the RPZ for a runway.**

Figure 4-10. *EMAS system at New York's John F. Kennedy International Airport.* (Figure courtesy Port Authority of New York and New Jersey)

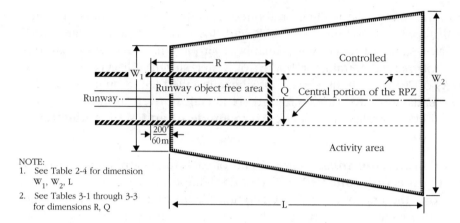

Figure 4-11. *Configuration of the RPZ for a runway.* (See Source FAA Advisory Circular 150/5300-13 Airport Design for specific dimensional requirements).

A runway's imaginary surfaces

While the runway itself is located entirely within the airfield of the airport, wide expanses of area extending far beyond the runway, but associated with the runway, exist for the protection of airspace used by aircraft approaching the airport.

Aircraft landing to or taking off from a runway need an area free of obstructions to operate safely. Within the Federal Aviation Administration's FAR Part 77—Objects Affecting Navigable Airspace—a series of *imaginary surfaces* is defined. These surfaces are referenced by airport management, land use planners, and the FAA itself, for determining whether or not natural terrain or manmade structures would be obstructions to the safe navigation of an aircraft operating on approach to a runway.

The imaginary surfaces defined in FAR Part 77 are the primary surface, the horizontal surface, the conical surface, the approach surface, and the transitional surface. The dimensions of each imaginary surface are defined in FAR Part 77 as follows (Fig. 4-12).

Primary surface   The primary surface is a surface longitudinally centered on a runway. When the runway has a specially prepared hard surface, the primary surface extends 200 feet beyond each end of that runway, but when the runway has no specially prepared hard surface, or planned hard surface, the primary surface ends at each end of that runway. The elevation of any point on the primary surface is the same as the elevation of the nearest point on the runway centerline. The width of a primary surface is:

- 250 feet for utility runways having only visual approaches.
- 500 feet for utility runways having nonprecision instrument approaches.

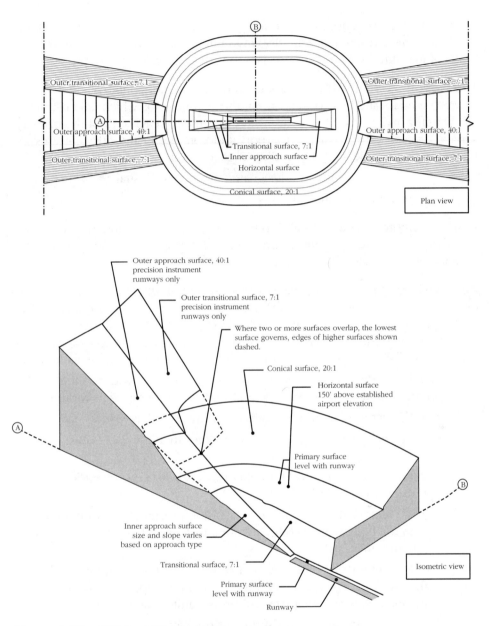

Figure 4-12. *FAR Part 77—Imaginary Surfaces.* (Source: FAA)

For other than utility runways the width is:

- 500 feet for visual runways having only visual approaches.
- 500 feet for nonprecision instrument runways having visibility minimums greater than three-fourths of a statute mile.

- 1,000 feet for a nonprecision instrument runway having a nonprecision instrument approach with visibility minimums as low as three-fourths of a statute mile, and for precision instrument runways.

Horizontal surface   The horizontal surface is a horizontal plane 150 feet above the established airport elevation, the perimeter of which is constructed by swinging arcs of specified radii from the center of each end of the primary surface of each runway of each airport and connecting the adjacent arcs by lines tangent to those arcs. The radius of each arc is:

- 5,000 feet for all runways designated as utility or visual.
- 10,000 feet for all other runways.

Conical surface   The conical surface extends outward and upward from the periphery of the horizontal surface at a slope of 20 to 1 for a horizontal distance of 4,000 feet.

Approach surface The approach surface is longitudinally centered on the extended runway centerline and extends outward and upward from each end of the primary surface. An approach surface is applied to each end of each runway on the basis of the type of approach available or planned for that runway end. The dimensions of the approach surface are determined as follows:

- The inner edge of the approach surface is the same width as the primary surface and it expands uniformly to an outer edge width of:
  - 1,250 feet for that end of a utility runway with only visual approaches.
  - 1,500 feet for that end of a runway other than a utility runway with only visual approaches.
  - 2,000 feet for that end of a utility runway with a nonprecision instrument approach.
  - 3,500 feet for that end of a nonprecision instrument runway other than utility, having visibility minimums greater than three-fourths of a statute mile.
  - 4,000 feet for that end of a nonprecision instrument runway, other than utility, having a nonprecision instrument approach with visibility minimums as low as three-fourths statute mile.
  - 16,000 feet for precision instrument runways.
- The approach surface extends for a horizontal distance of:
  - 5,000 feet at a slope of 20 to 1 for all utility and visual runways.
  - 10,000 feet at a slope of 34 to 1 for all nonprecision instrument runways other than utility.
  - 10,000 feet at a slope of 50 to 1 with an additional 40,000 feet at a slope of 40 to 1 for all precision instrument runways.

Transitional surface   Transitional surfaces extend outward and upward at right angles to the runway centerline and the runway centerline extends at a slope of 7 to 1 from the sides of the primary surface and from the sides of the approach surfaces. Transitional surfaces for those portions of the precision approach surface that project through and beyond the limits of the conical surface extend a distance of 5,000 feet measured horizontally from the edge of the approach surface and at right angles to the runway centerline.

These imaginary surfaces, along with the safety areas, RPZ, and of course the runways themselves, are the key elements of the runway component of an airport's airfield and adjacent vicinity.

Taxiways

The major function of taxiways is to provide access for aircraft to travel to and from the runways to other areas of the airport in an expeditious manner. **Taxiways** are identified as *parallel taxiways, entrance taxiways, bypass taxiways, or exit taxiways*.

A **parallel taxiway** is aligned parallel to an adjacent runway, whereas exit and entrance taxiways are typically oriented perpendicular to the runway, connecting the parallel taxiway with the runway. Entrance taxiways are located near the departure ends of runways; exit taxiways are located at various points along the runway to allow landing aircraft to efficiently exit the runway after landing. Bypass taxiways are located at areas of congestion at busy airports. They allow aircraft to bypass other aircraft parked on the parallel or entrance taxiways in order to reach the runway for takeoff.

Parallel taxiways are typically identified by alphabetic designators. The specific letter used to designate a given taxiway is arbitrary, although some airports use specific letters to identify their field. For example, a taxiway on the north side of an airfield might be designated taxiway N, and the taxiway on the south side of the field would be designated taxiway S. Other airports simply designate parallel taxiways in alphabetic order from one end of the airfield to the other.

Entrance, exit, and bypass taxiways connect the parallel taxiway with the runway. Entrance taxiways are located at the end of the runways, near the threshold. Exit taxiways are often located along the runway, and bypass taxiways are located adjacent to entrance taxiways, to allow for better aircraft sequencing onto a runway. Entrance, exit, and bypass taxiways are typically designated by the associated parallel taxiway, along with a number identifying the specific taxiway. For example, a series of entrance, exit, and bypass taxiways associated with parallel taxiway N may be numbered consecutively in series as N1, N2, N3, and so on.

Taxiways are planned with the following principles in mind:

1. Aircraft that have just landed should not interfere with aircraft taxiing to take off.

2. Taxi routes should provide the shortest distance between aircraft parking areas and runways.

3. At busy airports, taxiways are normally located at various points along runways so that landing aircraft can leave the runways as quickly as possible.

4. A taxiway designed to permit higher turnoff speeds reduces the time a landing aircraft is on the runway. Such taxiways are called *high-speed exit taxiways* and are typically aligned at a 30- to 45-degree angle connecting the runway with the parallel taxiway.

5. When possible, taxiways are planned so as not to cross an active runway.

The widths of taxiways are planned according to the type of aircraft in use. Specifically, the wingspan of the design aircraft is used as the primary planning characteristic for taxiway widths. Taxiway widths range from 25 feet for the smallest general aviation aircraft to 100 feet for aircraft with the largest wingspans.

Taxiway markings

All taxiways should have centerline markings and runway holding position markings whenever they intersect a runway. Taxiway edge markings are present whenever there is a need to separate the taxiway from a pavement that is not intended for aircraft use or to delineate the edge of the taxiway.

The **taxiway centerline** is a single continuous yellow line, 6 to 12 inches in width. This provides a visual cue to permit taxiing along a designated path. Centerlines along with properly planned taxiway widths are intended to ensure safe aircraft taxiing without the risk of hitting obstructions with aircraft wingtips.

Taxiway edge markings are used to define the edge of the taxiway. They are primarily used when the taxiway edge does not correspond with the edge of the pavement. There are two types of markings depending upon whether aircraft are permitted to cross the taxiway edge. **Continuous markings**, consisting of a continuous double yellow line, with each line being at least 6 inches in width spaced 6 inches apart, are used to define the taxiway edge from the shoulder or some other abutting paved surface not intended for use by aircraft. *Dashed markings* are used when there is an operational need to define the edge of a taxiway on a paved surface where the adjoining pavement to the taxiway edge is intended for use by an aircraft, for example, an aircraft apron or parking area. Dashed taxiway edge markings consist of a broken double yellow line, with each line being at least 6 inches in width, spaced 6 inches apart. These lines are 15 feet in length with 25-foot gaps.

Similar to runway shoulder markings, *taxiway shoulder markings* are sometimes used to further define the edge of the taxiway from adjacent unusable pavement. Taxiway shoulder markings are yellow lines running perpendicular to the centerline of the taxiway.

Runway holding position markings define the boundary between entrance, exit, and bypass taxiways and the runway. Runway holding position markings consist of a set of four lines, two solid lines and two dashed lines. The two dashed lines are located on the runway side of the boundary and the two solid lines are located on the taxiway side of the boundary. The lines are painted on top of a black background and are extend into the shoulder area of the taxiway. These recent enhancements were created to facilitate awareness of the boundary location. At controlled airports (i.e., airports with an active control tower), aircraft are not permitted to cross over the hold line from the taxiway to the runway without explicit permission from air traffic control. At uncontrolled airports, pilots are encouraged to stop, ensure that the runway environment is clear, and announce when they cross from the taxiway to the runway. Aircraft exiting the runway to the taxiway may cross over the hold line without explicit permission. An illustration of taxiway hold line markings is found in Fig. 4-13.

Enhanced taxiway centerline markings are found at the approaches from the parallel taxiways to the runway holding position markings, beginning 150 feet from the runway holding position markings. These enhanced markings consist of a series of long dashed lines on either side of the existing taxiway centerline, all on a painted black background. These enhanced markings were recently implemented at airports to provide pilots with an increased visual cue that they are approaching the boundary between the taxiway and the runway (Fig. 4-14).

Taxi shoulder markings are sometimes configured with paved shoulders. When such shoulders exist, they are marked with a series of yellow stripes leading from the solid yellow **taxiway edge markings**, which define the ends

Figure 4-13. *Runway holding position markings, yellow markings on black background.*

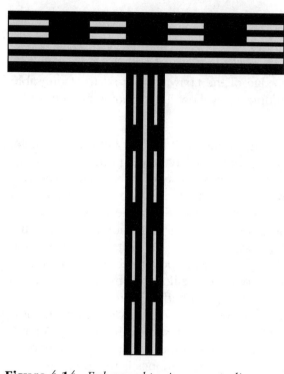

Figure 4-14. *Enhanced taxiway centerline markings leading up to runway holding position markings.*

of the useable taxiway pavement, to the end of the shoulder pavement. These markings provide guidance to pilots and ground vehicle operators that the shoulder is not to be used for normal operations. An illustration of taxi shoulder markings is found in Fig. 4-15.

At some busy airports that have complex taxiway systems and are prone to low-visibility conditions, the airport management may implement a **Surface Movement Guidance Control System (SMGCS)** plan. As part of this plan, points along taxiway routes may be marked with *geographic position markings*. These markings are used to identify the location of taxiing aircraft during low-visibility operations. They are positioned to the left of the taxiway centerline in the direction of taxiing. The geographic position marking is a circle composed of an outer black ring contiguous to a white ring with a pink circle in the middle. When installed on asphalt or other dark-colored pavements, the white ring and the black ring are reversed; that is, the white ring becomes the outer ring and the black ring becomes the inner ring. The marking is designated with either a number or a letter. The number corresponds to the consecutive position of the marking along the defined taxi route (Fig. 4-16).

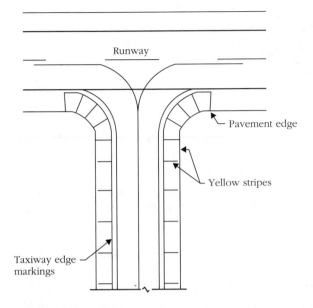

Figure 4-15. *Taxiway edge markings and shoulder markings.*

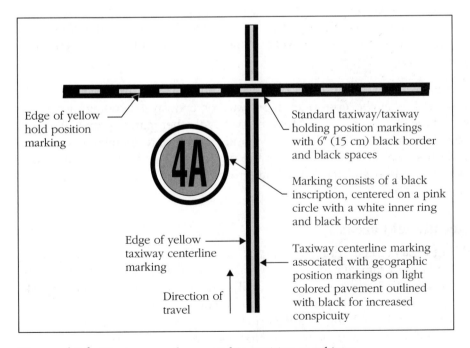

Figure 4-16. *Taxiway and geographic position markings.* (Source: FAA AIM)

Other airfield markings

On an airfield there typically exists a variety of markings in addition to those found on runways and taxiways. The primary purposes of these additional markings are to identify aircraft holding areas and other critical locations on the airfield and to provide guidance for ground service vehicles using the airfield.

Vehicle roadway markings are used when necessary to define a pathway for ground vehicle operations on or crossing areas that are also intended for the movement of aircraft. These markings consist of a white solid line to delineate each edge of the roadway and a dashed line to separate lanes within the edges of the roadway. In lieu of the solid lines, *zipper markings* may be used to delineate the edges of the vehicle roadway (Fig. 4-17).

VOR receiver checkpoint markings allow a pilot to check aircraft instruments with the signal emitted from the airport VOR (should there be one on the field) navigational aid. The marking consists of a painted circle with an arrow in the middle. The arrow is aligned in the direction of the VOR checkpoint azimuth (Fig. 4-18).

Nonmovement area boundary markings delineate the airfield's *movement area,* that is, the area under air traffic control. These markings are yellow and are located on the boundary between the movement and nonmovement areas. The nonmovement area boundary markings consist of two yellow lines (one solid and one dashed) 6 inches in width. The solid line is located on the nonmovement area side and the dashed yellow line is located on the movement area side (Fig. 4-19).

Markings of permanently closed runways and taxiways are identified with yellow crosses at the end of each runway or taxiway extending along the length of the pavement in 1,000-foot intervals. All other markings and lights are removed. Temporarily closed runways and taxiways are identified by single yellow crosses placed on the ends of the pavement (Fig. 4-20). For temporary markings, a raised and lighted yellow cross may be used in lieu of painted markings.

Other airfield areas

Holding areas (commonly referred to as *run-up areas*) are located at or very near the ends of runways for pilots to make final checks and await final clearance for takeoff. These areas are generally large enough so that other aircraft can bypass an aircraft still performing run-up checks or awaiting air traffic control clearance. The holding area is normally designed to accommodate two or three aircraft and allow enough space for one aircraft to bypass.

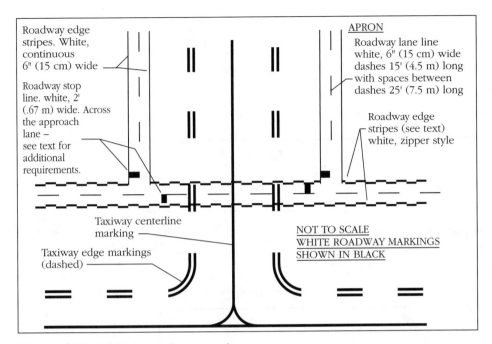

Figure 4-17. *Vehicle roadway markings.* (Source: FAA AIM)

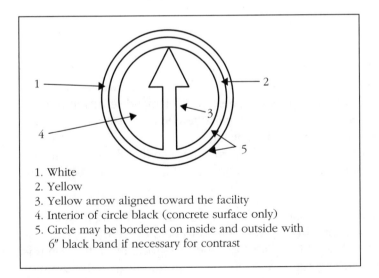

1. White
2. Yellow
3. Yellow arrow aligned toward the facility
4. Interior of circle black (concrete surface only)
5. Circle may be bordered on inside and outside with
   6″ black band if necessary for contrast

Figure 4-18. *VOR receiver checkpoint markings.* (Source: FAA AIM)

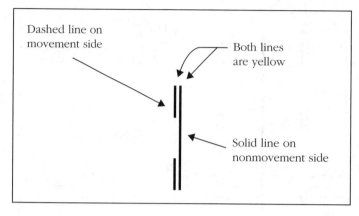

Figure 4-19. *Nonmovement area boundary markings.* (Source: FAA AIM)

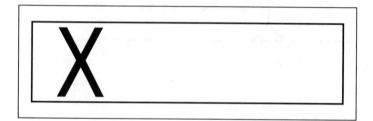

Figure 4-20. *Temporarily closed runway or taxiway.* (Source: FAA AIM)

Holding bays are apron areas located at various points off taxiways for temporary parking of aircraft. At some airports where peak demand results in full occupancy of all aircraft gate positions, aircraft will often be routed to a holding bay until a gate becomes available. Some holding bays are located as close as 250 feet from the active runway. During peak hours, aircraft are held in the bay until they are given a takeoff position; they then move to the holding area (Fig. 4-21).

An aircraft landing at a large commercial service airport might have to negotiate a mile or more of taxiways to reach the aircraft **apron** or *parking area*. Pilots usually are in possession of a map of the airfield to help move from one position to another on the airfield. If a pilot loses the way, local air traffic controllers will assist in providing *progressive* directions. In addition, a "follow me" truck might be sent to lead the pilot onto the parking apron area.

Yellow lines painted on the concrete parking apron adjoining the taxiways lead the pilot to the final positioning. Linemen will greet the incoming aircraft and direct the pilot with appropriate parking signals.

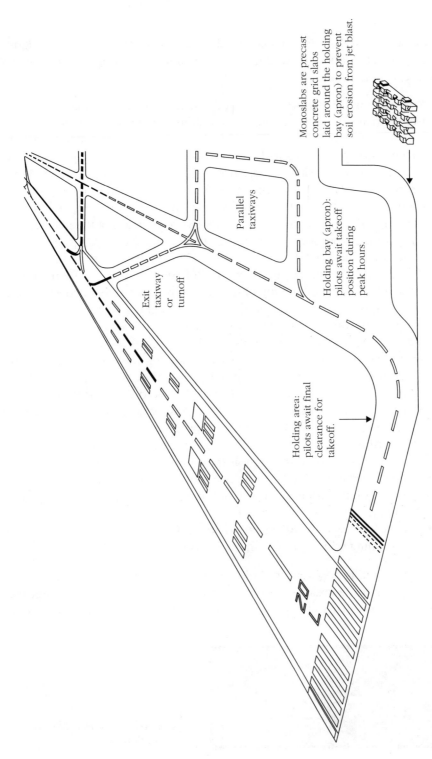

Parallel
taxiways

Exit
taxiway
or
turnoff

Holding bay (apron):
pilots await takeoff
position during
peak hours.

Holding area:
pilots await final
clearance for
takeoff.

Monoslabs are precast
concrete grid slabs
laid around the holding
bay (apron) to prevent
soil erosion from jet blast.

Figure 4-21. *Location of bolding areas and bolding bays.* (Source: FAA)

Airfield signage

There are six types of signs installed on airfields: mandatory instruction signs, location signs, direction signs, destination signs, information signs, and runway distance remaining signs.

Mandatory instruction signs have a red background with a black-outlined white inscription and are used to denote an entrance to a runway or critical area and areas where an aircraft is prohibited from entering. Typical mandatory instruction signs and applications include runway holding position signs, runway approach area holding position signs, ILS critical area holding position signs, and no-entry signs. *Runway holding position signs* are located at the holding positions on taxiways that intersect a runway or on runways that intersect other runways. The inscription on the sign contains the designation of the intersecting runway. The runway numbers on the sign are arranged to correspond to the respective runway threshold. For example, 9 - 27 indicates that the threshold for runway 9 is to the left and that the threshold of runway 27 is to the right. On taxiways that intersect the beginning of the takeoff runway, only the designation of the takeoff runway may appear on the sign; all other signs will have the designation of both runway directions (Fig. 4-22). If the sign is located on a taxiway that crosses the intersection of two runways, the designations for both runways will be shown on the sign along with arrows showing the approximate alignment of each runway. In addition to showing the approximate runway alignment, the arrow indicates the direction to the threshold of the runway whose designation is immediately next to the arrow. A runway holding position sign on a taxiway will be installed adjacent to holding position markings on the taxiway pavement. On runways, holding position marking will be located only on the runway pavement adjacent to the sign, if the runway is normally used for land and hold short operations (LAHSO). At commercial service airports with more than one runway on the airfield, runway holding position signs are also painted onto the taxiway surface in front of the runway holding position lines. These **surface painted runway holding position signs** provide added guidance to pilots in determining which runway they are about to enter.

At some airports, it is necessary to hold an aircraft on a taxiway located in the approach or departure area for a runway so that the aircraft does not interfere with operations on that runway. In these situations, a *runway approach area*

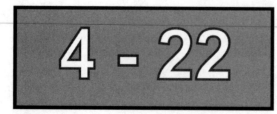

Figure 4-22. *Runway holding position sign.* (Source: FAA)

holding sign with the designation of the approach end of the runway followed by a dash (-) and the letters APCH will be located at the holding position on the taxiway (Fig. 4-23).

At some airports, when an Instrument Landing System (ILS) is being used, it is necessary to hold an aircraft on a taxiway at a location other than the holding position. In these situations an ILS critical area holding position sign, with the inscription ILS is used (Fig. 4-24).

A *no-entry sign* prohibits an aircraft from entering an area. The sign is identified by a white rectangular horizontal bar surrounded by a white ring on a red background (similar to "Do Not Enter" signs on automobile roads). Typically, this sign would be located on a taxiway intended to be used in only one direction or at the intersection of vehicle roadways with runways, taxiways, or aprons where the roadway may be mistaken as a taxiway or other aircraft movement surface (Fig. 4-25).

Figure 4-23. *Runway approach area holding sign.*
(Source: FAA AIM)

Figure 4-24. *ILS critical area holding position sign.*
(Source: FAA AIM)

Figure 4-25. *No-entry sign.*
(Source: FAA AIM)

Location signs are used to identify either a taxiway or runway on which the aircraft is located. Other location signs provide a visual cue to pilots to assist them in determining when they have exited an area. Location signs include taxiway location signs, runway location signs, runway boundary signs, and ILS Critical Area Boundary Signs.

Taxiway location signs are marked with a yellow inscription on a black background. The inscription is the designation of the taxiway on which the aircraft is located. These signs are installed along taxiways either by themselves or in conjunction with direction signs or runway holding position signs (Fig. 4-26).

Runway location signs have black backgrounds with yellow inscriptions and a yellow inner border. The inscription is the designation of the runway on which the aircraft is located. These signs are intended to complement the information available to pilots through their magnetic compasses and are typically installed where the proximity of two or more runways to one another could confuse pilots as to which runway they are on (Fig. 4-27).

Figure 4-26. *Taxiway location sign.* (Source: FAA AIM)

Figure 4-27. *Runway location sign.* (Source: FAA AIM)

Runway boundary signs have yellow backgrounds with black inscriptions with a graphic depicting the pavement holding position markings associated with the runway boundary (Fig. 4-28). This sign, which faces the runway and is visible to the pilot exiting the runway, is located adjacent to the holding position marking on the pavement. The sign is intended to provide pilots with another visual cue that they can use as a guide in deciding when they are "clear of the runway."

ILS critical area boundary signs have yellow backgrounds with black inscriptions and a graphic depicting the ILS pavement holding position marking. The marking is defined by two horizontal black bars adjoined by three sets of closely spaced sets of two vertical bars (Fig. 4-29). The sign is intended to provide pilots with another visual cue that they can use as a guide in deciding when they are "clear of the ILS critical area."

Figure 4-28. *Runway boundary sign.* (Source: FAA)

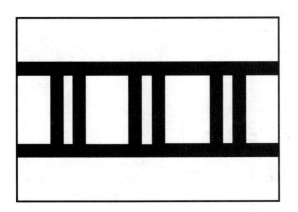

Figure 4-29. *ILS critical area boundary sign.* (Source: FAA AIM)

Direction signs have a yellow background with a black inscription. The inscription identifies the designations of the intersecting taxiways leading out of the intersection that a pilot would normally be expected to turn onto or hold short of. Each designation is accompanied by an arrow indicating the direction of the turn (Fig. 4-30). Direction signs are normally located on the left prior to an intersection. When used on a runway to indicate an exit, the sign is located on the same side of the runway as the exit. The taxiway designations and their associated arrows on the sign are arranged clockwise starting from the first taxiway on the pilot's left.

Destination signs also have a yellow background with a black inscription indicating a destination on the airfield. These signs always have an arrow showing the direction of the taxiing route to that destination. When the arrow on the destination sign indicates a turn, the sign is located prior to the intersection. Destinations commonly shown on these types of signs include runways, aprons, terminals, military areas, civil aviation areas, cargo areas, international areas, and fixed-base operators. An abbreviation may be used as the inscription on the sign for some of these destinations (Fig. 4-31).

When the inscription for two or more destinations having a common taxiing route is placed on a sign, the destinations are separated by a "dot" and one arrow would be used. When the inscription on a sign contains two or more

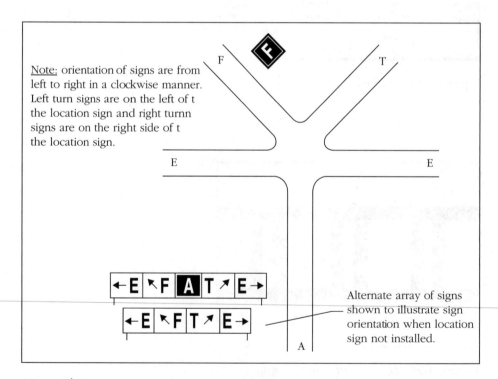

Figure 4-30. *Direction signs.* <small>(Source: FAA AIM)</small>

Figure 4-31. *Destination sign (to military ramp).* (Source: FAA AIM)

destinations having different taxiing routes, each destination will be accompanied by an arrow and will be separated from the other destinations on the sign with a vertical place message divider.

Information signs have a yellow background with a black inscription. They are used to provide the pilot with information on such things as areas that cannot be seen from the control tower, applicable radio frequencies, and noise abatement procedures. The airport operator determines the need, size, and location for these signs.

Runway distance remaining signs have a black background with a white numeral inscription. They may be installed along one or both sides of a runway. The number on the signs indicates the distance (in thousands of feet) of landing runway remaining (Fig. 4-32). The last sign, that is, the sign with the numeral 1, will be located at least 950 feet from the runway end.

Airfield lighting

To allow pilots to safely depart, arrive, and move about an airport's airfield during nighttime hours or periods of reduced visibility, nearly all commercial service and most general aviation airports are equipped with airfield lighting systems. These systems may include runway lighting, taxiway lighting, and approach lighting systems, as well as various other lighting systems ranging from an airport's beacon to ramp and perimeter lighting systems. Each of these systems is available in varying configurations and intensities.

Runway lighting

Runway lighting is extremely important for nighttime aircraft operations or in poor visibility weather conditions. Runway lighting systems include approach lighting systems, visual glideslope indicators, runway end identifiers, runway edge light systems, and in-runway lighting systems. As their names imply, approach lighting systems aid aircraft in properly aligning with the runway on approach to landing, and in-runway lighting systems aid aircraft in landing and takeoff operations on and in the immediate vicinity of the runway (Fig. 4-33).

Figure 4-32. *Runway distance remaining sign.* (Source: FAA AIM)

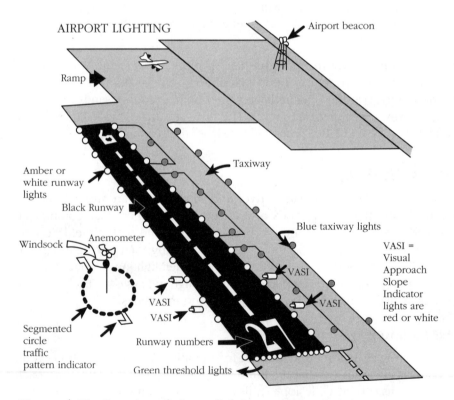

Figure 4-33. *Overview of airport lighting systems.* (Source: NASA)

Approach lighting systems   Approach lighting systems (ALS) provide the basic means for aircraft to identify runways when operating in poor weather conditions and when operating under IFR. ALS are a configuration of signal lights starting at the landing threshold and extending back from the runway, called the approach area, a distance of 2,400 to 3,000 feet for precision instrument runways and 1,400 to 1,500 feet for nonprecision instrument runways. Some systems include sequenced flashing lights, which appear to the pilot as a ball of light traveling toward the runway at high speed (Fig. 4-34).

The following approach lighting systems are in use at civil airports in the United States (Fig. 4-35):

ALSF-1:   Approach light system 2,400 feet in length with sequenced flashing lights in ILS Cat-I configuration (see further in this section for a full description of ILS).

ALSF-2:   Approach light system 2,400 feet in length with sequenced flashing lights in ILS Cat-II configuration.

SSALF: Simplified short-approach light system with sequenced flashing lights.

SSALR: Simplified short-approach light system with runway alignment indicator lights.

MALSF: Medium-intensity approach light system 1,400 feet in length with sequenced flashing lights.

MALSR: Medium-intensity approach light system 1,400 feet in length with runway alignment indicator lights.

LDIN: Lead-in-light system, which consists of one or more series of flashing lights installed at or near ground level that provides positive visual guidance along an approach path, either curing or straight, where special problems exist with hazardous terrain, obstructions, or noise abatement procedures.

RAIL: Runway alignment indicator lights. Sequenced flashing lights which are installed in combination with other light systems.

ODALS: Omnidirectional approach lighting system consisting of seven omnidirectional flashing lights located in the approach area of a non-precision runway. Five lights are located on the runway centerline with the first light located 300 feet up from the threshold and extending at equal intervals up to 1,500 feet from the threshold. The other two lights are located, one on each side of the runway threshold, at a lateral distance of 40 feet from the runway edge, or 75 feet from the runway edge when installed on a runway equipped with a Visual Approach Slope Indicator (VASI).

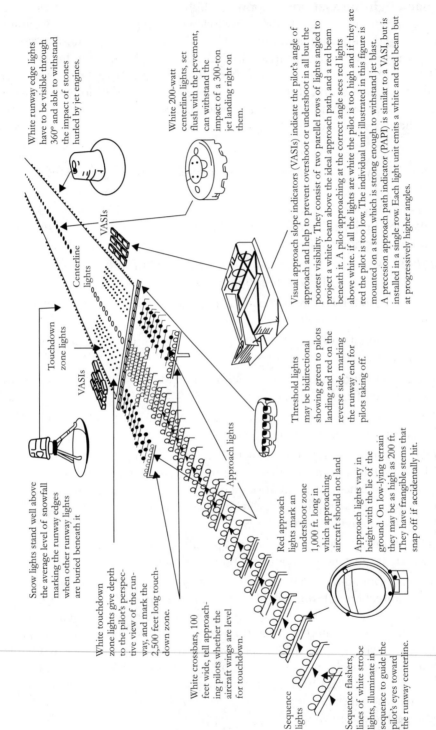

White runway edge lights have to be visible through 360° and able to withstand the impact of stones hurled by jet engines.

White 200-watt centerline lights, set flush with the pevement, can withstand the impact of a 300-ton jet landing right on them.

VASIs

Centerline lights

Touchdown zone lights

VASIs

Snow lights stand well above the average level of snowfall marking the runway edges when other runway lights are buried beneath it

White touchdown zone lights give depth to the pilot's perspective view of the runway, and mark the 2,500 feet long touchdown zone.

White crossbars, 100 feet wide, tell approaching pilots whether the aircraft wings are level for touchdown.

Approach lights

Threshold lights may be bidirectional showing green to pilots landing and red on the reverse side, marking the runway end for pilots taking off.

Red approach lights mark an undershoot zone 1,000 ft. long in which approaching aircraft should not land

Approach lights vary in height with the lie of the ground. On low-lying terrain they may be as high as 200 ft. They have frangible stems that snap off if accidentally hit.

Sequence lights

Sequence flashers, lines of white strobe lights, illuminate in sequence to guide the pilot's eyes toward the runway centerline.

Visual approach slope indicators (VASIs) indicate the pilot's angle of approach and help to prevent overshoot or undershoot in all but the poorest visibility. They consist of two parellel rows of lights angled to project a white beam above the ideal approach path, and a red beam beneath it. A pilot approaching at the correct angle sees red lights above white. if all the lights are white the pilot is too high and if they are red the pilot is too low. The individual unit illustrated in this figure is mounted on a stem which is strong enough to withstand jet blast.
A precision approach path indicator (PAPI) is similar to a VASI, but is installed in a single row. Each light unit emits a white and red beam but at progressively higher angles.

Figure 4-34. *Precision approach lighting systems.* (Source: FAA)

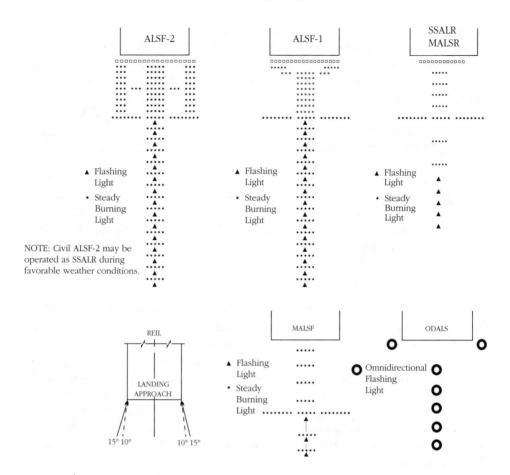

Figure 4-35. *Approach lighting systems.* (Source: FAA AIM)

Visual glideslope indicators Visual glideslope indicators are lighting systems located adjacent to runways on the airfield to assist aircraft with visually based vertical alignment on approach to landing. The five most common visual glideslope indicators are visual approach slope indicators (VASI), precision approach path indicators (PAPI), tricolor systems, pulsating systems, and alignment of elements systems.

The **visual approach slope indicator (VASI)** is a system of lights so arranged to provide visual descent guidance information during an aircraft's approach to a runway. These lights are visible from 3 to 5 miles during the day and up to 20 miles or more at night. The visual glide path of the VASI provides safe obstruction clearance within 10 degrees of the extended runway centerline and to 4 nautical miles from the runway threshold.

VASIs may consist of 2, 4, 6, 12, or 16 light units arranged in bars referred to as near, middle, and far bars. Most VASIs have two bars, near and far, and may consist of 2, 4, or 12 light units. Some VASIs consist of three bars, near, middle, and far, which provide an additional visual glide path to accommodate aircraft with high cockpits. This installation may consist of either 6 or 16 light units. VASIs consisting of 2, 4, or 6 light units are located on one side of the runway, usually to the left. Where the VASI consists of 12 or 16 light units, the units are located on both sides of the runway.

Two-bar VASIs provide one visual glide path that is normally set at 3 degrees of slope. Three-bar VASIs provide two visual glide paths. The lower glide path is provided by the near and middle bars and is normally set at 3 degrees of slope, whereas the upper glide path, provided by the middle and far bars, is normally set one-quarter of a degree higher. The higher glide path is typically used only by aircraft with higher cockpit heights to assure proper threshold crossing height (Fig. 4-36).

The basic principle of the VASI is that of color differentiation between red and white. Each light unit projects a beam of light having a white segment in the upper part of the beam and red segment in the lower part of the beam. The light units are arranged so that the pilot using the VASIs during an approach will see the combination of lights associated with their height relative to the approach path. For example, on a two-bar VASI, the glide slope is associated with the pilot seeing red lights emanating from the far bar and white lights from the near bar. If the aircraft on approach is above the glide path, both bars would be seen as white. If the aircraft is below the glide path, both bars would be seen as having red lights (Fig. 4-37).

The **precision approach path indicator (PAPI)** uses light units similar to the VASI, but they are installed in a single row of their two or four light units. These systems have an effective visual range of about 5 miles during the day and up

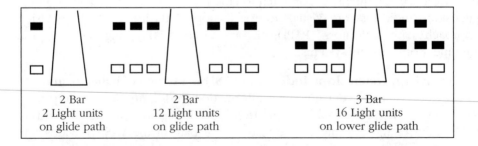

Figure 4-36. *Various VASI configurations: dark = red; white = white.*

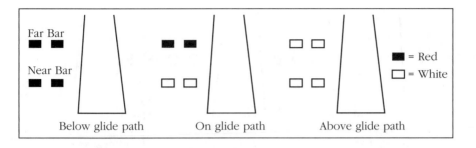

Figure 4-37. *Two-bar VASI.* (Source: FAA AIM)

to 20 miles at night. The row of light units is normally installed on the left side of the runway. As with the VASI, the light units on the PAPI are equipped with red and white beams that project various degrees of glide path to the runway. The PAPI is said to be more precise than a VASI because it allows the pilot to judge approximately how many degrees above or below the glide path the aircraft is on approach by the number red versus white lights observed. For example, on a four-light PAPI, observing two red and two white lights denotes on glide path, three red and one white light denotes slightly below (approximately 0.2 degrees) glide path, four red lights denote 0.5 or more degrees below glide slope, and so on (Fig. 4-38).

Tricolor visual approach slope indicators normally consist of a single light unit projecting a three-color visual approach path to the runway. The below glide path indication is red, the slightly below and above glide path indications are amber, and the on glide path indication is green. These types of indicators have a useful range of approximately 1/2 to 1 mile during the day and up to 5 miles at night depending on visibility conditions (Fig. 4-39).

Pulsating visual approach slope indicators normally consist of a single light unit projecting a two-color visual approach path to the runway. The on glide path indication is a steady white light. The slightly below glide path indication is a steady red light. If the aircraft descends further below the glide path, the red light begins to pulsate. The above glide path indication is a pulsating white light. The pulsating rate increases as the aircraft gets further above or below the desired glide slope. The useful range of the system is about 4 miles during the day and up to 10 miles at night (Fig. 4-40).

Alignment of elements systems are installed on some small general aviation airports. They are low-cost systems consisting of three painted plywood panels, normally black and white or fluorescent orange. Some of these systems are lighted for night use. The useful range of these systems is approximately three-quarters of a mile. To use the system, the pilot positions the aircraft so

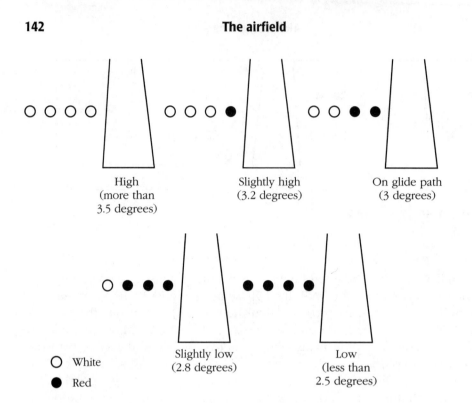

Figure 4-38. *PAPI (precision approach path indicator).* (Source: FAA AIM)

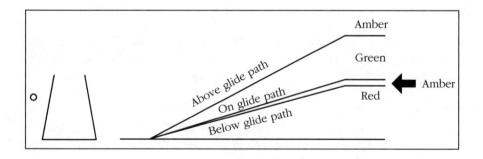

Figure 4-39. *Tricolor VASI.* (Source: FAA AIM)

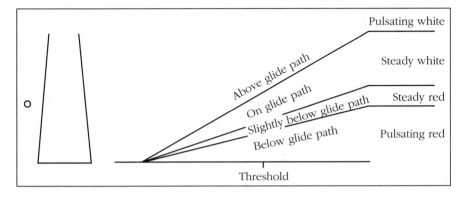

Figure 4-40. *Pulsating VASI.* (Source: FAA AIM)

the elements are in alignment. If the pilot is above the glide path, the center panel will appear to be above the outer two panels. If the pilot is below the glide path, the center panel will appear to be below the outer two panels (Fig. 4-41).

Runway end identifier lights   Runway end identifier lights (REILs) are installed at many airfields to provide rapid and positive identification of the approach end of a runway. The system consists of a pair of synchronized flashing lights located laterally on each side of the runway threshold. REILs may be either omnidirectional or unidirectional facing the approach area. They are effective for identifying a runway surrounded by a preponderance of other lighting, a runway that lacks contrast with surrounding terrain, or a runway during reduced visibility.

Runway edge light systems   Runway edge lights are used to outline the edges of runways during periods of darkness or reduced visibility. These light systems are classified according to the intensity or brightness they are capable of producing. Runway edge light systems include:

HIRL—high-intensity runway lights

MIRL—medium-intensity runway lights

LIRL—low-intensity runway lights

The HIRL and MIRL systems typically have variable intensity controls, whereas the LIRLs normally have one intensity setting.

Runway edge lights are white, except on instrument runways where yellow lights replace white on the last 2,000 feet or half the runway length, whichever is less, to form a caution zone for landings. The lights marking the ends of the runway emit red light toward the runway to indicate the end of the runway to

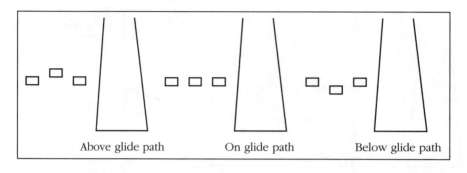

Figure 4-41. *Alignment of elements system.* (Source: FAA AIM)

a departing aircraft and emit green outward from the runway end to indicate the threshold to landing aircraft.

In-runway lighting   Lighting systems integrated into the runway pavement include runway centerline lighting systems (RCLS), touchdown zone lights (TDZL), taxiway lead-off lights, and land and hold short lights. These lighting systems are intended to aid aircraft on approach, on takeoff, and for taxiing on and off the runway (Fig. 4-42).

Runway centerline lighting systems (RCLS) are installed on some precision instrument runways to facilitate landing under reduced visibility conditions. They are located along the runway centerline and are spaced at 50-foot intervals. When viewed from the landing threshold, the runway centerline lights are white until the last 3,000 feet of the runway. The white lights begin to alternate with red for the next 2,000 feet, and for the last 1,000 feet of runway, all centerline lights are red.

Touchdown zone lights (TDZL) are installed on some precision approach runways to indicate the touchdown zone when landing under adverse visibility conditions. They consist of two rows of transverse light bars disposed symmetrically about the runway centerline. The system consists of steady-burning white lights that start 100 feet beyond the landing threshold and extend to 3,000 feet beyond the landing threshold or to the midpoint of the runway, whichever is less.

Taxiway lead-off lights extend from the runway centerline to a point on an exit taxiway to expedite movement of aircraft from the runway. These lights alternate green and yellow from the runway centerline to the runway holding position.

Land and hold short lights are used to indicate the hold short point on certain runways that are approved for **land and hold short operations (LAHSO).**

Figure 4-42. *In-runway lighting at Daytona Beach International Airport.*
(Photo: S. Young)

Land and hold short lights consist of a row of pulsing white lights installed across the runway at the hold short point. Where installed, the lights will be on anytime LAHSO is in effect. These lights will be off when LAHSO is not in effect.

Taxiway lighting

Many airports are equipped with taxiway lighting to facilitate the movement of aircraft on the airfield at night or in poor visibility conditions. Taxiway lighting includes taxiway edge lights, taxiway centerline lights, clearance bar lights, runway guard lights, and stop bar lights.

Taxiway edge lights are used to outline the edges of taxiways during periods of darkness or restricted visibility conditions. These lights emit blue light. **Taxiway centerline lights** are located along taxiway centerlines in a straight line on straight portions, on the centerline of curved portions, and along designated taxiing paths in portions of runways, aircraft ramp, and parking areas. Taxiway centerline lights are steady burning and emit green light.

Clearance bar lights are installed at holding positions on taxiways in order to increase the conspicuity of the holding position in low-visibility conditions. They may also be installed to indicate the location of intersecting taxiways during periods of darkness. Clearance bars consist of three in-pavement steady-burning yellow lights.

Runway guard lights are installed at intersections of runways and taxiways. They are primarily used to enhance the conspicuity of taxiway/runway intersections during low-visibility conditions but may be used in all weather conditions. Runway guard lights consist of either a pair of elevated flashing yellow lights installed on either side of the taxiway, or a row of in-pavement yellow lights installed across the entire runway, at the runway holding position marking.

Stop bar lights are used to confirm instructions from air traffic controllers' clearance to enter or cross an active runway in low-visibility conditions. A stop bar consists of a row of red, unidirectional, steady-burning in-pavement lights installed across the entire taxiway at the runway holding position, and elevated steady-burning red lights on each side. A controlled stop bar is operated in conjunction with the taxiway centerline lead-on lights which extend from the stop bar toward the runway. Following clearance to proceed, the stop bar is turned off and the lead-on lights are turned on. The stop bar and lead-on lights are automatically reset by a sensor or backup timer.

Other airfield lighting

In addition to the lighting located on runways and taxiways, lighting is required to identify potential obstructions to aircraft operations on, and in the vicinity of, an airport's airfield.

Obstruction lights are implemented to warn pilots of the presence of obstructions during daytime and nighttime conditions. They may be lighted in any of the following combinations:

> *Aviation red obstruction lights* are beacons that flash a red omnidirectional beam of light at a rate of 20 to 40 flashes per minute during daytime hours and burn steadily at night.

> *Medium-intensity flashing white obstruction lights* may be used during daytime and twilight with automatically selected reduced intensity for nighttime operation. When this system is used on structures 500 feet or less in height, other methods of marking and lighting the structure may be omitted. However, if the structure is greater than 500 feet, the top of the obstruction must be marked with aviation orange and white paint.

> *High-intensity white obstruction lights* are flashing high-intensity white lights during daytime with reduced intensity for twilight and nighttime operation. When this type of system is used, the marking of structures with red obstruction lights and aviation orange and white paint may be omitted.

A combination of flashing aviation red beacons and steady-burning aviation red lights for nighttime operations and flashing high-intensity white lights for daytime operations is *dual lighting* of obstructions. With dual lighting, aviation orange and white markings may be omitted.

Finally, airfield lighting is used to identify the location and type of an airport. The airport's *aeronautical light beacon* is considered a visual navigational aid for aircraft. The beacon displays flashes of white and/or colored light to indicate the location of an airport or heliport. The light used may be a rotating beacon or one or more flashing lights.

Airport and heliport beacons have a vertical light distribution to make them most effective from 1 to 10 degrees above the horizon; however, they typically can be seen well above and below this peak spread. The beacon may be an omnidirectional capacitor-discharge device, or it may rotate at a constant speed, which produces the visual effect of flashes at regular intervals. Flashes may be one or two colors alternately. The total number of flashes are:

1. 24 to 30 per minute for beacons marking airports.

2. 30 to 45 per minute for beacons marking heliports.

The color or color combination displayed by a particular beacon indicates the type of airport it is identifying:

1. Alternating white and green: Lighted land airport

2. Alternating white and yellow: Lighted water airport

3. Alternating green, yellow, and white: Lighted heliport

4. Flashing white: Unlighted airport

Military airport beacons flash alternatively white and green, but are differentiated from civil beacons by dual-peaked (two quick) white flashes between the green flashes.

The *airport code beacon,* which can be seen from all directions, is used to identify airports and landmarks. The code beacon flashes the three- or four-character airport identifier in International Morse Code six to eight times per minute. Green flashes are displayed for land airports; yellow flashes indicate water airports.

Airport beacons are activated during nighttime hours, and during hours of reduced visibility. Specifically, airport beacons are lit during daytime hours when the ground visibility is less than 3 miles and/or the cloud ceiling is less than 1,000 feet.

Navigational aids (NAVAIDS) located on airfields

Various types of **navigational aids (NAVAIDS)** are in use today to aid aircraft both to fly between locations and to approach an airport for landing, particularly in poor weather conditions. Often these aids are located on airport airfields and, hence, airport management and planners should be aware of how they operate and where they may be placed in relation to other facilities on the airfield.

Nondirectional radio beacons (NDB)

The **nondirectional radio beacon,** or **NDB,** is the oldest of the radio signal–based Navigational Aids Used on Airfields. NDBs emit low- or medium-frequency radio signals whereby the pilot of an aircraft properly equipped with an automatic direction finder (ADF) can determine bearings and "home" in on the station. NDBs normally operate in the frequency of 190 to 535 kilohertz (kHz) and transmit a continuous carrier with either 400 or 1,020 hertz (Hz) modulation. NDBs are considered to be nonprecision navigational aids, and thus runways approached by aircraft utilizing an NDB for navigational aid are equipped with nonprecision instrument markings.

An NDB is normally mounted on a 35-foot-high pole. An NDB may be located on or adjacent to the airport, at least 100 feet clear of metal buildings, power lines, or metal fences to avoid radio signal interference (Fig. 4-43).

Very-high-frequency omnidirectional range radio beacons (VOR)

The **VOR** is the most common ground-based electronic navigational aid found in the United States today. The VOR transmits a set of very-high-frequency navigational signals, which, when identified by navigation instruments in aircraft, determine the location of the VOR from the aircraft with respect to magnetic north. VORs operate within the 108.0 to 117.95 MHz frequency band

Figure 4-43. *NDB nondirectional beacon.* (Source: FAA)

and have a power output necessary to provide coverage within their assigned operational service volume. The standard VORs found on airports are called TVORs, and have a typical operational service volume of 25 nautical miles in radius from the airport. VORs are considered to be nonprecision navigational aids, and thus runways approached by aircraft utilizing a VOR for navigational aid are equipped with nonprecision instrument markings.

When located on an airport's airfield, a VOR should be located at least 500 feet from the centerline of any runway and 250 feet from the centerline of any taxiway. If the airport has intersecting runways, the VOR should be located near the intersection to provide accurate navigational guidance for approach to both runways (Fig. 4-44).

VOR signals are susceptible to distortion caused by reflections off other objects. As such, VORs should be located at least 1,000 feet from any structures and trees. Metal fences should be at least 500 feet away from the antenna and overhead power lines should be at least 1,000 feet from the antenna.

Instrument Landing Systems (ILS)

The most common navigational aid used by aircraft for both lateral and vertical guidance on approach to runways is the **Instrument Landing System (ILS)** (Fig. 4-45). The ILS is designed to provide an approach path for exact alignment and descent of an aircraft on approach to a runway. By virtue of the fact that the ILS provides both lateral and vertical guidance, it is considered a *precision approach* system, and is associated with precision approach markings on its associated runway. The ILS has been the standard precision approach

Figure 4-44. *VOR beacon on the airfield at Washington Reagan National Airport.* (Photo courtesy Seth Young)

FAA Instrument Landing Systems

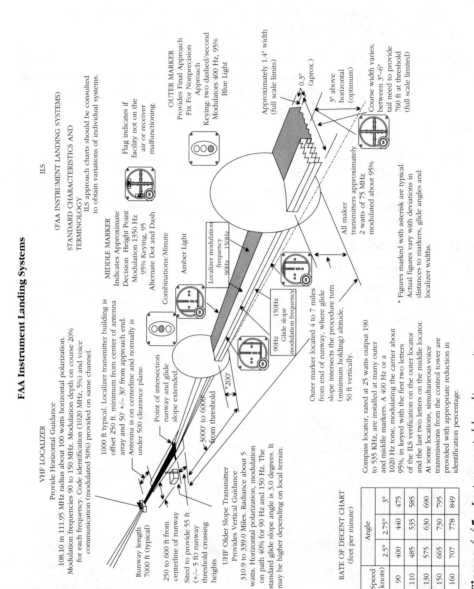

Figure 4-45. *Instrumental landing system.* (Source: FAA AIM)

navigational aid since its introduction in the United States in 1941. The ILS provides guidance by radio beams that define a straight-line path to the runway at a fixed slope of approximately 3 degrees, beginning 5 to 7 miles from the runway threshold. All aircraft approaching the airport under ILS guidance must follow this path in single file.

The ground equipment that comprise an ILS consists of two highly directional transmitting systems and, along the approach, up to three marker beacons. The directional transmitters are known as the *localizer* and *glide slope.*

The **localizer** transmitter operates on one of 40 ILS channels within the frequency range of 108.10 to 111.95 MHz. Signals provide the pilot with course guidance to the runway centerline. The localizer antenna is sited on the extended runway centerline 1,000 to 2,000 feet beyond the far end of the runway. An area of radius 250 feet along with a rectangular area extending from the antenna down the runway at lengths ranging from 2,000 to 7,000 feet and widths ranging from 500 to 600 feet is called the *ILS critical area,* and must be left free of any objects. In addition, when an aircraft is on approach using the ILS as guidance, no other vehicles or aircraft are allowed in or over the ILS critical area.

The **glide slope transmitter** transmits UHF frequencies on one of the 40 ILS channels within the frequency range 329.15 to 335.00 MHz radiating in the direction of the approach. The glide slope transmitter is located between 750 feet and 1,250 feet from the approach end of the runway (down the runway) and offset 250 to 650 feet from the runway centerline. The glide slope antenna may be located on either side of the runway, but preferably the side offering the least possibility of signal reflections from buildings, power lines, and other objects.

In addition to the localizer and glide slope, the ILS is typically equipped with marker beacons to assist pilots in identifying their location on approach, known as the outer marker (OM), middle marker (MM), and in the cases of advanced (i.e., Cat-II or III) ILS systems, an inner marker (IM).

The **outer marker** is typically located from 5 to 7 miles from the end of the runway threshold. A vertically emitted radio signal activates a rapidly flashing blue light on the aircraft's marker beacon receiver when the aircraft passes overhead. The outer marker also produces an audio signal, two Morse code dashes per second, at a low tone to further alert the pilot to the aircraft's position on approach.

The **middle marker** indicates a position approximately 3,500 feet from the runway threshold. This is typically the location where an aircraft on approach will be at an altitude of approximately 200 feet above the elevation of the landing area. This marker's audio signal is a series of alternating Morse code dots and dashes at a high tone. For most ILS systems, the pilot should be able to visually identify the runway either by pavement or with the assistance of an associated approach lighting system. If the pilot cannot visually see the runway at this altitude, called the *design height,* the pilot must declare a *missed approach* and abort the landing.

For advanced ILS systems, the decision height may be less than 200 feet. The **inner marker** identifies the location on the approach of the designated decision height in this case. The inner marker's audio signal is a series of Morse code dots at a high tone.

An ILS allows an aircraft to approach a runway for landing under varying reduced cloud ceiling and visibility conditions, depending on the sophistication of the ILS equipment. The sophistication of an ILS is identified by its category. Table 4-2 identifies the lowest cloud ceiling and visibility allowed at an airport for an appropriately equipped aircraft to land.

ILS systems may also be accompanied by **runway visual range (RVR)** facilities. RVR facilities provide a measurement of horizontal visibility, that is, how far ahead the pilot of an aircraft should be able to see high-intensity runway edge lights or contrasting objects. RVR installations consist of a projector/receiver unit or multiple units located along and ahead of the runway threshold. The number of RVR facilities required depends on the ILS category system installed. ILS CAT-I and CAT-II systems that allow approaches with visibilities between 1,600 and 2,400 feet require one RVR, called a *touchdown RVR,* located 750 to 1,000 feet from the runway threshold, normally behind the ILS glide slope antenna. CAT-II systems that allow approaches with visibilities less than 1,600 feet require an additional RVR, called a *rollout RVR,* located 750 to 1,000 feet from the rollout end of the runway. ILS CAT-III systems and CAT-II systems on runways longer than 8,000 feet require a *midpoint RVR,* located within 250 feet of the longitudinal midpoint of the runway. All RVRs are located adjacent to the runway using the ILS.

Air traffic control and surveillance facilities located on the airfield

At many airports, especially those that experience a high level of operational activity, air traffic control and surveillance procedures are located on the airfield to control and facilitate the safe and efficient movement of aircraft to, from, and around the airport's airfield.

Air traffic control towers

Perhaps the most prominent air traffic control facility located on an airfield is the **air traffic control tower (ATCT).** From control towers, air traffic control personnel control flight operations within the airport's designated airspace [typically within a 5-mile radius of the airport, from the ground to 2,500 feet above ground level (AGL)] and the operation of aircraft and vehicles on the airfield's movement area.

The typical ATCT site ranges from 1 to 4 acres, which includes the facility, associated administration buildings, and vehicle parking. ATCT sites must meet the specific requirements. There must be maximum visibility of the local airspace, including local air traffic patterns, approaches to all runways or landing areas, and to all runway and taxiway surfaces. Furthermore, the ATCT must not derogate the signal generated by any existing or planned electronic navigational aid or other air traffic control facility.

Airport surveillance radar

Airport surveillance radars (ASR) are radar facilities located on the airport airfield used to control air traffic. ASR antennas scan through 360 degrees to present an air traffic controller with the location of all aircraft within 60 nautical miles of the airport. The site for an ASR antenna is flexible, subject to certain location and clearance guidelines.

The location of an ASR antenna should be as close to the ATCT control room as practical. Typical distances between the ASR antenna and the ATCT range between 12,000 and 20,000 feet. Antennas should be located at least 1,500 feet from any building or object that might cause signal reflections and at least one-half mile from other electronic equipment. ASR antennas are typically elevated from 25 to 85 feet off the ground, so as to maintain proper line-of-site clearance.

Airport surface detection equipment

Surveillance and control of aircraft movement on the airport surface is normally accomplished largely by visual means, but during periods of low visibility caused by conditions such as rain, fog, and night, the surface movement of aircraft and service vehicles are drastically reduced. To improve the safety and efficiency of ground movement operations in low visibility, controllers take advantage of two radar-based systems employed at the busier airports. These systems are called **airport surface detection equipment (ASDE-3)** and **Airport Movement Area Safety Systems (AMASS).**

ASDE-3 is a high-resolution ground-mapping radar that provides surveillance of taxiing aircraft and service vehicles at the highest activity airports. AMASS enhances the function of the ASDE-3 radar by providing automated alerts and warnings to potential runway incursions and other hazards. AMASS can visually and aurally prompt ATCT controllers to respond to situations that potentially compromise safety. Because of the high cost of implementation, ASDE-3 and

AMASS have been limited to the very busiest airports. A less expensive ASDE system, known as ASDE-X is at airports where surface surveillance radar would be beneficial at less implementation expense.

The ideal location for ASDE equipment is on the top of air traffic control towers, so as to provide line-of-sight coverage for the entire airfield. If not on an ATCT, ASDE equipment may be placed on a free-standing tower up to 100 feet tall and located within 6,000 feet of the ATCT.

Weather reporting facilities located on airfields

Many airports are equipped with automated weather reporting facilities to provide pilots with up-to-date meteorological information including cloud height, visibility, wind speed and direction, temperature, dew point temperature, and precipitation information. Two of the most common systems are the Automated Weather Observing System (AWOS) and the Automated Surface Observing System (ASOS).

The **Automated Weather Observing System (AWOS)** is a suite of sensors which provide a minute-to-minute update that is usually provided to pilots by a radio on a frequency between 118.0 and 136.0 MHz. Six different AWOS types are available with varying weather reporting capabilities. Table 4-3 lists the different types of AWOS systems and their capabilities.

The **Automated Surface Observing System (ASOS)** is another automated observing system sponsored by the FAA, National Weather Service (NWS), and the Department of Defense (DOD). ASOS provides weather observations that include air and dew point temperature, wind, air pressure, visibility, sky conditions, and precipitation. A total of 882 airports are currently equipped with ASOS or AWOS of varying capabilities (Fig. 4-46).

Table 4-3. AWOS Capabilities

System	Capabilities
AWOS I	Wind speed, wind gust, wind direction, variable wind direction, temperature, air pressure, and density altitude
AWOS II	AWOS I capabilities, visibility, and variable visibility
AWOS III	AWOS II capabilities, sky conditions, cloud heights, and type
AWOS III-P	AWOS III capabilities, present thunder, and precipitation identification
AWOS III-T	AWOS III capabilities, thunderstorm, and lightning detection
AWOS III-P-T	AWOS III capabilities, present weather, and lightning detection

Figure 4-46. *ASOS-3 system.* (Source: FAA)

Wind indicators

Perhaps the simplest system located on airfields that report meteorological conditions are *wind indicators*. Three of the most common wind indicators include wind socks, wind tees, and tetrahedrons. These systems provide vital information at airports where no other sources of weather information are provided so that pilots may appropriately determine the appropriate runway to use for takeoff and landing. At airports where other sources of weather information are provided, wind indicators give the pilot supplemental information of possibly highly variable wind changes while on approach or takeoff.

A **wind sock** is a hollow flaglike object that depicts approximate wind direction and speed. As air flows into the wind sock, it becomes oriented so that it is pointing away from the source of the wind, that is, in the *downwind direction,* toward the approach end of the runway of appropriate use. In addition, the stronger the wind, the straighter the extension of the wind sock.

A **wind tee** is similar to that of a typical weather vane. The wind tee points into the direction of the source of the wind. The typical wind tee is designed in the form of an aircraft to illustrate the suggested direction of operations on the basis of wind direction. The wind tee does not provide any information regarding the speed of the wind.

A **tetrahedron** is a landing indicator typically located near a wind direction indicator. The tetrahedron may swing around with the small end point into the wind, or it may be manually positioned to depict recommended landing direction. The tetrahedron is usually large and painted in such a manner that makes it easily visible to aircraft on approach to the airport.

At airports without control towers, wind direction indicators are usually placed on airfields surrounded by a segmented circle. A segmented circle is a set of markings that depict the runway configuration and recommended traffic patterns of aircraft at the airfield. The segmented circle provides further assistance to pilots with relation to suggesting the proper runway and traffic procedures to use when utilizing the airfield.

Security infrastructure on airfields

Airport facilities require protection from acts of vandalism, theft, and potential terrorist attack. To provide a measure of protection, unauthorized persons must be precluded from having access to all airfield facilities. At most airports where air traffic control facilities, approach lighting systems, and other navigation and weather aids are present, *perimeter fencing* around the airfield is strongly recommended. In addition, security procedures should be established for the protection of the airfield and its facilities.

Access to the airfield from the perimeter is typically regulated by some means of *controlled access.* At smaller airports controlled access measures may be limited to simple padlocks securing access gates adjoining the perimeter fence. Other access controls include the use of identification cards and number combinations to open electronically secured access points. Further details regarding airport security infrastructure may be found in Chap. 8 of this text.

Concluding remarks

The facilities that are located on an airport's airfield comprise a wide variety of technologies that together accommodate the operation of aircraft between the airport and the local airspace. Proper planning and management of the airfield and associated facilities are a necessary component of successful airport operations.

Key terms

airside

airfield

airspace

landside

terminal

ground access

ARFF (aircraft rescue and fire fighting) facilities

runways

primary runway

crosswind runways

parallel primary runways

MSL (mean sea level)

flexible pavement

rigid pavement

runway designators

runway centerlines

runway threshold markings

relocated threshold

displaced threshold

Runway aiming points

runway touchdown zone markings

runway side stripes

ALS (approach lighting systems)

VASI (visual approach slope indicator)

PAPI (precision approach path indicator)

REIL (runway end identifier lights)

LAHSO (land and hold short operations)

taxiways

SMGCS (Surface Movement Guidance Control System)

airport and heliport beacons

mandatory instruction sign

location signs

direction sign

destination sign

NAVAID (navigational aid)

NDB (nondirectional radio beacon)

VOR (VHF omnidirectional range radio) beacon

ILS (Instrument Landing System)

ASR (airport surveillance radar)

ASDE (airport surface detection equipment)

AWOS (Automated Weather Observing System)

ASOS (Automated Surface Observing Systems)

wind sock

wind tee

tetrahedron

Questions for review and discussion

1. What are the four components that make up an airport?
2. How are runways identified on an airfield?
3. What is the difference between a displaced threshold and a relocated threshold?
4. What are the differences between visual, nonprecision instrument, and precision instrument runways?
5. What is SMGCS? How does it help airfield operations?
6. How are taxiways identified on an airfield?
7. What types of taxiways exist on an airfield?
8. What are the imaginary surfaces described in FAR Part 77 for? What are the dimensions of these surfaces?
9. How does an ILS work?
10. What makes the GPS system so different from other navigational aid technologies?

11. What types of technologies exist on airfields to aid aircraft on approach to landing?

12. What types of airport beacons exist? What do the different lighting combinations mean?

13. What are the two sets of flight rules under which airports may operate?

14. What are the different types of signs that are located on an airfield? How are they marked? What do they mean?

15. What are some of the facilities on the airfield that help detect and communicate wind and weather information?

Suggested readings

Ashford, Norman, and Paul H. Wright. *Airport Engineering*. New York: Wiley-Interscience Publications, Wiley, 1979.

Deem, Warren H., and John S. Reed. *Airport Land Needs*. San Francisco: Arthur D. Little, 1966.

FAR/AIM 2010, United States Department of Transportation, 2010.

Horonjeff, Robert. *Planning and Design of Airports,* 4th ed. New York: McGraw-Hill, 1994.

5

Airspace and air traffic management

Outline

- Introduction
- Brief history of air traffic control
- The present-day air traffic control management and operating infrastructure
  - The FAA's Air Traffic Organization (FAA ATO)
  - Air Traffic Control System Command Center
- The basics of air traffic control
  - Visual flight rules (VFR) versus instrument flight rules (IFR)
    - Airspace classes
    - TRACONs
  - Air traffic control towers
  - Victor Airways and Jet Ways
  - Special-use airspace
  - Flight service stations
  - Terminal Area Air Traffic Control Procedures
  - Traditional and Modern "NextGen" Procedures
- Current and future enhancements to air traffic management
  - En route Navigation
  - Modernized Approaches to Airports
  - Airport Surface Movement Management

Objectives

The objectives of this section are to educate the reader with information to:

- Discuss the history of the U.S. air traffic control system.
- Identify the various classes of U.S. airspace.

- Discuss the hierarchical air traffic control management structure.
- Discuss the goals of national airspace system modernization.
- Describe some of the technologies used to modernize air traffic control.
- Understand how air traffic control affects airport management.

Introduction

Whether at a small general aviation airport or at the largest of commercial airline hubs, every aircraft that operates into or out of an airport will either directly interact with, or at least be wary of, the complex hierarchy of organizations, facilities, and regulations that control the nation's and the world's airspace.

In the United States, the Federal Aviation Administration (FAA) owns and operates the facilities that make up the nation's air traffic control system. In most of the rest of the world, the airspace above many nations is controlled by that nation's local government, and supervised by the International Civil Aviation Organization (ICAO). In recent years, many regions of the world have shifted to private or corporate ownership and operation of air traffic control. Examples include Air Services Australia, National Air Traffic Services, Ltd. (serving the United Kingdom), and NavCanada.

Brief history of air traffic control

The roots of today's **air traffic control (ATC)** systems began in the 1920s when pilots relied on scattered radio stations and rotating light beacons to fly from one airport to the next. At the end of the 1920s the federal government introduced the first radio-based navigational aids, known as the "four-course radio range." As its name implies, the four-course radio range transmitted radio waves in four directions, typically, north, south, east, and west. By interpreting the reception of each signal on navigational radios in the aircraft, pilots could judge where they were in the airspace, and report their location to local air traffic "controllers" typically located at each airport.

The first national air traffic control center originated at Newark Airport in Newark, New Jersey, as a privately operated venture formed by cooperative airline companies in October 1935. On July 8, 1936, the Department of Commerce assumed operation of air traffic responsibilities. The duties of air traffic control at the time were to receive routine position reports from aircraft and monitor their respective courses along each of their planned routes. At the control center, each aircraft was physically identified by the creation of a "flight strip," then known as "shrimp boats" and placed on a large map of the U.S. airspace. If two aircraft seemed like they were destined to converge, the control center would radio each pilot to warn aircraft of traffic in close range and perhaps make suggestions for slight course deviations to avoid a collision. Through the 1930s and 1940s the Department of Commerce opened a series of air traffic control centers throughout the country (Fig. 5-1).

By 1950, the technology of air traffic control was significantly enhanced by the introduction of **radar** into the ATC environment. The CAA began to deploy the first **airport surveillance radar (ASR-1)** systems (Fig. 5-2). ASR antennas had the capability of identifying aircraft locations at frequencies as high as 7 seconds. These locations would be reported as "blips" on radar scopes monitored by air traffic controllers. Subsequent advances in radar technology included *transponder encoding technology,* which allowed air traffic controllers the ability to identify not only the location of the aircraft, but also its altitude, speed, and even the aircraft's itinerary information, such as its originating airport and planned destination. With these technologies in place, and the aircraft in the skies flying at faster speeds, air traffic control adapted the concept of *positive control* for aircraft flying in higher altitudes, in poor visibility weather conditions, and around high traffic areas at low altitudes near the busiest airports. Under positive control, the air traffic controller determines the appropriate altitude, direction, and speed at which the aircraft should travel. If a pilot wishes

Figure 5-1. *The first air traffic control center, Newark, N.J., circa 1940.* (Source: FAA)

Figure 5-2. *Early airport surveillance radar.* (Source: FAA)

to deviate from course, altitude, or speed, permission must be granted by air traffic control before any deviations may be made.

The advent of computer technology allowed many of the tasks of air traffic controllers to be automated. In 1967, IBM delivered a prototype computer to the air traffic control center in Jacksonville, Florida (Fig. 5-3). This system, known as the Automated Radar Traffic System (ARTS), serves as the foundation of today's advanced air traffic control technologies.

By the mid-1970s, the FAA had achieved a semiautomatic air traffic control system based on a marriage of radar and computer technology. By automating certain routine tasks, the system allowed controllers to concentrate more effectively on the vital task of providing aircraft separation. Despite its effectiveness,

Figure 5-3. *Automated Radar Traffic (ARTS) System at Jacksonville Center circa 1970.* (Source: FAA)

however, the system required further enhancement to keep pace with the increased volumes of air traffic, particularly since the deregulation of commercial air carriers in 1978.

To meet the challenge of traffic growth, the FAA unveiled the **National Airspace System (NAS) Plan** in January 1982. The new plan called for more advanced systems for en route and terminal-level air traffic control, modernized flight service stations, and improvements in ground-to-air surveillance and communication.

While preparing the NAS Plan, the FAA faced a strike by air traffic controllers. The **Professional Air Traffic Controllers Organization (PATCO)** led a strike in August 1981. The federal government deemed the strike illegal and dismissed over 11,000 strike participants and decertified PATCO as a result of the strike.

In February 1991, the FAA enhanced the NAS plan with the more-comprehensive **Capital Improvement Plan (CIP).** The new plan outlined a program for further enhancement of the air traffic control system, including higher levels of automation as well as new radar, communication, and weather forecasting systems. Programs included the deployment of Terminal Doppler Weather Radar Systems with the ability to warn pilots and controllers of meteorological hazards, with specific attention paid to lighting and thunderstorm activity.

The present-day air traffic control management and operating infrastructure

In the United States, the air traffic control system is operated and managed in a hierarchical structure, ranging from control towers, which monitor and control the movement of aircraft at and around individual airports, to one system command center, which oversees approximately 5,000 aircraft currently in flight, at any given point in time, over the entire United States.

The FAA's Air Traffic Organization (FAA ATO)

The FAA's Air Traffic Organization (ATO) is the office within the FAA primarily responsible for the safe and efficient movement of air traffic within the NAS. The ATO is responsible for administering the rules, regulations, and procedures for air traffic operations, including en route and oceanic operations and terminal area operations. The nation's workforce air traffic controllers fall under the authority of the ATO. In addition, the ATO has the responsibility of ensuring that the air traffic management infrastructure supporting the NAS is well maintained and operational. The ATO also develops rules, procedures, and guidance for operating within the NAS, and develops training procedures for those operating within the air traffic environment.

The ATO supervises each level of the air traffic control operational hierarchy, described in the subsequent paragraphs.

Air Traffic Control System Command Center

At the top of the air traffic control operational hierarchy is the **Air Traffic Control System Command Center (ATCSCC).** The ATCSCC provides macro-level management of every aircraft currently in the national airspace system, as well as those aircraft with itineraries planned hours into the future. The ATCSCC in its current form was established in 1994 and currently resides in Herndon, Virginia. The role of the ATCSCC is to manage the flow of air traffic within the continental United States. The ATCSCC regulates air traffic when weather, equipment, runway closures, or other impacting conditions place stress on the National Airspace System. In these instances, traffic management specialists at the ATCSCC take action to modify traffic demands in order to reduce potential delays and unsafe situations in the air. Some of the strategies used by ATCSCC include the implementation of speed restrictions on aircraft, and imposition of ground delay programs, known as *ground holds,* on aircraft. Under a ground delay program, aircraft destined for an airport with potential delays upon arrival time will be held at its originating airport in order to avoid congestion and delays on route.

The ATCSCC's Airport Reservation Office (ARO) processes all requests for aircraft operating under instrument flight rules (IFR) at designated high-density traffic airports and allots reservations on a first-come first-served basis. As of 2010, four airports were considered high-density airports. They are John F. Kennedy International Airport and LaGuardia Airport in New York, Chicago's O'Hare International Airport, and Ronald Reagan Washington National Airport in Washington, D.C. The ARO also allocates reservations to and from airports with above-normal traffic demand because of special events and circumstances, such as major sporting events, evacuations due to extreme weather, and the like. By implementing a Special Traffic Management Program (STMP), the ARO controls the number of operations generated by an event, allowing for a limited number of reservations in specific time intervals (Fig. 5-4).

ATCSCC employs an Enhanced Traffic Management System (ETMS) to predict, on a national and local level, traffic surges, gaps, and volume based on current and anticipated airborne aircraft. ETMS specialists evaluated the projected flow of traffic into airports and airspace *sectors* and then implemented the least restrictive action necessary to ensure that traffic demand does not exceed system capacity.

The ATCSCC is also responsible for issuing *notices to airmen (NOTAMs)* to provide the most up-to-date information regarding the status of the National Airspace System. Examples of NOTAM information include runway closures, malfunctions to navigational aids, missile and rocket launches, and any areas restricted because of national security issues.

Although the ATCSCC has ultimate control of every aircraft in the system, it is not the job of the ATCSCC to monitor and control the flights of individual

Figure 5-4. *FAA Air Traffic Control System Command Center (ATCSCC), Herndon, Virginia.* (Source: FAA)

aircraft. The tasks of controlling individual aircraft are divided among the lower levels of the ATC hierarchy, specifically the Air Route Traffic Control Centers (ARTCCs), Terminal Radar Approach Control (TRACON) facilities, and air traffic control towers (ATCTs).

The basics of air traffic control

Aircraft flying between airports within the United States operate under varying levels of air traffic control, depending on the location and altitude at which they are traveling and the weather conditions while in flight. In many areas of the United States, particularly at low altitudes around unpopulated areas, aircraft may fly under no direct control by ATC. In contrast, in poor weather conditions, around busy air traffic areas, and at high altitudes, aircraft must fly under *positive control,* where altitude, direction, and speed of aircraft are dictated by air traffic controllers. Air traffic control operating rules are found in Federal Aviation Regulations Subchapter E—"Airspace"—and Subchapter F—"Air Traffic and General Operating Rules"—including:

FAR Part 71	Designation of Airspace Areas
FAR Part 73	Special-Use Airspace
FAR Part 91	General Operating and Flight Rules
FAR Part 93	Special Air Traffic Rules
FAR Part 95	IFR Altitudes
FAR Part 97	Standard Instrument Procedures

Within these regulations, the airspace within the NAS, and the rules for safely operating within the ATC functions of the NAS, are provided.

Visual flight rules (VFR) versus instrument flight rules (IFR)

One factor that determines the level of control an aircraft will be subject to depends, in part, on the type of flight rules the aircraft is operating under. The flight rules, in turn, depend, in part, on the weather conditions during flight. Under weather conditions where the visibility is sufficient to see and avoid other aircraft, and the pilot can keep the aircraft sufficiently clear of clouds, the pilot may operate under **visual flight rules (VFR).** When visibility is insufficient or a pilot's route takes the aircraft through clouds, the aircraft must fly under **instrument flight rules (IFR).** While flying under VFR, there are often times when positive control by ATC is unnecessary; under IFR, positive control is mandated. In the airport environment, VFR rules typically apply when the cloud ceiling at the airport is greater than 1,000 feet above the elevation of the airfield and the visibility is greater than 3 statute miles. When the cloud ceilings or visibility is lower, then the airport is said to be operating under IFR.

Airspace classes

The visibility and cloud clearance criteria determining whether or not an aircraft must fly under IFR versus VFR depends largely on the class of airspace through which the aircraft will be flying. The airspace class of any given location in the United States is defined by the FAA and identified by pilots by referencing air traffic control maps, called *sectionals, terminal area charts,* or *aeronautical charts.* It is important for airport management, as well, to identify the class of airspace under which their airport lies, for it certainly has an impact on aircraft operations at the airport. Since 1993, airspace has been classified as Class A, Class B, Class C, Class D, Class E, or Class G airspace (Fig. 5-5).

Class A airspace, known as **Positive Control Airspace** prior to 1993, is located continuously throughout the continental United States, including the waters surrounding the continental United States out to 12 miles from the coastline, and Alaska, beginning at an altitude of 18,000 feet above sea level (MSL) up to 60,000 feet MSL (known as FL 600). Unless otherwise authorized, all aircraft operating in Class A airspace must operate under IFR.

Class A airspace is controlled by ATC at Air Route Traffic Control Centers (ARTCCs). There are 21 ARTCCs in the United States, each controlling one of 20 contiguous areas in the continental United States, and the area surrounding Alaska.

Class B airspace, known as **Terminal Radar Service Areas (TRSA)** prior to 1993, surrounds the nation's busiest airports (in terms of commercial passenger enplanements or IFR operations). The configuration of each Class B airspace area is specific to each area, but typically consists of a surface area and two or more layers of controlled airspace. The shape of Class B airspace is often described as an "upside down wedding cake." Generally, Class B airspace

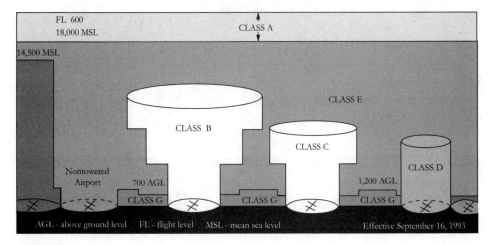

Figure 5-5. *Airspace classifications.* (Source: FAA)

centers on the busiest airport in the area, extending from the surface to 10,000 feet MSL. Aircraft must be granted permission to fly within Class B airspace. Aircraft flying under VFR must be able to remain clear of clouds while in Class B airspace. All aircraft flying in Class B airspace fly under the control of ATC.

Class B airspace is identified by thick dark blue lines, and altitude designations on aeronautical charts (Fig. 5-6). As of 2010, 39 airports in the United States are centered on Class B airspace, and are thus classified as Class B airports.

Figure 5-6. *Class B airspace surrounding Seattle-Tacoma International Airport.*
(Source: NOAA) (Not for navigational use)

Class C airspace, known as **Airport Radar Service Areas (ARSA)** prior to 1993, surrounds those airports that serve moderately high levels of IFR operations or passenger enplanements. Class C is generally considered areas of moderate air traffic volumes, but not as busy as Class B airspace. Class C airspace is usually centered on an airport of moderately high volumes of traffic, ranging from the surface to 4,000 feet above the airport's elevation within 5 miles of the airport, and from 1,200 feet above the surface to 4,000 feet above the surface from 5 to 10 miles from the airport. Class C airspace is also in the form of an inverted wedding cake. When in Class C airspace, each aircraft must establish two-way radio communications with the ATC facility providing air traffic services prior to entering the airspace, and thereafter maintain those communications while in the airspace. To fly under VFR, there must exist at least 3 miles of visibility and aircraft must be able to remain at least 500 feet below, 1,000 feet above, and 2,000 feet horizontally from any clouds. ATC will control aircraft flying under both VFR and IFR to maintain adequate separation from other aircraft under IFR. Aircraft flying VFR are responsible to see and avoid any other traffic. Class C airspace is identified by solid magenta rings and altitude designators on aeronautical charts (Fig. 5-7). A total of 123 U.S.-operated airports are considered Class C airspace airports.

TRACONs

Class B and Class C airspace, as well as some airspace extending beyond the limits of Class B and Class C airspace, is typically serviced by a **Terminal Radar Approach Control (TRACON) facility.** There are approximately 160 TRACON facilities located within the United States controlling air traffic within approximately a 30-mile radius of the busiest airport in the area, from altitudes under 15,000 feet MSL, with the exception of the areas immediately surrounding the airport, which are typically controlled by an **air traffic control tower (ATCT).** The primary objectives of TRACON controllers are to facilitate the transition of aircraft to and from the local airport's airspace into an aircraft's en route phase of flight, and to coordinate the typically high volumes of air traffic flying within the area. TRACON facilities operate strictly by monitoring aircraft by radar and hence may not necessarily be located on airport property, although many are.

Class D airspace, known as Airport Traffic Areas or **Control Zones** (CZ) prior to 1993, surround those airports not in Class B or Class C airspace but do have an air traffic control tower in operation. Class D airspace is generally a cylindrical area, 5 miles in radius from the airport, ranging from the surface to 2,500 above the elevation of the airport. Unless authorized, each aircraft must establish two-way radio communications with the ATC facility providing air traffic services prior to entering the airspace and thereafter maintain those communications while in the airspace. While IFR traffic is controlled by ATC to maintain adequate separation in the airspace, VFR traffic generally is not,

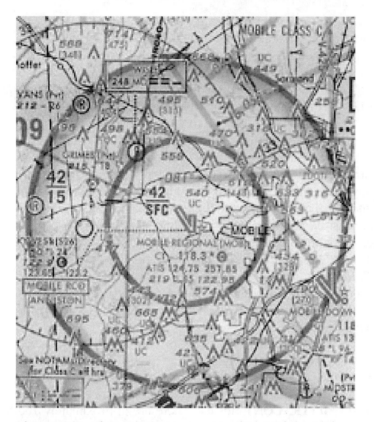

Figure 5-7. *Class C airspace surrounding Mobile, Alabama, International Airport.* (Source: National Aeronautical Charting Office, FAA) (Not for navigational use)

except when performing runway operations (takeoffs or landings). In order to operate VFR in Class D airspace, pilots must have at least 3 miles of visibility and be able to remain at least 500 feet below, 1,000 feet above, and 2,000 feet horizontally from clouds.

Class D airspace is identified by a dashed blue ring and altitude designator on aeronautical charts.

Airports operating under Class B, Class C, and Class D airspace almost always have an operational control tower (ATCT) monitoring operations on and within 5 miles of the airport. The ATCT typically controls all surface movement on the airport, as well as any air traffic departing, landing, or overflying the airport from the surface to 2,500 feet above the ground (AGL). ATCTs located at airports within Class B, Class C, and some Class D airspaces may be equipped with airport surveillance radar (ASR) to facilitate air traffic controllers in identifying and adequately managing the flow of potentially high volumes of traffic within the local airspace. ATCTs in Class B and Class C airspace may also be equipped with airport surface detection equipment (ASDE) to aid in controlling the movement of aircraft on the airfield itself.

Air traffic control towers

As of 2010, nearly 500 airports in the United States were equipped with air traffic control towers. Many of these ATCTs are directly managed by the FAA, and nearly half are operated by private companies at smaller airports. These airports are part of the FAA's Contract Tower Program, which provides funding to airports to construct and support the operation of federal contract towers (FCTs). Services provided to airports under the Contract Tower Program are identical to that of Federal ATCTs, with the exception that they do not control traffic under IFR, but tend to have operating costs approximately half their federal counterparts. Under the federal Contract Tower Program, "low-density airports" are eligible to participate in the program.

Class E airspace, known as **General Controlled Airspace** prior to 1993, generally exists in the absence of Class A, B, C, or D airspace extending upward from the surface to 18,000 feet MSL within 5 miles of airports without control towers. In other areas, Class E airspace generally exists from 14,500 feet MSL to 18,000 feet MSL over the contiguous United States, including the waters within 12 miles off the coast, and Alaska. In addition, federal airways, known as Victor Airways, and Jet Routes, which generally exist from 700 or 1,200 feet above the ground (AGL) are considered Class E airspace. Only aircraft operating under IFR receive positive control in Class E airspace. VFR traffic is responsible to see and avoid all traffic. All aircraft operating under VFR must have at least 3 miles of visibility and be able to remain at least 500 feet below, 1,000 feet above, and 2,000 feet horizontally from clouds at altitudes below 10,000 feet and must have at least 5 miles of visibility and remain 1,000 feet above, 1,000 feet below, and 1 mile clear of clouds at or above 10,000 feet MSL.

Class G airspace, known as **Uncontrolled Airspace** prior to 1993, encompasses the airspace in the absence of Class A, B, C, D, or E airspace. This limited area typically reaches from the surface to 14,500 feet MSL in areas that are not part of federal airways, and from the surface to 700 or 1,200 feet AGL in areas that are part of federal airways. Many remote airfields lie under Class G airspace, and hence have the very basic minimum of air traffic control services, if any at all. Aircraft flying in Class G airspace receive air traffic control assistance only if the workload on air traffic controllers permits. Aircraft flying under IFR generally do not operate in Class G airspace. Aircraft flying under VFR are responsible to see and avoid all other aircraft, must have at least 1 mile of visibility, and be able to remain clear of clouds when flying in daylight conditions below 1,200 feet AGL. At night, when operating under 1,200 feet AGL,

VFR aircraft must have at least 3 miles of visibility, and be able to remain 500 feet below, 1,000 feet above, and 2,000 feet horizontally from clouds. When operating at altitudes above 1,200 feet AGL but less than 10,000 feet MSL, aircraft operating in Class G airspace require at least 1 mile of visibility during the day, and 3 miles of visibility at night, and be at least 500 feet below, 1,000 feet above, and 2,000 feet horizontally clear of clouds. When operating at or above 10,000 feet MSL, and 1,200 feet AGL (at areas of high ground elevation, it is possible to be flying at greater than 10,000 feet MSL and less than 1,200 feet AGL), aircraft in Class G airspace must have at least 5 miles of visibility and remain at least 1,000 feet below, 1,000 feet above, and 1 mile horizontally clear of clouds.

All of the above visibility and cloud clearance requirements have been designated by ATC in the name of safety for all users of the national airspace system.

Victor Airways and Jet Ways

Whether flying by VFR or IFR rules, aircraft flying within the airspace system have traditionally been encouraged to fly on designated corridors known as Federal Air Routes. At low altitudes, the air routes are known as Victor Airways, named so because they are typically defined by a direct line from one VOR navigation facility to another. Victor Airways are typically 8 nautical miles in width and generally range from 1,200 **above the ground (AGL)** up to but not including 18,000 feet **above sea level (MSL).** Victor Airways are identified by light blue lines and designators [denoted by a V followed by a route number (e.g., V123)] on low-altitude aeronautical charts. At altitudes between 18,000 feet MSL and 45,000 feet MSL, routes are known as Jet Routes. Jet Routes are identified as magenta lines and designators [denoted by a J followed by a route number (e.g., J4)] on high-altitude aeronautical charts. In addition to Victor Airways and Jet Routes, an increasing number of Federal Air Routes, known as T-Routes, are being created with the application of GPS-based navigation, and these are discussed in detail later in this chapter.

Special-use airspace

ATC designates certain areas of airspace as special-use airspace (SUA), designed to segregate flight activity related to military and national security needs from other airspace users. There are six different kinds of SUA: prohibited areas, restricted areas, military operations areas, alert areas, warning areas, and controlled firing areas.

Prohibited areas are established over security-sensitive ground facilities such as the White House, certain military installations, and presidential homes and retreats. All aircraft are prohibited from flight operations within a prohibited area unless specific prior approval is obtained from the FAA or the local controlling agency.

Restricted areas are established in areas where ongoing or intermittent activities occur that create unusual hazards to aircraft, such as artillery firing, aerial firing, and missile testing. Restricted areas differ from prohibited areas in that most restricted areas have specific hours of operation. Entry during restricted hours requires specific permission from the FAA or the local controlling agency.

Since September 11, 2001, the FAA and TSA have collaborated in designating and issuing **temporary flight restrictions (TFR)** that identify restricted areas for a period of time for reasons of national security. TFRs have been issued to restrict aviation activity around sporting events, military base activities, or other areas deemed to be security sensitive or potential terrorist targets for given periods of time. It is of utmost importance to pilots and airport managers alike to be aware of any TFRs that may be issued.

Military operations areas (MOA) are established to contain certain military activities, such as air combat maneuvers, intercepts, and acrobatics. Civilian flights are allowed within an MOA even when the area is in use by the military. ATC will provide separation services for IFR traffic within MOAs.

Alert areas contain a high volume of pilot training or an unusual type of aerial activity, such as helicopter activity near oil rigs, which could present a hazard to other aircraft. There are no special requirements for operations within alert areas other than heightened vigilance by pilots.

Warning areas contain the same kind of hazardous flight activity as restricted areas, but are located over domestic and international waters. Warning areas generally begin 3 miles offshore.

Controlled firings areas contain civilian and military activities that could be hazardous to nonparticipating aircraft, such as rocket testing, ordnance disposal, and blasting. They are different from prohibited and restricted areas in that radar or a ground lookout is used to indicate when an aircraft is approaching the area, at which time all activities are suspended.

Flight service stations

Much of the information available to both pilots and airport management regarding the most up-to-date information about air traffic control policies, SUA, NOTAMs, and the itineraries of pilots flying under IFR is provided by ATC flight service stations (FSS). The FSS also broadcast weather reports, process flight plans for aircraft flying under both VFR and IFR, and issue airport advisories.

There are 15 FSS in the United States. In addition, there exist 61 automated flight service stations (AFSS) in the United States, which are equipped with the latest weather, traffic, and communications technologies.

Terminal Area Air Traffic Control Procedures

A significant element of air traffic management is the coordination of aircraft arriving to, and departing from, airports. To facilitate these movements, particularly for aircraft operating under IFR, FAA ATO publishes standard approach and/or departure procedures at most public-use airports in the United States. In general, these are known as **Terminal Instrument Procedures (TERPS)**. Procedures that are used during a departure from an airport are known as **Standard Instrument Departures (SIDs)**, also known as **Departure Procedures (DPs)**, and those that are used during an approach to an airport are known as **instrument Approach Procedures (IAPs)**. These procedures are nearly always developed for one or more runways at one or more airports in a particular vicinity, are created with particular consideration for surrounding terrain and other obstacles, and often reference NAVAID facilities located on or near the airport. These procedures may be considered the primary arrival and departure routes to and from airports, and as such airport management should be keenly aware of any procedures that exist for the airport.

Traditional and Modern "NextGen" Procedures

Traditional terminal air traffic control procedures have been created by referencing traditional ground-based analog NAVAIDS such as those described in Chap. 4 of this text. The most common traditional approach procedures include those that reference NDBs, VORs, and ILS systems. Traditional DPs most commonly reference VORs located some distance from the departing airport. Figures 5-8, 5-9, 5-10, and 5-11 illustrate examples of DPs, and NDB, VOR, and ILS approach procedures, respectively. NDB- and VOR-based approach procedures are known as **nonprecision approaches**, as they rely solely on lateral guidance to direct aircraft to an airport. ILS approaches are considered **precision approaches** as they employ both lateral and vertical guidance in directing aircraft on approach. Precision approaches often provide approach accessibility during poorer weather conditions than their nonprecision counterparts.

In the early 2000s new TERPS procedures that referenced GPS-defined waypoints, rather than ground-based NAVAIDS stations, began to be published. These procedures have adopted the term **Area Navigation (RNAV)** procedures. In the latter part of the decade, a proliferation of a higher accuracy RNAV procedure, known as the **Required Navigation Performance (RNP)** procedure, has greatly improved access to airports that have previously been limited

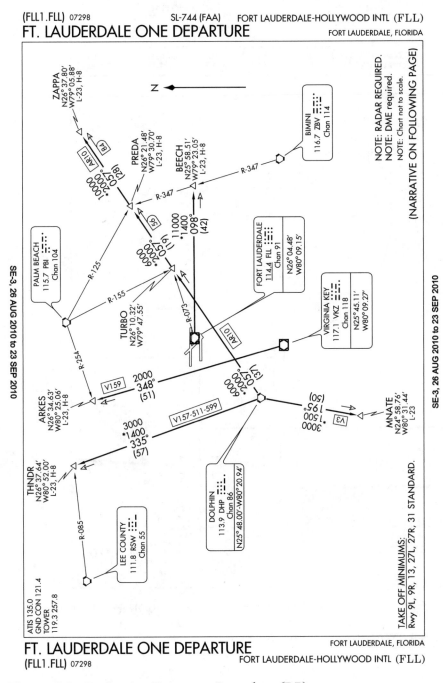

Figure 5-8. *Instrument Departure Procedure (DP).*

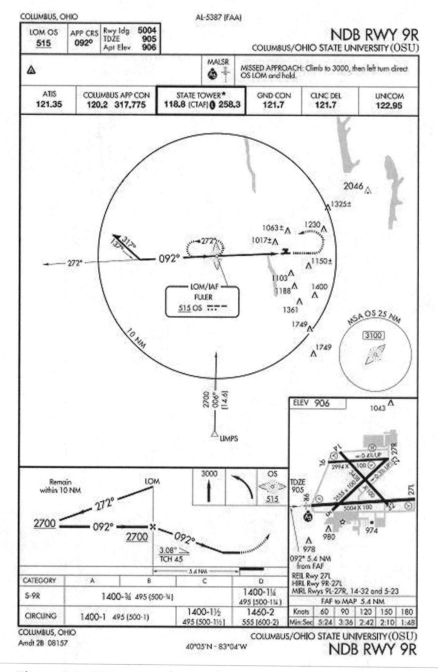

Figure 5-9. *NDB approach procedure.*

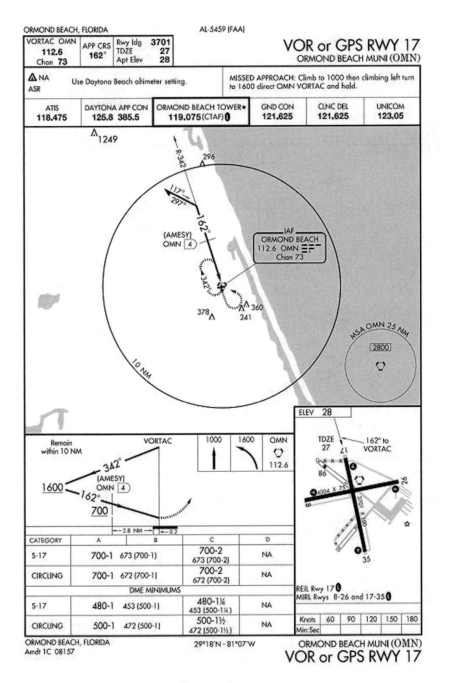

Figure 5-10. *VOR approach procedure.*

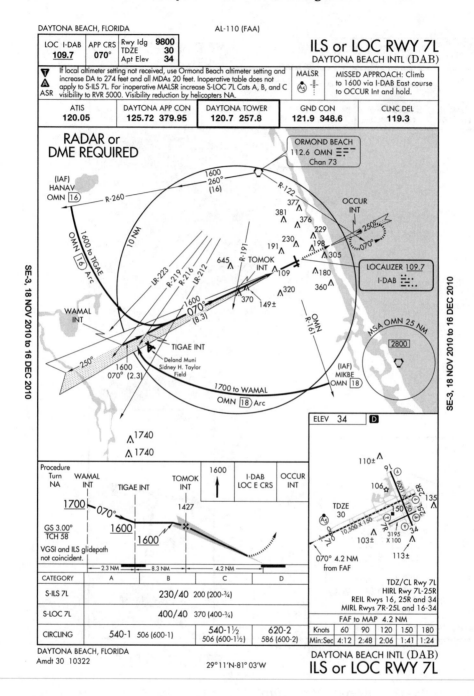

Figure 5-11. *ILS approach procedure.*

due to terrain, airspace restrictions, or conflicts with approaches to other airports. Both RNAV and RNP procedures are continuously being published at airports of all sizes and activity levels, often replacing or *overlaying* on existing traditional procedures. Figures 5-12 and 5-13 provide illustrated examples of RNAV and RNP procedures, respectively.

While a complete description of these procedures is outside the scope of this text, it should be noted that each of these procedures have been designed primarily for approach to a single runway on each airport (although they can be used to access any airport on the airfield under certain weather conditions). These procedures have the following effects on airport operations.

1. They define the minimum weather conditions under which aircraft will be able to approach the airport. These **minimums** include the lowest height of the cloud ceiling and the minimum visibility allowed for safe landing.

2. They are created based on existing terrain and manmade obstacles, and are often affected by new vertical construction within the vicinity of the airport.

3. They affect the physical layout of the airport, from the markings on runways to the size of safety areas.

The procedures are published and maintained by FAA ATO, which does not typically have direct communication with the airport management community. However, it is important for airport management to be aware of these procedures, because these procedures have a direct effect on how the airport and the surrounding land uses are operated, managed, and planned.

Current and future enhancements to air traffic management

The last decade of the twentieth century witnessed the nexus of two critical avenues that has led to the revolution of air traffic management in the early twenty-first century. The overwhelming growth in air transportation demand in the 1990s and early 2000s led to historical levels of congestion and delays at the nation's airports, much attributed to an antiquated air traffic control system whose capacity was being severely tested, and created a need for the modernization of air traffic control. The development and proliferation of advanced navigation and communication technologies, including the Global Positioning System (GPS) and digital-based communications (such as Internet and cell-based voice and data transfer) are now being applied to modernize the air traffic management paradigm. This modernization effort has come to be known as the Next-Generation Air Traffic Management System, or **NextGen.**

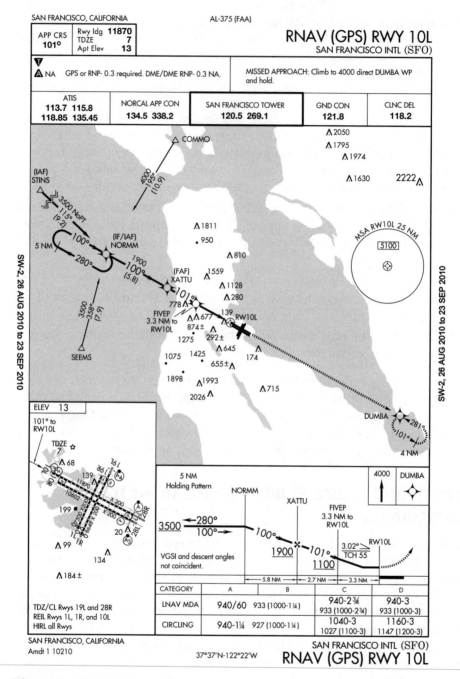

Figure 5-12. *RNAV procedure.*

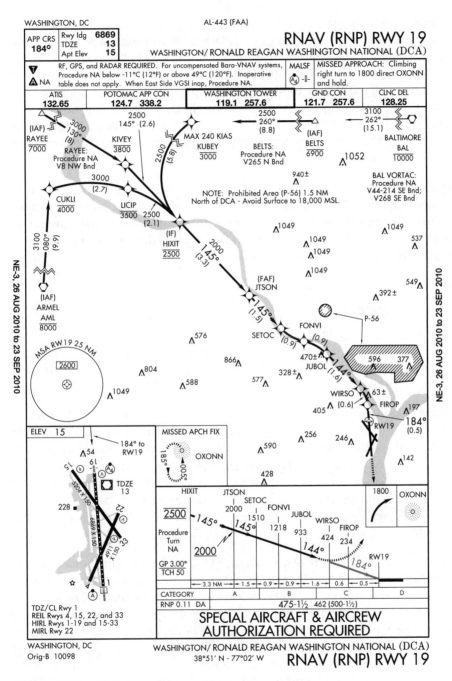

Figure 5-13. *RNP procedure.*

The NextGen modernization effort began with the Congress creating the Joint Planning and Development Office, or **JPDO**, a consortium of seven government organizations with the interests and resources to collectively and cooperatively set the agenda for air traffic modernization. The JPDO was established in 2003 as part of the Vision 100—Century of Aviation Reauthorization Act and includes the U.S. Departments of Transportation, Defense, Homeland Security, Commerce, the FAA, NASA, and the White House Office of Science and Technology Policy. The mission of the JPDO is to unify the efforts of government and industry to form the framework for NextGen.

NextGen is a widely complex system of improvements and enhancements to air traffic management. The system includes enhancement technologies intended to improve the efficiency and safety of **en route navigation, approaches to airports, airport surface movements, flight planning, weather information, and overall communications** among all users of the national airspace system.

En route Navigation

The central technology involved in improving the efficiency of en route navigation under NextGen is known as **Automated Dependent Surveillance—Broadcast (ADS-B)**. ADS-B is, in part, an application of the **Global Positioning System (GPS).**

GPS was developed and is maintained by the U.S. Department of Defense, primarily for the military and activities associated with national defense. In July 1995, GPS gained full operational capability for civilian use, although with reduced accuracy. Since 1995, GPS technology has been applied at increasing volumes, particularly in seafaring transportation, aviation, automobiles, and most recently portable smart phones. GPS applications have become the standard technologies for identifying and navigating between locations around the world, and are an integral component of NextGen.

GPS consists of three segments, space, control, and user. The *space segment* consists of 24 NAVSTAR satellites (21 in active use and 3 spare) placed in circular, geosynchronous orbits, 10,900 nautical miles above the earth. The satellites are positioned so that at least five satellites are always "in view" to a user no matter where that user is on the earth. The satellites continuously broadcast navigation signals, identifying their positions, which are used by GPS receivers to calculate position information at the receiver location. Five monitor stations, three uplink antennas, and a master control station located at Falcon Air Force Base in Colorado Springs, Colorado, make up the *control segment*. The stations track all GPS satellites and calculate precise orbit locations. From this information, the master control station issues updated navigation messages for each satellite, thus maintaining the most accurate position information possible.

The *user segment* includes antennas, receivers, and processors that use the position and time signals broadcast from GPS satellites to calculate precise position, as well as speed, direction of travel, and time. Measurements collected simultaneously from three satellites provide accurate two-dimensional information, usually in terms of latitude and longitudinal positions. A minimum of four satellites providing measurements allows for three-dimensional information, latitude, longitude, and elevation above sea level. Database information contained in GPS receivers correlate this basic position information with referenced points in the database, such as airports, roads, and other landmarks of interest. GPS units with appropriate software technology have the ability to build, save, and navigate according to user-defined routes connected by points, known as *fixes* associated with position locations.

Initially, GPS navigation in the NAS was approved only as an aid to navigation under VFR conditions. This was due to the fact that the position accuracy of GPS was degraded by the DOD, under a program called *selective availability (SA),* for reasons of national security. SA limited the accuracy of GPS readings to between 300 and 1,000 feet, which, among other things, precluded aircraft from navigation with sufficient accuracy to fly and make precision approaches to airports under IFR conditions. SA was turned off by presidential order in May 2000, resulting in position accuracy errors approaching 100 feet. As a result, along with the development of more advanced GPS receivers, pilots gained the ability to fly under IFR and make nonprecision approaches using GPS, effectively legitimizing the concept of what was originally known as **free flight**, that is, the ability to navigate directly between two points using GPS rather than having to rely on ground-based NAVAIDS which often resulted in circuitous routings. In addition to free flight, the FAA continues to enhance the federal airways system by creating T-routes in addition to traditional Victor airways and jet routes. T-routes reference GPS-based defined waypoints, rather than land-based VOR radio beacons. T-routes often offer more direct routing for aircraft and also avoid traditional congestion points in the airspace. The development of differential GPS systems facilitates further GPS position accuracy, sufficient to allow precision approaches using GPS under IFR conditions.

The **Wide Area Augmentation System (WAAS)** is one such augmentation to GPS. WAAS includes integrity broadcasts, differential corrections, and additional ranging signals. The primary objective of WAAS is to provide the accuracy, integrity, availability, and continuity required to support all phases of flight. In doing so, WAAS is designed to allow GPS to be used for en route navigation and nonprecision approaches throughout the NAS, as well as for making precision (equivalent to ILS CAT I) approaches to selected airports. WAAS allows a pilot to determine a horizontal and vertical position to within 25 feet. The wide area of coverage for this system includes the entire United States and some outlying areas.

WAAS consists of a network of ground reference stations that monitor GPS signals. Data from these reference stations are linked to master stations, where the validity of the signals from each satellite is assessed and wide area corrections provide a direct verification of the integrity of the signal from each satellite in view.

The last of 25 initial WAAS reference stations was installed in June 1998. Operational and testing activities in preparation for initial WAAS system commissioning was completed in July 1999. As of 2010, nearly all aviation applied GPS receivers come equipped with WAAS capabilities, further encouraging the use of GPS navigation for operations under both VFR and IFR.

To aid in the efficiency of managing en route navigation, ADS-B will enable transmission of GPS position information, aircraft identification, altitude, velocity vector, and intent information to other users of the system, including other aircraft. This is a fundamental improvement from the current radar-based air traffic control system. Under the current system, such aircraft information is only known by the aircraft itself and the air traffic controller via the radar scope. Aircraft must rely on the air traffic controller to make them aware of other aircraft in the vicinity. With ADS-B, aircraft will be able to "communicate" with each other, allowing pilots to know the positions of any other aircraft in the vicinity.

It is important to note that previous traffic awareness technology also exists. The **Traffic Alert and Collision Avoidance System (TCAS)** was one such technology that had begun to be implemented in the late 1990s. TCAS-enabled aircraft provided the pilot with enhanced traffic surveillance information, using transponder technology. Part of the TCAS capability is a display showing the pilot the relative positions and velocities of aircraft up to 40 miles away. The instrument sounds an alarm when it determines that another aircraft will pass too closely to the subject aircraft. TCAS provides a backup to the air traffic control system's regular separation processes. ADS-B provides an enhancement to older systems such as TCAS in particular because ADS-B offers the ability for two-way data communications between aircraft. This technology offers the possibility for aircraft not only to be aware of other traffic in the area, but also to self-correct to avoid any hazardous reduction in separation. Furthermore, ADS-B may be applied to allow aircraft to route themselves, and even sequence each other on a route, or an approach to an airport that is optimally safe and efficient. The applications of ADS-B form the backbone of the new paradigm that is NextGen air traffic management. Such possibilities were demonstrated in 2005 under the NASA-sponsored research program on Small Aircraft Transportation Systems (SATS). As illustrated in Figure 5-14, the conceptual designs created under the SATS program in the early 2000s have evolved into the modern "glass-cockpits" found in an ever-growing number of aircraft. Systems such as ADS-B are vital components of NextGen and are expected to have a profound impact on the future of air traffic management.

Figure 5-14. *Conceptual cockpit designs created under the SATS program in the early 2000's have evolved into the glass-cockpits of a growing number of modern aircraft.* (Figure courtesy NASA)

Modernized Approaches to Airports

A number of visual and electronic landing aids at or near airports assist pilots in locating the runway, particularly during IMC. Approach procedures have traditionally been based on the type and accuracy of landing aids available, geography, traffic, and other factors. As navigational technologies improve, operating procedures for approach are modified and enhanced commensurate with the characteristics of the new technology.

The application of GPS, ADS-B, and other technologies are being applied not only to creating a safer and more efficient en route air traffic management system but also to create a more efficient system for aircraft approaching to and departing from the nation's airports. This is a highly critical aspect of NextGen, as many of the capacity, safety, and even environmental issues associated with the current air traffic control system occur at the nation's busiest, most congested, and environmentally challenged airports.

One of the applications of NextGen expected to have some of the greatest impacts to approaches to airports is the transition from the use of approach procedures using ground-based NAVAIDS, such as an ILS or VOR, to **GPS-based area navigation** (known as **RNAV**) procedures. While an older term historically used to define navigation by the use of triangulating among a number of ground-based radio beacons, RNAV has come to be known as applying GPS to create a series of waypoints along which a pilot would fly that would safely align the pilot with an airport's runway. More flexible than an approach defined by a few ILS radio signals, RNAV procedures allow for custom-designed approach routings to avoid local terrain, buildings, or even noise-sensitive areas. Furthermore, multiple RNAV approaches may be created for a given runway, which theoretically can tremendously increase the capacity of the local airspace around an airport. Since creating an RNAV procedure is relatively inexpensive as compared to installing ILS systems, RNAV approach procedures may be created for multiple runways at a given airport, and even at airports without any ground-based NAVAIDS, opening up access to thousands of more runways for commercial and general aviation airports during IFR conditions. An enhancement to RNAV is known as **Required Navigation Performance (RNP)**. RNP procedures allow for extremely precise approach routing to airports, allowing aircraft to fly through curved narrow defined corridors to avoid terrain or sensitive land uses. Two examples of how RNP procedures have improved local airspace performance to airports are found in Juneau, Alaska, and Washington, DC (see Fig. 5-13).

The application of ADS-B navigation technology is also paving the way for more efficient and direct approaches to airports through what is known as **continuous descent approaches (CDA)**. Rather than requiring aircraft to descend and stay at lower altitudes in preparation for an approach to an airports, which is both inefficient from an aircraft performance perspective and also creates serious air traffic and environmental impacts, CDA allows aircraft to make steady continuous descents to an airport. From an airport management perspective, this is expected to create greater capacity in the local airspace, as well as significantly reduce the environmental noise and air quality impacts of aircraft flying level at lower altitudes in preparation for their approaches to an airport.

Airport Surface Movement Management

In addition to providing enhanced technologies for the management of aircraft within the airspace, NextGen recognizes that improvements in safety and efficiency of the system must include enhancements to the way aircraft movements are controlled on the airfield. Applications of ADS-B, as well as enhanced ground-based radar systems, are being implemented to achieve this mission.

One such technological implementation is known as **Airport Surface Detection Equipment Model X (ASDE-X).** ASDE-X combines radar, satellites, and ADS-B technology to provide highly accurate positioning and movement information of aircraft and ground vehicles maneuvering around an airfield. ASDE-X is being applied at airports with complex airfield environments as a method of surface traffic control. The application of ASDE-X systems has the primary intention of improving safety on airfields by mitigating the risk of runway incursions, that is, the unauthorized entry to a runway of an aircraft or ground vehicle, which may result in a collision between an aircraft and another vehicle. In addition to safety enhancements, ASDE-X systems are being applied to most efficiently move aircraft and vehicles around the airfield, with the intention of reducing delays caused by airfield congestion. As of 2010, more than 40 of the United States' busiest commercial service airports have implemented ASDE-X systems.

Concluding remarks

The FAA's complex system of management hierarchies, facilities, policies, and technologies that make up air traffic control plays a vital role in the management of the civil aviation system in general and the operation of airports in particular. Although ultimate strategic and daily operational decisions on the control of air traffic lie with government or contracted air traffic controllers, as well as pilots of commercial and general aviation aircraft, knowledge of the air traffic control system, with particular respect to local airspace classifications, the presence of particular navigational aids, and policies for aircraft operations on and within the vicinity of the airport, is vital to the overall efficient management of the airport. The modernization of the national airspace and air traffic management under the NextGen paradigm is an incredibly complex set of solutions that affects every part of the system, including airports. The sheer complexity of the system presents challenges toward system-wide implementation. As recently as 2010, however, increasing elements of NextGen modernization are being implemented into and around airports and regions of airspace around the nation. A system-wide completion of NextGen is estimated to occur in the year 2025. Until then, a series of incremental enhancements to technology and procedures will continue to occur, and contribute to improving the safety and efficiency of the movement of aircraft within the national airspace system.

Key terms

ATC (air traffic control)
ATO (FAA Air Traffic Organization)
radar
ASR (airport surveillance radar)
NAS (National Airspace System) Plan

PATCO (Professional Air Traffic Controllers Organization)

NATCA (National Air Traffic Controllers Association)

CIP (Capital Improvement Plan)

ATCSCC (Air Traffic Control System Command Center)

VFR (visual flight rules)

IFR (instrument flight rules)

NOTAM (notices to airmen)

Class A airspace: Positive Control Airspace

Class B airspace: Terminal Radar Service Areas (TRSA)

Class C airspace: Airport Radar Service Areas (ARSA)

TRACON (Terminal Radar Approach Control)

ATCT (air traffic control tower)

Class D airspace: Control Zones

Class E airspace: General Controlled Airspace

Class G airspace: Uncontrolled Airspace

AGL (above ground level)

MSL (above mean sea level)

prohibited areas

restricted areas

TFR (temporary flight restrictions)

MOA (military operations areas)

alert areas

warning areas

controlled firings areas

TERPS (Terminal Instrument Procedures)

STARS (Standard Terminal Arrival Route Procedures)

SIDs (Standard Instrument Departures)

DP (Departure Procedure)

Nonprecision approaches

Precision approaches

NextGen (Next-Generation Air Transportation System)

JPDO (Joint Planning and Development Office)

RNAV (Area Navigation)

RNP (Required Navigation Performance)

ADS-B (Automated Dependent Surveillance - Broadcast)

ATM (air traffic management)

CPDLC (Controller-to-Pilot Data Link Communications)

NEXCOM (Next-Generation Air-to-Ground Communications)

GPS (Global Positioning System)

WAAS (Wide Area Augmentation System)

TCAS (Traffic Alert and Collision Avoidance System)

ADS (Automated Dependent Surveillance)

STARS (Standard Terminal Automation Replacement System)

ITWS (Integrated Terminal Weather System)

NWS (National Weather Service)

TDWR (Terminal Doppler Weather Radar)

NEXRAD (Next-Generation Weather Radar)

LLWAS (Low-Level Windshear Alert System)

ASOS (Automated Surface Observing System)

WARP (Weather and Radar Processors)

CDM (Collaborative Decision Making)

FSM (Flight Schedule Monitor)

RVSM (Reduced Vertical Separation Minima)

Questions for review and discussion

1. How did the implementation of radar affect the nation's air traffic control system?

2. How is the current administrative operating structure of air traffic control organized in the United States?

3. What is the ATCSCC? What purpose does the ATCSCC serve in the ATC systems?

4. What are the different classes of airspace that exist in the current NAS? How do these classes vary in location and air traffic control regulations?

5. What purposes to ATCTs serve at airports?

6. How do contract towers differ from federal air traffic control towers?

7. What are the different types of special-use airspace? How does each affect the movement of aircraft in their vicinity?

8. What are TFRs? Why have TFRs been implemented? How do TFRs affect airspace and the air traffic control system?

9. What is NextGen? How does NextGen expect to change aircraft management through the NAS?

10. How does GPS work?

11. What technologies exist that enhance the capabilities of GPS?

12. What are the technological enhancements to communications associated with NAS modernization?

13. What is Automated Dependent Surveillance Broadcast (ADS-B)?

14. What types of weather reporting technologies are being developed to improve the safety and efficiency of aircraft traveling in the NAS?

15. What are some of the air traffic management strategies being developed to enhance aircraft movement through the NAS?

Suggested readings

Airport and Air Traffic Control System. Washington, D.C.: U.S. Congress, Office of Technology Assessment, January 2002.

Airport Capacity and Operations. Washington, D.C.: Transportation Research Board, 1991.

Airport Capacity Enhancement Plan. Washington, D.C.: FAA, December 2002.

Airport System Capacity: Strategic Choices. Washington, D.C.: Transportation Research Board, 1990.

Airport System Development. Washington, D.C.: U.S. Congress, Office of Technology Assessment, August 1984.

Capital Investment Plan. Washington, D.C.: FAA, December 1990.

Improving the Air Traffic Control System: An Assessment of the National Airspace System Plan. Washington, D.C.: Congressional Budget Office, August 1983.

Nagid, Giora. "Simultaneous Operations on Closely Spaced Parallel Runways Promise Relief from Airport Congestion," *ICAO Journal* (Montreal, Canada) April 1995, pp. 17–18.

National Airspace System Plan, rev. ed. Washington, D.C.: FAA, April 2001.

Nolan, M. *Fundamentals of Air Traffic Control,* New York: McGraw-Hill, 1994.

Parameters of Future ATC Systems Relating to Airport Capacity/Delay. Washington, D.C.: Federal Aviation Administration, June 1978

NextGen Joint Planning and Development office (http://www.jpdo.gov)

6

Airport operations management under 14 CFR Part 139

Outline

- Introduction
- Part 139 airport classifications
- Inspections and compliance
- Specific areas of airport management of importance to airports found in 14 CFR Part 139
  - Pavement management
    - Runway surface friction
  - Aircraft rescue and firefighting (ARFF)
  - Snow and ice control
    - Timing
    - Equipment and procedures
    - Ice accumulation
    - Aircraft deicing
  - Bird and wildlife hazard management
    - Bird hazards
- Self-inspection programs
  - Ramp/apron–aircraft parking areas
  - Taxiways
  - Runways
  - Fueling facilities
  - Buildings and hangars
  - Components of a safety self-inspection program
- SMS—safety management systems for airports

Objectives

The objectives of this section are to educate the reader with information to:

- Understand the requirements under 14 CFR Part 139 to operate airports serving commercial air carrier operations.
- Describe the different types of airfield pavements, their potential failures, and various types of maintenance programs.
- Describe the major items included in a snow and ice control plan.
- Identify the areas of concern with respect to safety inspection programs.
- Understand the aircraft rescue and fire fighting requirements for a given airport.
- Discuss approaches to mitigating bird and wildlife hazards.
- Be aware of safety management systems.

Introduction

The effective management of the facilities that exist on and around an airport's airfield is vital to the safety and efficiency of aircraft operations. Because of this, airport operations management represents many of the defining issues concerning airport planners and managers. For airports serving, or intending to serve, air carrier operations, the Federal Aviation Administration's Regulations 14 CFR Part 139—Certification of Airports—defines specific policies, activities, and standards for airfield operations management that are required for compliance.

Initially, 14 CFR Part 139 applied to land airports that served any scheduled or unscheduled passenger air carrier operation that was conducted with aircraft having a seating capacity of more than 30 passengers. In 2004, the application of 14 CFR Part 139 was expanded to airports serving scheduled air carrier service on aircraft with a seating capacity of more than nine passengers. An air carrier operation is defined as a takeoff or landing of an aircraft operating under FAR Part 121—Operating Requirements: Domestic, Flag, and Supplemental Operations. During the period 15 minutes before such an operation through 15 minutes after the operation is completed, the airport hosting the air carrier operation must be in compliance with the requirements of 14 CFR Part 139. There are approximately 550 airports in the United States that are certificated under 14 CFR Part 139. All airports that receive federal funding, however, are required to be operated and maintained in a safe and serviceable condition in accordance with minimum standards prescribed by other federal, state, and local agencies. In addition, all airports are subject to comply with regulations specific to the community to which they belong.

Part 139 airport classifications

14 CFR Part 139 classifies airports into four classes, depending on the types of air carrier operations served at airports. This classification is based on the size and scheduling of the aircraft serving air carrier operations at the airport. The following table summarizes the classification of Part 139 airports.

Type of Air Carrier Operation	Class I	Class II	Class III	Class IV
Scheduled large air carrier aircraft (at least 31 seats)	X			
Unscheduled large air carrier aircraft (at least 31 seats)	X	X	–	X
Scheduled small air carrier aircraft (more than 9 but less than 31 seats)	X	X	X	–
Number of airports in United States by Part 139 Class	380	50	35	75

Class I airports may serve scheduled air carrier operations on aircraft with seating capacity of more than 9 passengers, including scheduled and unscheduled service on aircraft with seating capacity of more than 30 passengers.

Class II airports may serve scheduled air carrier operations on aircraft with seating capacity of more than 9 passengers but only unscheduled service on aircraft with seating capacity of more than 30 passengers.

Class III airports may only serve scheduled service on aircraft with seating capacity of more than 9 but less than 31 seats, and do not serve any aircraft with seating capacity of more than 30 seats.

Class IV airports may only serve unscheduled air carrier service.

The FAA defines a **scheduled operation** as "any common carriage passenger-carrying operation for compensation or hire conducted by an air carrier for which the air carrier or its representatives offers in advance the departure location, departure time, and arrival location."

The FAA defines an **unscheduled operation** as "any common carriage passenger-carrying operation for compensation or hire, using aircraft designed for at least 31 passenger seats, conducted by an air carrier for which the departure time, departure location, and arrival location are specifically negotiated with the customer or the customer's representative."

Much of the regulations described in 14 CFR Part 139 are designed to be broad ranging and generic in nature, so as to be applicable to any civil-use airport. Each individual airport is required, then, to create operational procedures specific to its unique environment that comply with the regulations listed in 14 CFR Part 139. For airports to be in compliance with 14 CFR Part 139, a comprehensive list of operational procedures are required to be compiled into an **Airport Certification Manual (ACM).** Airports with an approved ACM and a completed on-site FAA inspection are issued an **Airport Operating Certificate (AOC),** which certifies the airport's compliance with 14 CFR Part 139.

14 CFR Part 139 provides a listing of the specific areas of airfield operations that must meet particular compliance standards. **Table 6-1** provides a summary list of the 29 elements of the ACM that are required for Part 139 compliance. It should be noted that Class I, II, and III airports must comply with all of the 29 elements listed, while Class IV airports are exempt from some requirements.

The elements listed in Table 6-1 may be considered to fit 3 categories:

Administration and record keeping: All certified airports are required to have formal administrative and record keeping policies, such as a formal description of the airport's airfield including all movement areas, safety areas, surrounding obstructions, and any other areas on or around the airport that may be significant to emergency operations; the airport's organizational chart; lines of responsibility; and personnel training procedures.

Operations and maintenance: All certified airports are required to have formal procedures for the safe operation of the airport, and procedures for controlling the movement on and around the airfield infrastructure, including paved and nonpaved areas, lights, markings, and signage, and for Class I, II, and III airports, wildlife hazard management and snow and ice control.

Safety and emergency operations: All certified airports are required to have a formal emergency plan, a description of public protection, and aircraft rescue and fire fighting requirements.

An airport's ACM is written to describe to the FAA how the airport will specifically address these elements. Clearly, an airport's ACM is unique to that airport, as no two airports are exactly alike. The Federal Aviation Administration assesses an airport's compliance with airport operation regulations predominantly on the basis of the airport's unique ACM. Because of this, the ACM is considered to be one of the most personal and important documents created by airport management.

Inspections and compliance

Each year, the FAA conducts inspections of each 14 CFR Part 139 certified airport. In doing so, the FAA inspects to see whether the airport is operating in accordance with its own ACM. According to the FAA, a Part 139 inspection consists of the following phases:

Preinspection review: This includes a review of office airport files and airport certification manual.

Table 6-1 Required Airport Certification Manual Elements

Manual Elements	Airport Certificate Class			
	Class I	Class II	Class III	Class IV
1. Lines of succession of airport operational responsibility	X	X	X	X
2. Each current exemption issued to the airport from the requirements of this part	X	X	X	X
3. Any limitations imposed by the Administrator	X	X	X	X
4. A grid map or other means of identifying locations and terrain features on and around the airport that are significant to emergency operations	X	X	X	X
5. The location of each obstruction required to be lighted or marked within the airport's area of authority	X	X	X	X
6. A description of each movement area available for air carriers and its safety areas, and each road described in §139.319(k) that serves it	X	X	X	X
7. Procedures for avoidance of interruption or failure during construction work of utilities serving facilities or NAVAIDS that support air carrier operations	X	X	X	
8. A description of the system for maintaining records, as required under §139.301	X	X	X	X
9. A description of personnel training, as required under §139.303	X	X	X	X
10. Procedures for maintaining the paved areas, as required under §139.305	X	X	X	X
11. Procedures for maintaining the unpaved areas, as required under §139.307	X	X	X	X

(continued)

197

Table 6-1 Required Airport Certification Manual Elements (continued)

Manual Elements	Airport Certificate Class			
	Class I	Class II	Class III	Class IV
12. Procedures for maintaining the safety areas, as required under §139.309	X	X	X	X
13. A plan showing the runway and taxiway identification system, including the location and inscription of signs, runway markings, and holding position markings, as required under §139.311	X	X	X	X
14. A description of, and procedures for maintaining, the marking, signs, and lighting systems, as required under §139.311	X	X	X	X
15. A snow and ice control plan, as required under §139.313	X	X	X	
16. A description of the facilities, equipment, personnel, and procedures for meeting the aircraft rescue and firefighting requirements, in accordance with §§139.315, 139.317, and 139.319	X	X	X	X
17. A description of any approved exemption to aircraft rescue and firefighting requirements, as authorized under §139.111	X	X	X	X
18. Procedures for protecting persons and property during the storing, dispensing, and handling of fuel and other hazardous substances and materials, as required under §139.321	X	X	X	X
19. A description of, and procedures for maintaining, the traffic and wind direction indicators, as required under §139.323	X	X	X	X
20. An emergency plan as required under §139.325	X	X	X	X
21. Procedures for conducting the self-inspection program, as required under §139.327	X	X	X	X

	Col 1	Col 2	Col 3
22. Procedures for controlling pedestrians and ground vehicles in movement areas and safety areas, as required under §139.329	X	X	
23. Procedures for obstruction removal, marking, or lighting, as required under §139.331	X	X	X
24. Procedures for protection of NAVAIDS, as required under §139.333	X	X	X
25. A description of public protection, as required under §139.335	X	X	X
26. Procedures for wildlife hazard management, as required under §139.337	X	X	X
27. Procedures for airport condition reporting, as required under §139.339	X	X	X
28. Procedures for identifying, marking, and lighting construction and other unserviceable areas, as required under §139.341	X	X	X
29. Any other item that the Administrator finds is necessary to ensure safety in air transportation	X	X	X

In-briefing with airport management: Organize inspection time schedule and meet with different airport personnel.

Administrative inspection of airport files, paperwork, etc. This also includes updating the Airport Master Record (FAA Form 5010) and reviewing the Airport Certification Manual (ACM), Notices to Airmen (NOTAM), airfield self-inspection forms, etc.

Movement area inspection: Check the approach slopes of each runway end; inspect movement areas to find out the condition of the pavement, markings, lighting, signs, abutting shoulders, and safety areas; watch ground vehicle operations; ensure that the public is protected against inadvertent entry and jet or propeller blast; check for the presence of any wildlife; check the traffic and wind direction indicators.

Aircraft rescue and fire fighting inspection: Conduct a timed-response drill; review aircraft rescue and fire fighting personnel training records, including annual live-fire drill and documentation of basic emergency medical care training; check equipment and protective clothing for operation, condition, and availability.

Fueling facilities inspection: Inspection of fuel farm and mobile fuelers; check airport files for documentation of their quarterly inspections of the fueling facility; review certification from each tenant fueling agent about completion of fire safety training.

Night inspection: Evaluate runway/taxiway and apron lighting and signage, pavement marking, airport beacon, wind cone, lighting, and obstruction lighting for compliance with Part 139 and the ACM. A night inspection is conducted if air carrier operations are conducted or expected to be conducted at an airport at night or the airport has an instrument approach.

Post inspection briefing with airport management: Discuss findings; issue Letter of Correction noting violations and/or discrepancies if any are found; agree on a reasonable date for correcting any violations, and give safety recommendations.

According to the FAA, if it finds that an airport is not meeting its obligations, the FAA often imposes an administrative action. It can also impose a financial penalty for each day the airport continues to violate a Part 139 requirement. In extreme cases, FAA might revoke the airport's certificate or limit the areas of an airport where air carriers can land or takeoff. Clearly it is in the best interests of airports that desire to serve commercial service operations to maintain administration and operations in accordance with an approved ACM to remain in compliance with 14 CFR Part 139.

Furthermore, while not a requirement for operation, it is recommended that airports not operating under Part 139 certification implement the elements that are found in ACMs as they do aid in effectively managing a safe airport facility.

Specific areas of airport management of importance to airports found in 14 CFR Part 139

While it is not the intention of this text to simply replicate all of the content found in 14 CFR Part 139, a number of specific areas of airport management that are emphasized in 14 CFR Part 139 are described in this chapter. It is strongly recommended that the reader supplement this chapter with a reading of 14 CFR Part 139, and for those working within airport management, a review of the airports ACM.

The areas described in further detail in this chapter are Pavement Management, Aircraft Rescue and Firefighting (ARFF), Snow and Ice Control, Wildlife Hazard Management, and Self-Inspection Programs. It should be stressed that these subjects are of importance and interests to all of airport management, not solely those operating under 14 CFR Part 139.

Pavement management

For most aircraft, the presence of strong, level, dry, and well-maintained pavement surfaces are required for safe movement to, from, and around an airport's airfield. Thus, the inspection, maintenance, and repair of the runways, taxiways, and apron areas as part of an airfield pavement management program are of utmost importance to airport management.

14 CFR Part 139, Section 139.305, covers some specific characteristics that define the minimum quality standards for airfield pavements, including:

- Pavement edges shall not exceed 3 inches difference in elevation between abutting pavement sections and between full-strength pavement and abutting shoulders.
- Pavement surfaces shall have no hole exceeding 3 inches in depth or any hole the slope of which from any point in the hole to the nearest point at the lip of the hole is 45 degrees or greater as measured from the pavement surface plane, unless, in either case, the entire area of the hole can be covered by a 5-inch diameter circle.
- Pavement shall be free of cracks and surface variations that could impair directional control of air carrier aircraft.
- Mud, dirt, sand, loose aggregate, debris, foreign objects, rubber deposits, and other contaminants shall be removed promptly and as

completely as practicable, with exceptions for snow and ice removal operations.

- Any chemical solvent that is used to clean any pavement area shall be removed as soon as possible, with exceptions for snow and ice removal operations.
- The pavement shall be sufficiently drained and free of depressions to prevent ponding that obscures markings or impairs safe aircraft operations.

Runways are typically paved using one of two sets of materials. Runways may be constructed of **flexible (asphalt)** or **rigid (concrete)** materials. Concrete, a rigid pavement that can remain useful for 20 to 40 years, is typically found at large commercial service airports and former military base airfields. Runways made of rigid pavements are typically constructed by aligning a series of concrete slabs connected by joints that allow for pavement contraction and expansion as a result of the loading of aircraft on the pavement surface, and as a result of changes in air temperature. Runways constructed from flexible pavement mixtures are typically found at most smaller airports. Flexible pavement runways are typically much less expensive to construct than rigid pavement runways. The life of asphalt runways typically lasts between 15 and 20 years, given proper design, construction, and maintenance.

Because of its flexible material characteristics, asphalt paving requires no visible joints or seams. Asphalt might be less expensive to install than concrete, but generally requires much more maintenance in the long run. Much depends on the preparation and grading of the underlying ground, known as the **subgrade,** as well as vigilance and prompt attention to maintenance needs. Moisture is the primary enemy. If water does not drain off the surface and away from the pavement edging quickly, it will filter to the underlying layers of the pavement and weaken it to the point where the overlying layers sag and break open. Potholes then appear as heavy rains wash away loose material.

After years of use, as well as merely by exposure to atmospheric conditions, asphalt runways begin to lose their elasticity. When this occurs, cracks begin to appear on the pavement surface, which allows moisture to penetrate and further weaken the pavement. Ultimately, the pavement is no longer able to support heavy loads. This is known as pavement failure. The life of a pavement may be prolonged in part by patching weakened areas and filling cracks to reduce further moisture penetration.

Concrete runways or taxiways are usually found at large airports with high volumes of air carrier traffic because of their relatively high load-bearing capabilities and resistance to the destructive effects of weather. Concrete also resists deterioration from oil or fuel spillage better than asphalt, and for this reason is generally used for parking ramps and around hangars at all types of airports.

Concrete, being a rigid material that expands and contracts with temperature change, is laid down in slabs separated by contraction and expansion joints. The joints are filled with flexible binder, which either compresses and extrudes or shrinks as the concrete contracts or expands. In colder temperatures, as the concrete contracts, the joints might separate enough to admit material that is essentially incompressible, such as sand or water when frozen.

When incompressible materials infiltrate the joints in concrete, tremendous pressures are generated during later expansion of the slabs, and the concrete might fracture in the joint area. This is known as **spalling.** The fractured edges permit precipitation to seep underneath the pavement surface, causing the subgrade to be washed away. This leads to empty foundation under the concrete slabs, which in turn causes the slabs to become misaligned and perhaps break.

Incompressible material in the expansion joints can also cause the slabs to pop out, that is, to rise and slide over adjacent slabs. It can also cause slabs to buckle upward, cracking the surface and opening up areas for moisture to seep underneath the pavement surface. Considerable amounts of concrete surface can be destroyed in a relatively short time because of poorly maintained expansion joints. Even if the concrete slabs are misaligned only to a small degree, they present a hazard. Landing gear, particularly nose wheels, can be significantly damaged, because irregular surfaces can blow tires and wrench airplanes out of control.

Periodic on-the-ground inspections can easily spot joint openings, surface cracks, and other problems before the runway becomes a hazard to aviation operations. Specific runway conditions that are considered hazards include alligatoring of asphalt surfaces, pavement cracking, rutting, raveling, and the creation of potholes.

The following symptoms provide evidence of potential pavement failures:
- Ponding of water on or near pavement
- Building up of soil or heavy turf at pavement edges, preventing water runoff
- Clogged or overgrown ditches
- Erosion of soil at pavement edges
- Open or silted-in joints
- Surface cracking or crumbling
- Undulating or bumpy surfaces

A number of actions can be taken to repair the distresses that occur in concrete and asphalt pavements. The determining factor in selecting an action is the degree to which the pavement has deteriorated. Pavements that have little deterioration generally require moderate maintenance, whereas pavements that are more extensively deteriorated require *rehabilitation* or *reconstruction*. The FAA defines **pavement maintenance** as "any regular or recurring work necessary, on a continuing basis, to preserve existing pavement facilities in good condition, any work involved in the care or cleaning of existing pavement facilities, and incidental or minor repair work on existing pavement facilities." Pavement maintenance involves, for example, sealing of small surface cracks.

The FAA defines **pavement rehabilitation** as the "development required to preserve, repair, or restore the financial integrity" of the pavement. Adding an additional layer of asphalt on the surface of a runway with the goal of restrengthening the pavement would be considered a rehabilitation. The FAA typically provides airport funding through the AIP program for pavement rehabilitation projects. Pavement maintenance projects are generally not eligible for AIP funding.

Though approaches to repairing pavements may differ, some experts note that appropriately timed maintenance and rehabilitation forestalls the need to replace the pavement entirely, termed **pavement reconstruction,** which is a far more expensive process. An appropriate maintenance program can minimize pavement deterioration. Similarly, rehabilitation can extend the time needed until the pavement must be replaced.

A proper pavement management program evaluates the present condition of a pavement and predicts its future condition through the use of a pavement condition index. By projecting the rate of deterioration, a life cycle cost analysis can be performed for various alternatives, and the optimal time of application of the best alternative is determined.

During the first 75 percent of its life, a pavement's performance is relatively stable. It is during the last 25 percent of its life that pavement begins to deteriorate rapidly. The challenge of pavement management programs is to predict as accurately as possible when that 75 percent life cycle point will be reached for a particular piece of pavement so its maintenance and rehabilitation can be scheduled at the appropriate times.

The longer a pavement's life can be stretched until it must be rehabilitated, the lower the overall life cycle cost of the pavement will be. According to the FAA's own estimates, the total costs for ignoring maintenance and periodically rehabilitating poor pavement can be up to four times as high as the cost for maintaining the same piece of pavement in good condition.

An accurate and complete evaluation of the existing pavement system is one of the key factors contributing to the success of a maintenance project. Major strides have been made in this area with the development and application of **nondestructive testing (NDT).**

One of the most effective and valuable of the nondestructive techniques is *vibratory* or *dynamic testing*. This technique measures the strength of the pavement system by subjecting it to a vibratory load and measuring the amount the pavement responds or *deflects* under this known load. Of the many devices available to perform these tests, one of the most popular is the Road Rater. This maneuverable device can perform an accurate test in as little as 12 seconds.

Pavement evaluations in the past normally included large numbers of expensive and time-consuming destructive tests, including cores, borings, and test pits. Selected on a visual or random basis, the locations at which these tests were performed yielded results of varying degrees of success. Unfortunately, taking enough tests to provide reasonable assurance that the results were meaningful also meant high costs and excessive pavement closure time.

Taking advantage of the economy and speed of vibratory testing, it is possible to saturate the pavement system with tests to determine a very minimum number of locations at which destructive tests can be performed for a complete and accurate evaluation of a pavement and its components. Additionally, the results frequently point out other factors contributing to pavement weakness such as drainage deficiencies.

Rigid concrete pavements may also be examined with this technique to evaluate the likelihood of voids, extent of pumping, load transfer qualities, and the degree to which cracked sections reseat themselves under traffic loadings. These considerations determine the amount and time of remedial or preventive maintenance appropriate to prepare the concrete pavement system so that it will perform adequately after it is overlayed.

The testing of the airport surface will reveal maintenance needed to upgrade the pavement system to the specifications as outlined in the airport manager's long-range plan. Products and methods for pavement maintenance are constantly changing. Environmental conditions require different applications of similar materials and methods of construction. Solutions to the problem of pavement maintenance take on specific characteristics, primarily on the basis of the specific environment and activity levels individual airports experience.

Maintenance in general substantially reduces the need for extensive repairs or replacement of deteriorating airport surfaces. The ultimate solution to pavement difficulties is to discover and repair minor damage before major damage evolves. Airport operations personnel normally make daily inspections of pavement surfaces, note emerging problems, and call for technical assistance.

Periodically, an inspection by a civil engineer is made to check on the more subtle forms of pavement distress.

Runway surface friction

One of the more important characteristics of runway pavements, in particular, is surface friction. Surface friction allows aircraft to safely accelerate for takeoff, and to decelerate after landing. Lack of sufficient surface friction will result in aircraft skidding, slipping, and general loss of control on the runway surface.

Runway pavement surface friction is threatened by normal wear, moisture, contaminants, and pavement abnormalities. Repeated traffic movements wear down the runway surface. Wet weather can create **dynamic** or **viscous hydroplaning.** Dynamic hydroplaning is a condition where landing gear tires ride up on a cushioning film of water on the runway surfaces. Viscous hydroplaning occurs when a thin film of oil, dirt, or rubber particles mixes with water and prevents tires from making sure contact with pavement. Contaminants, rubber deposits, and dust particles accumulate over a period of time and smother the surface. The pavement itself might have depressed surface areas that are subject to ponding during periods of rainfall.

The most effective and economical method of reducing hydroplaning is **runway grooving.** One-quarter-inch grooves spaced approximately 1¼ inches apart are made (generally with diamond blades) in the runway surface. These safety grooves help provide better drainage on the runway surface, furnish escape routes for water under the tire footprint to prevent dynamic hydroplaning, and offer a means of escape for superheated steam in reverted rubber skids. Grooving also assists in draining surface areas that tend to pond, reducing the risks of spray ingesting, fluid drag on takeoff, and impacting spray damage. Unfortunately, the grooves become filled with foreign matter and must be cleaned periodically. The removal of rubber deposits and other contaminants includes use of high-pressure water, chemical solvents, and high-velocity impact techniques.

The *high-pressure water method* is based on high-pressure water jets aimed at the pavement surface to blast contaminants off the pavement surface. The technique is environmentally clean and removes deposits in a minimum of time. High-pressure water equipment operates between 5,000 and 8,000 psi and is capable of pressures exceeding 10,000 psi. The high-pressure water method of runway surface cleaning may be used only in temperatures greater than 40 degrees Fahrenheit, where the risk of icing is minimized.

Chemical solvents have also been used successfully to remove contaminants from both concrete and asphalt runways. Chemicals must meet environmental standards. Acid-based chemicals are used on concrete runways and alkaline chemicals on asphalt.

The *high-velocity impact method* consists of throwing abrasive particles at high velocity at the runway surface. This technique blasts contaminants from the surface and can be adjusted to produce the desired surface texture. The abrasive material is propelled mechanically from the peripheral tips of radial blades in a high-speed, fanlike wheel. This reconditioning operation may be carried out during all temperature conditions and seasons except during rain, or in standing water, slush, snow, or ice.

Regardless of the type of pavement used on an airfield's runways, taxiways, and apron areas, a prescribed plan for pavement inspection, maintenance, and rehabilitation is essential for the safe operation and movement of aircraft at the airport.

Aircraft rescue and fire fighting (ARFF)

Although the incidents of fires and emergencies occurring at an airport are rare, when they do occur, especially on an aircraft, the fire fighting and rescue capabilities at the airport may mean the difference between life and death for pilots, passengers, and other airport personnel. Because of this, aircraft rescue and fire fighting (ARFF) services are strongly recommended at all airports and are required to be present at all airports operating under 14 CFR Part 139. For those airports not operating under 14 CFR Part 139, an agreement with local municipal rescue and firefighting agencies is necessary for safe operations.

The characteristics of aircraft fires are different from those of other structures and equipment because of the speed at which they develop and the intense heat they generate. Because of this, 14 CFR Part 139 designates specific ARFF requirements based on the type of aircraft that typically use any given airport.

14 CFR Part 139.315 designates the *ARFF index* of an airport based on the length (from nose to tail) of air carrier aircraft that use the airport and the average number of daily departures of air carrier aircraft. ARFF index is determined by the longest aircraft that serves the airport on an average of five or more departures per day. Index determination based on aircraft length is as follows:

Index A: Aircraft less than 90 feet in length

Index B: Aircraft more than 90 feet but less than 126 feet in length

Index C: Aircraft more than 126 feet but less than 159 feet in length

Index D: Aircraft more than 159 feet but less than 200 feet in length

Index E: Aircraft greater than 200 feet in length

The index system is based on an area that must be secured to effect evacuation or protection of aircraft occupants should an accident involving fire occur. The protected area is equal to the length of the aircraft, multiplied by a 100-foot width, consisting of 40 feet on each side of the fuselage plus a 20-foot allowance for fuselage width. The indexing system was based on this critical area concept, expressed in aircraft length, to provide a more equitable protection to all aircraft using the airport.

ARFF uses combinations of water, dry chemicals, and **aqueous film-forming foam (AFFF)** to fight aircraft-based and other airfield fires. 14 CFR Part 139.317 describes the required ARFF equipment and agents to be present at the airport, based on the airport's ARFF index. These minimum requirements are as follows:

Index A airports require one ARFF vehicle carrying at least:

 1. 500 pounds of sodium-based dry chemical, halon 1211, or clean agent

or

 2. 450 pounds of potassium-based dry chemical and 100 pounds of water and AFFF for simultaneous water and foam application

Index B airports require either of the following:

 1. One vehicle carrying at least 500 pounds of sodium-based dry chemical, halon 1211, or clean agent, and 1,500 gallons of water, and AFFF for foam production

or

 2. Two vehicles, with one vehicle carrying the agents required for Index A and one vehicle carrying enough water and AFFF so that the total quantity of water for foam production carried by both vehicles is at least 1,500 gallons

Index C airports require either:

 1. Three vehicles, with one vehicle carrying the agents required for Index A, and two vehicles carrying enough water and AFFF so that the total quantity of water for foam production carried by all three vehicles is at least 3,000 gallons

or

 2. Two vehicles, with one vehicle carrying the requirements for Index B, and one vehicle carrying enough water for foam production by both vehicles is 3,000 gallons

Index D airports require three vehicles, including:

 1. One vehicle carrying the agents required for Index A

 2. Two vehicles carrying enough water and AFFF so that the total quantity of water for foam production carried by all three vehicles is at least 4,000 gallons

Index E airports require three vehicles, including:

1. One vehicle carrying the agents required for Index A
2. Two vehicles carrying enough water and AFFF so that the total quantity of water for foam production carried by all three vehicles is at least 6,000 gallons

14 CFR Part 139 indicates a minimum response time of the first vehicle to an incident, defined by the ability to reach the midpoint of the runway farthest from the vehicle's assigned post, of 3 minutes from when an alarm is sounded, with all other vehicles required to the scene within a minimum of 4 minutes.

Until the 1960s, airport fire fighting equipment consisted of little more than modified versions of the gear used by municipal fire services. Today, nearly every major airport is equipped with rapid intervention vehicles (RIVs) able to reach runways within 2 minutes of an alarm. Heavy-duty vehicles are designed to cross rough ground to reach a distant runway or go into rough terrain, where many accidents tend to occur (Fig. 6-1).

RIVs are fast trucks that carry foam, water, medical and rescue equipment, and lights for use in fog and darkness. Their crews begin holding operations to contain the fire and clear escape routes. Heavy-duty foam tenders follow. They are large, but fast and maneuverable, and carry about 10 times more foam than the RIV. Turret-mounted foam guns swivel to project the foam up to 300 feet (Fig. 6-2).

Most airport ARFF facilities are based on the quick delivery of foam extinguishing agents to the scene of an accident. Foam is the general selection because the two main ingredients, foam concentrate and water, can be brought to the scene and applied to an existing fire most efficiently.

Foam smothers the flames and cools the surrounding area to prevent further outbreak of fire. Water is only really effective as a coolant. Dry powder (either sodium or potassium bicarbonate base) is most effective on localized fires in wheels or tires, or in an electrical apparatus. Foam does have some limitations as an extinguishing agent. It must be applied in large quantities in what the National Fire Protection Association describes as a "gentle manner so as to form an impervious fire-resistance blanket" when dealing with large flammable liquid spills, such as fuel and hydraulic fluids.

The foam blanket, once applied, can be broken by wind, clear water streams, turbulence, or even the "heat baking" generated by residual heat in metals or burned-over surfaces. Applying a good blanket of foam and keeping the blanket intact is a primary concern within ARFF procedures.

The light rescue unit illustrated on the left can carry 300 pounds of dry powder sodium bicarbonate in two units pressurized by carbon dioxide. Each discharge nozzle can eject powder at the rate of 3 pounds per second over a range of 39 feet.

The heavy-duty fire tender illustrated on the right can discharge over 10,000 gallons of water or foam a minute through its monitor and over 1,000 gallons through each of its two hand-lines, while moving forward or backward.

The tank holds 200 gallons of foam concentrate and is designed so that the base slopes down to a sump.

The four-person four door cap is made of double-skinned insulated aluminum.

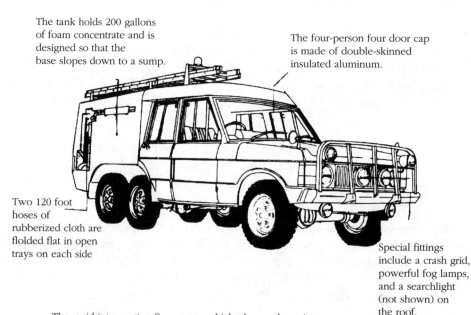

Two 120 foot hoses of rubberized cloth are flolded flat in open trays on each side

Special fittings include a crash grid, powerful fog lamps, and a searchlight (not shown) on the roof.

The rapid intervention fire rescue vehicle shown above is designed to accelerate to 70 mph as fast as a sports car, despite the weight of foam and dquipment carried. It carries 240 gallons of a concentrated ready-mixed water and foam solution, and first aid and rescue equipment. It is used to contain the fire and keep aircaraft escape routes open until the main fire fighting force arrives. This versatile chassis can be fitted with stretchers and other special equipment, for use as an ambulance.

Figure 6-1. *Typical ARFF vehicles.*

Figure 6-2. *Advanced ARFF "Striker" vehicle.* (Photo courtesy Oshkosh, Inc.)

Training is a key ingredient to the overall effectiveness of ARFF. There are two basic challenges to airport management in this regard, initial training and maintaining fire fighting readiness and efficiency. To keep ARFF personnel and equipment in top working order, intensive in-service training programs should be developed. 14 CFR Part 139 suggests that any ARFF training curriculum contain instruction in the following areas:

- Airport familiarization
- Aircraft familiarization
- Rescue and fire fighting personnel safety
- Emergency communications systems at the airport, including fire alarms
- Use of the fire hoses, nozzles, turrets, and other appliances required for compliance
- Application of the types of extinguishing agents required for compliance
- Emergency aircraft evacuation assistance
- Fire fighting operations
- Adapting and using structural rescue and fire fighting equipment for aircraft rescue and fire fighting
- Aircraft cargo hazards
- Familiarization with firefighters' duties under the airport emergency plan

Furthermore, at least one ARFF person on duty must be trained in emergency medical care, covering the following areas:

- Bleeding
- Cardiopulmonary resuscitation
- Shock
- Primary patient survey
- Injuries to the skull, spine, chest, and extremities
- Internal injuries
- Movement of patients
- Burns
- Triage

14 CFR Part 139 requires all ARFF personnel to participate in at least one live-fire drill every 12 months. Many airports conduct real-time accident drills, which include ARFF, as well as all other elements of airport management and local community services, every 3 years.

Snow and ice control

In many areas in the northern and mountainous regions of the United States, the removal of snow and ice from airfield pavements represents a significant portion of an airport's overall operations budget. How effective this expenditure is depends on the ability of management to plan and execute an efficient snow and ice control plan (SICP).

14 CFR Part 139.313, states specifically that all airports operating under 14 CFR Part 139 where snow and icing conditions regularly occur shall prepare, maintain, and carry out an SICP. The SICP shall include instructions and procedures for:

- Prompt removal or control, as completely as practical, of snow, ice, and slush on each pavement area
- Positioning snow on movement area surfaces so that all air carrier aircraft propellers, engine pods, rotors, and wingtips will clear any snowdrift and snowbank as the aircrafts' landing gear traverses any full-strength portion of the pavement area
- Selection and application of approved materials for snow and ice control to ensure that they adhere to snow and ice sufficiently to minimize engine ingestion
- Timely commencement of snow and ice control operations
- Prompt notification of all air carriers using the airport when any portion of the pavement area normally available to them is less than satisfactorily cleared for safe operation by their aircraft

A typical SICP tends to include:

1. A brief statement of the purpose of the program.

2. A listing of the personnel and organizations (airport and other) responsible for the SICP: Many airports hire additional personnel during the winter months or utilize personnel from the streets and sanitation departments on an emergency basis.

3. Standards and procedures to be followed: There are a number of excellent sources that airport management uses in preparation of this aspect of their SICP. They include the *Air Transportation Snow Removal Handbook*, published by the ATA, and FAA Advisory Circular 150/5200-30C, Airport Winter Safety and Operations. The AAAE also sponsors an annual International Aviation Snow Symposium at which workshops are held covering all aspects of snow removal.

4. Training: Because the airport snow removal program requires special skills, a training program is normally an integral part of the plan. This includes classroom training in such areas as airport orientation, snow removal standards and procedures, use of various types of equipment, aircraft characteristics (capabilities and limitations), description of hazards and problem areas at the airport, communications, and safety procedures. On-site training includes a review of operational areas and hazards, test runs with equipment to accustom operators to area dimensions and maneuvering techniques, and communications practice while on the job.

In addition, SICPs should include air traffic control communications, safety considerations, inspection standards, and notice to airmen (NOTAM) responsibilities.

Timing

Knowing when to implement the SICP in order to maintain safe operations and avoid unnecessary repetition of certain activities is critically and generally learned through experience. Weather forecasts including the following information can be helpful in this regard:

- Forecasted beginning of any snowfall
- Estimated duration, intensity, and accumulation
- Types of precipitation expected
- Anticipated wind directions and velocities during the snowfall
- Temperature ranges during and after the snowfall
- Cloud coverage following the snowfall

Snow removal is generally geared to the operational limitations of the most critical aircraft using the airport. Large jet aircraft have a takeoff limitation of

1/2 inch of heavy wet snow or slush, and 1 inch of snow of medium moisture content. This means that removal operations must get underway before such conditions occur, and must continue without interruption until the end of the snowfall event and snow removal has progressed to the point where aircraft operations may be carried on with safety.

Specifically, airports under 14 CFR Part 139 should include in their ACM SICP a set of areas on the airfield on which snow removal will take priority areas. These "priority 1" areas typically include one or more primary active runways and a set of taxiways and ramp areas most critical to safe aircraft operations during winter weather conditions. The following table identifies the FAA's recommended times to clear priority 1 runway surfaces of at least 1 inch of snow to accommodate aircraft operations (ref. FAA AC 150/5200-30C):

	Clearance Times (hours)	
# Annual Operations	Commercial Service Airport	General Aviation Airport
40,000 or more	1/2	2
10,000 but less than 40,000	1	3
6,000 but less than 10,000	1 ½	4
Less than 6,000	2	6

The above table is to be used as a reference by airport management to aid in determining the appropriate number of snow removal vehicles and equipment for the airport.

Snow removal operations normally are started on the active runway and progress to other runways and taxiways. At the same time this work is proceeding, snow clearing from ramps, aircraft loading positions, service areas, and public facilities also takes place, because all of these areas are closely related in the overall operation of an airport facility.

Equipment and procedures

There are two basic methods of removing snow and ice: *mechanical* and *chemical*. Most removal is accomplished by mechanical means, because chemical methods are generally more expensive and less effective than those available for highway use, for example. Underground hot water and electrical heating systems are used around ramps areas at some large airports. Such systems are very expensive to construct and maintain, precluding them from implementation at most airports.

The three mechanical methods of snow removal include *plows, blowers/throwers,* and *brushes.* Snowplows available for airport use do not differ significantly from those used on automobile highways. Snowplow blades are available in steel, steel with special carbide steel cutting edges, rubber, and polyurethane edges. The carbide steel edge gives longer life than the traditional steel edges, cuts packed snow more effectively, and can be more effective in the removal of ice that is not bound to the pavement surface.

Rubber blades have a longer life than steel blades, make less noise and vibration (which contributes to the operator's comfort), and work well on slush, although not as well as steel on dry or packed snow. Rubber blades cost considerably more than steel blades, but generally last 5 to 10 times longer, depending on how carefully they are maintained.

The snowblower or snow thrower is the primary mechanical device for removal of hazardous snow accumulations such as windrows and snowbanks. Blowers are frequently used to clear taxiways, ramps, and parking areas prior to windrow removal on the runway.

Snow brushes are primarily used to clean up the residue left on the surface by plows or blowers. They are also used to clear surfaces of a light snow and to remove sand spread on the runway to improve friction. The brush is the only use of the three basic types of equipment that has year-round use at the airport. Runways and taxiways can be kept clean and free of debris with the brush, which prevents *foreign object debris (FOD)* from damaging aircraft propellers or turbine engines. The brush typically operates the slowest of the three types, and because of its relative ineffectiveness in removing and appreciable snow accumulation will not be useful as the initial attack machine on most snowfalls. Its use, however, can eliminate the problems caused by freezing residue on the surfaces and the concern that aircraft operators have regarding turbine ingestion and propeller erosion caused by loose, dry sand on the runway. Brushes are available with either steel or synthetic bristles. The steel bristle cuts ice more effectively, but the nylon or polypropylene bristle is more effective on very wet snow or slush.

Snow removal equipment is expensive, but losses in revenue sustained by an airport closed by snow may be far greater, if appropriate equipment is not acquired (Fig. 6-3).

For airfield pavements, different types of chemicals may be used to prevent or remove snow and use accumulations. Such chemicals include urea, acetate-based compounds, and sodium formate. Urea is a solid synthesized crystalline granular compound that is often used as fertilizer. Urea is effective in removing ice and snow in temperatures as low as 15 degrees Fahrenheit. Acetate-based compounds include potassium acetate (known as Cryotech), calcium magnesium acetate (CMA), or sodium acetate (Clearway 2). Acetate-based compounds are known to be effective in temperatures as low as 250 degrees Fahrenheit.

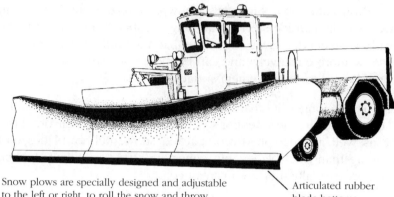

Snow plows are specially designed and adjustable
to the left or right, to roll the snow and throw
it to either side. The articulated rubber blades are
designed to clear runway lights. The plow illustrated
above can clear 21 feet per pass at up to 40 mph.

Articulated rubber
blade bottoms

A snow blower or thrower such as the one illustrated
above can clear windrows at speeds of 35 mph
and handle as much as 3,000 tons per hour.

A snow brush shown above can clear a heavy snowfall
in one operation. Angled up to 45 degrees, the 14 foot long brush scatters snow
up to 50 yards to either side. Variable speed control can be set to slow
speeds for sweeping surface dirt up to 550 rpm to clear heavy slush. By means
of an air deflector snow is thrown high to be carried away by the wind, or low to
avoid blowback.

Figure 6-3. *Airfield snow removal equipment.*

Snow removal normally begins as soon as there are traces of precipitation on the runways. Snow is usually allowed to accumulate to 1 inch on ramp areas before airport personnel or contractors are called in for removal. Accumulation is normally trucked to a snow dump in an outlying area of the airport.

Snow clearance operations on runways are normally carried out by a chain of four or five vehicles working in an echelon formation. First the plows move down the runways at speeds up to 35 mph and move the snow to the pavement edges (Fig. 6-4). This is normally followed by snowblowers or snow throwers, which disperse the windrowed snow into the open areas beyond the pavement. Snow brushes are used for cleaning snow from semiflush-type mixtures such as in-runway lighting installations, and for the removal of slush and very light snow accumulation.

Ice accumulation

Although snow accumulation can pose serious risks to the safety of airfield operations, ice is the most difficult problem to cope with, presenting the greatest hazards to aircraft operations. Many airports attempt to control such conditions through the use of sand. Unfortunately, dry sand spread on runways and taxiways is quickly removed by an aircraft engine blast, so a means of securing the sand to the ice is necessary. The most successful method uses conventional weed-burning equipment. The procedure is to apply sand to the icy surface by hydraulically powered and operated truck spreaders. These lay down a uniform

Figure 6-4. *Snowplows move the snow to the pavement edges where snowblowers disperse the windrowed snow.* (Photo courtesy FAA

layer of sand. This is immediately followed by flame-thrower type burner units, which heat the sand particles and melt the ice sufficiently to produce a coarse, sandpaperlike surface upon refreezing. This method provides sufficient surface friction for aircraft operations until thawing temperatures cause the sand particles to sink into the ice. This method of ice control normally is used when the ice thickness is 1/4 inch or more. The process has the additional advantage of dissipating some of the ice through evaporation and weakening the ice structure by a honeycombing effect that takes place when the open flame embeds the sand particles into the ice.

Polypropylene glycol and ethylene glycol are two liquid compounds approved for use to remove existing accumulations of ice or to prevent ice formation. These deicing chemicals work by lowering the freezing point of precipitation. These chemicals tend to be too expensive for use in pavement ice control, however, and thus are limited primarily to deicing the aircraft.

Aircraft deicing

The presence of ice or significant snow accumulation on an aircraft's wings or fuselage has potential significant adverse effects on the performance of aircraft in flight. Because of this, the removal of such accumulations is required prior to flight. This removal process is known as aircraft deicing.

Aircraft deicing is accomplished by spraying one of two types of heated aqueous solutions onto the aircraft. The heat of the solution and the force of the spray melt and remove the accumulation. The chemical properties of the solution act as an antifreeze to prevent significant accumulation prior to takeoff.

The two most common types of aircraft deicing fluids, known as Type I and Type II fluids, are distinguished by their relative viscosities. Type I fluid is a mixture of glycol and water that is heated to 180 degrees Fahrenheit. Applied to clean frozen precipitation on the aircraft, Type I fluid protects aircraft from snowfall for approximately 15 minutes, and from freezing rain for approximately 3 to 5 minutes.

Type II fluid is a thicker solution of glycol and water that uses a polymer as a thickening agent. Type II fluid is not heated prior to application. Type II fluid is used primarily during periods of heavy snowfall. Once Type II fluid is applied, the mixture adheres to the aircraft's outer surface, rather than running off. If significant ice accumulation exists on the aircraft, it must be removed using a Type I fluid. Type II fluids will prevent reaccumulation of snow and ice on aircraft for as long as 45 minutes.

The application of Type I and Type II fluids for deicing is performed in varying manners at different airports. Most airports have deicing vehicles that travel from aircraft to aircraft to perform deicing. Other airports have fixed deicing stations, to which aircraft taxi prior to takeoff. Overall, the application of deicing fluids is performed in 5 to 10 minutes, depending on the size of the aircraft.

The cost of Type I and Type II solutions is relatively high and the environmental effects of fluid runoff into the ground and water sources in the surrounding areas may be significant. Because of this, a process for limiting and controlling runoff is an important part of the overall deicing process.

Bird and wildlife hazard management

Birds and other wildlife striking aircraft in operation in the vicinity of an airport have the potential to cause serious damage to aircraft and loss of human life. Between 1990 and 2007, more than 82,000 aircraft collisions with wildlife, known as "wildlife strikes," were reported to the FAA (source ACI). In 2007 alone, more than 7,600 strikes were reported—more than quadruple the number of reported strikes in 1990. More than 98 percent of wildlife strikes involved birds, with the remaining involving a wide variety of land-based animals, from deer to coyotes, to alligators. While more than 85 percent of wildlife strikes have had minimal impact on flight, others have demonstrated the tremendous danger to aircraft.

The most recent widely publicized event occurred in January 2009, when a large flock of Canadian geese struck U.S. Airways flight 1549 causing a double engine failure shortly after takeoff from New York's LaGuardia Airport, and an emergency water landing of the aircraft in the Hudson River. Miraculously, there were no fatalities from this event. However, the event brought to the forefront the threat of wildlife to safe aircraft operations, particularly within the vicinity of airports.

14 CFR Part 139.337 directs airports to conduct a study and provide a wildlife management program for airports when any of the following events has occurred on or near the airport:

1. An air carrier aircraft experiences a multiple-bird strike or engine ingestion.
2. An air carrier aircraft experiences a damaging collision with wildlife other than birds.
3. Wildlife of a size or in numbers capable of causing either of the above events is observed to have access to any airport flight pattern or movement area.

Any wildlife hazard management program should be formulated and implemented on the basis of an ecological study of the environment.

Bird hazards

A flock of birds ingested into a jet engine at takeoff can cause a dangerous stall, and a single large bird hitting an engine with the force of a bullet might smash a fan blade, which can cost thousands of dollars to replace. Airport managers, as well as all other members of the aviation community, are aware of the hazards that can be caused by birds. 14 CFR Part 139 requires that airport operators must show that they have established instructions and procedures for the prevention or removal of factors at the airport that attract, or might attract, bird activity. Many airport managers call upon the expertise of ornithologists to help analyze bird activity at their particular location. The ornithologist can provide useful data such as identification of species, estimates of the number of birds involved, habitat and diet, migrating characteristics, tendency to fly in flocks, and flight patterns. Most allowable control techniques are intended to cause birds roosting or feeding at the airport to go elsewhere and overflights to use different routes away from the airport. There are a variety of control techniques available that can be used individually or in combination, including:

- Elimination of food sources through better planning and implementation of a regimen for vegetation management on the airport property.
- Elimination of habitat such as trees, ponds, building ledges, and other roosting areas: Proper water retention management, including better drainage and elimination of wetlands and low areas, is particularly important in discouraging bird population.
- Physical annoyance, such as noisemakers, high-pressure water from fire hoses, and decoys, such as papier-mâché owls to frighten birds.
- Chemical treatment to cause dispersal and movement of flocks or death: Effective insect control would also be a part of chemical treatment.
- Continual upgrading of scientific methods used in assessing the effectiveness of different bird control techniques.
- Better training and management of a team dedicated to bird hazard management.
- Use of firearms or other mechanical means of killing.

The use of trained birds of prey, such as falcons and hawks, complements a number of other measures enacted in recent years in the fight against bird strikes. Moreover, several airports have turned to border collies as an effective way to chase birds (Fig. 6-5).

Some of the listed techniques are not always feasible. If large numbers of birds are involved, the use of shotguns, for example, is generally ineffective. Chemical poisoning of nuisance birds is generally not allowed for environmental reasons because of possible harmful side effects from toxic agents and large concentrations of dead birds, both of which can pose significant public health hazards. Chemicals are available that, when mixed with food, cause birds to exhibit erratic behavior and emit distress cries. These alarming reactions result in dispersal of flocks and the movement of individual birds to different

Figure 6-5. *Border collies have been specifically trained to chase birds from aircraft flight paths.*
(Photo courtesy Lee County, Florida, Port Authority)

locations. These chemicals may be applied only by personnel licensed by the Environmental Protection Agency.

For other wildlife, the placement of fencing around the airport property will deter the movement of grazing animals to the airfield. For wildlife that reside on the airfield, which may range from foxes to turtles, careful observation of their movement patterns and routine airfield inspections are recommended to maintain aircraft operational areas free of these potential hazards.

Self-inspection programs

Clearly one of the most important concerns of airport management is operational safety. The Federal Aviation Act of 1958 and the requirements surrounding

FAR 139 were primarily established in the interest of promoting safety. To ensure that these regulations are continuously met, airport management should carry out a comprehensive safety inspection program. The frequency of overall inspections varies by airport, but certain facilities and equipment must be inspected as often as daily, if not hourly. Some of these facilities include runways, taxiways, and navigational aids. Other elements are normally inspected with a frequency commensurate with how critical they are to the overall safety of airport operations. The FAA's *Airport Certification Program Handbook* suggests the following general categories in which emphasis on elimination, improvement, or education should be placed:

1. Hazards created by weather conditions such as snow, ice, and slush on or adjacent to runways, taxiways, and aprons
2. Obstacles on and around airfield surfaces
3. Hazards that threaten the safety of the public
4. Hazards created by erosion, or broken or damaged facilities in the approach, takeoff, taxi, and apron areas
5. Hazards occurring on airports during construction activity, such as holes, ditches, obstacles, and so forth
6. Bird hazards adjacent to the airport
7. Inadequate maintenance personnel or equipment

In addition, the FAA's Advisory Circular 150/5200-18C—*Airport Safety Self-Inspection* establishes a checklist primarily designed for operators of airports. This list includes some of the more important items that are often overlooked and result in damage to aircraft and injury to people. The following examples taken from this source are not all-inclusive. They do, however, give a good idea of the areas of major concern to an airport manager, especially at a general aviation airport.

Ramp/apron–aircraft parking areas

1. Unsealed pavement cracks, weak or failing equipment, buildup of shoulders causing entrapment of water, poor drainage, and growth of vegetation are repaired.
2. Adequate aircraft parking and tie down areas are provided, well clear of taxiways, and prominently marked.
3. Areas are free of obstructions such as blocks, chocks, loose gravel, baggage carts, and improperly parked ground service vehicles.
4. Deadlines are provided for safe passenger loading and unloading, cargo handling, and aircraft servicing.
5. Fuel trucks and other airport vehicles are parked in specified areas away from aircraft.

6. Unauthorized vehicles are prohibited from entering the ramp area.

7. "NO SMOKING" signs are prominently displayed in all areas where aircraft are being fueled.

8. Fire extinguishers are provided and are in good working condition.

9. Adequate directional signage is provided.

10. Flood lights, power outlets, and grounding rods are all in good condition.

Taxiways

1. Unsealed pavement cracks, weak or failing pavement, buildup or erosion of shoulders, poor drainage are repaired.

2. The taxiway is free of weeds, foreign object debris (FOD), and other obstructions.

3. Shoulders are firm and are marked as necessary for easy reference.

4. Yellow centerlines are provided and are in good condition.

5. Hold lines are provided and clearly visible.

6. Unauthorized vehicles, people, or animals are prevented from occupying taxiways.

7. Necessary directional signs are provided and are so located as to be well clear of taxi areas.

8. Lighting systems are in good working order.

Runways

1. Runway lights and markers are clearly visible, are operated at correct brilliance, properly leveled and oriented, equipped with usable lamps of correct wattage, clear; clean lenses are in runway lights, clean green lenses are in threshold lights, and all lights are unobstructed by vegetation.

2. The threshold is properly marked and lighted.

3. Runway designators are well painted.

4. The ends of runways are flush with the surrounding ground (no lip).

5. Overrun areas are in good condition.

6. Shoulders are firm, clearly marked, and free from washouts, holes, or ditches.

7. The centerline (white) is well painted.

8. All approach areas are clear of obstruction. It should be noted whether views of ends of other runways are unobstructed by vegetation, trees, terrain, or other obstruction, and whether unauthorized vehicles or livestock have access to the runways or airfield.

9. Procedures for removal of disabled aircraft from runways are in place.

Fueling facilities

1. Fueling areas are clearly defined and are located away from aircraft parking areas.
2. Pumps are placarded to properly identify the type of fuel dispensed.
3. Grounding means are provided for all fueling operations.
4. Fire extinguishers are provided and are in good condition.
5. Fuel hose and nozzle units are stored in clean areas for protection from weather and contamination.
6. Fuel filters are regularly checked.
7. Tanks are regularly checked for water or contamination.
8. Locks are provided and used on fuel tank filler caps.
9. Fuel tank vents are regularly checked.
10. Fueling areas are kept clean and free of debris.
11. Rags are stored in closed containers.
12. Oil is kept in storage bins or closets.
13. Oil cans are kept in proper containers.
14. "NO SMOKING" signs are posted.
15. Stepladders are provided, properly stored, clean, and in good repair.

Buildings and hangars

1. All buildings and aircraft hangars are free of debris, trash, unusable aircraft parts, and other potentially hazardous objects of no practical use.
2. Fire protection with an adequate umber of fire extinguishers in good operational condition and with dates of service record are available. Fire and rescue equipment and first aid and emergency services are provided. Smoke detectors and emergency lighting are in working order.
3. All tools and unused equipment are properly stored.
4. Paints, oils, and other chemical compounds are stored in exclusive, preferably fireproof, areas.
5. "NO SMOKING" signs are properly posted.
6. Restricted area signs are properly posted.
7. Exit signs are posted.
8. Building identification is properly posted with signs or numbers.
9. Buildings are provided with appropriate locks on doors and windows commensurate with security needs.
10. The areas around buildings are clean, free of weeds, debris, and unsafe terrain.

Components of a safety self-inspection program

The FAA recommends that any airport safety self-inspection program should have the following four components:

Regularly scheduled inspection of physical facilities: for 14 CFR Part 139 certificated airports, daily scheduled inspections of such facilities are required, including nighttime inspection of all lighting systems.

Continuous surveillance inspection programs: of airport operations, construction, and maintenance activities.

Periodic inspections of the surrounding environment: including determination of any existing or planned construction that may become obstacles to safe air navigation, or any increase in wildlife activity.

Special condition inspections: during and after unusual events, such as weather events, high-volume traffic events, or other unusual conditions.

Any safety self-inspection program should include a record keeping activity of all inspections and their findings.

SMS—safety management systems for airports

The purpose of 14 CFR Part 139 has always been to ensure that the operation of an airfield that accommodates commercial service operations be as safe as possible. As has been detailed in this chapter, 14 CFR Part 139 focuses on the safety of the airfield's infrastructure, and the services that contribute to maintaining safe operations. However, to date there has been no formal requirement to develop a formal *culture* of safety. The development of safety management systems (SMS) at airports is one method that such a formal culture may be established.

While not yet formally included in 14 CFR Part 139, in 2010, the FAA's office of Airports issued a notice of proposed rulemaking (NPRM) that would require Part 139 airports to develop formal Safety Management Systems (SMS). This action has been taken by FAA following the rulemaking by ICAO in 2005 requiring all of its member states to have SMS at all their international airports.

The FAA defines SMS as a "formal, top-down business-like approach to managing safety risk, including systematic procedures, practices, and policies for the management of safety." SMS comprises four primary elements: safety policy, safety promotion, safety risk management, and safety assurance.

Safety policy is further defined by the FAA as the "fundamental approach to managing safety that is to be adopted within an organization." Safety policy includes the rules by which the airport maintains safe operations, the chain of command and communication procedures as it pertains to safety issues, and language that describes the overall mission and vision of a safe airport facility. Airports currently implementing SMS have often created a position of **Safety Manager** within the organization, whose position has the operational responsibility of overseeing the SMS at the airport.

Safety promotion is further defined by the FAA as "a combination of safety culture, training, and data sharing activities that supports the implementation and operation of SMS." Because airports can often be large complex organizations with a wide variety of employees, contractors, vendors, tenants, and of course passengers, promotion of the safety culture at an airport to all of its users is of vital importance. Safety promotion includes such activities as formal training and education, effective communication, and promotion of the idea of "continuous safety improvement."

Safety Risk Management (SRM) is further defined by the FAA as the "formal process within SMS composed of describing the system, identifying the hazards, assessing the risk, and controlling the risk." SRM is in fact the operational foundation of safety management systems.

The FAA describes five formal phases to the SRM process. They are:

Phase I: System Description: The first phase of SRM is to formally identify, inventory, and understand the airport operating environment. It is clearly important to completely understand how the airport operates if there is to be a successful SMS program. Most important elements of the system to be described include the airfield layout, the number and type of operations that occur on the airfield, surrounding infrastructure, the numbers and types of employees and service vehicles operating on and around the airfield, and the standard operating procedures for all operations on and around the airfield. This understanding is in addition to a full understanding of the airport's 14 CFR Part 139 ACM.

Phase II: Hazard Identification: The second phase of SRM is to identify any existing or potential hazards to the operation of the system. Hazard identification includes investigating all operations of the system and hypothesizing all potential avenues for system failure that may result in damage to people and property. Some examples of potential hazards include pavement condition (from cracking to **foreign object damage (FOD)**), the presence of certain types of wildlife, weather (such as reduced visibility, or wet icy conditions), equipment (e.g., operation of heavy machinery, or equipment in need of repair), human factors (ranging from staffing considerations to complacency), or any number of externalities caused by users of the airport (from pilots who may be unfamiliar with the airport environment to the general public and their lack of expertise in safe airport operations).

Phase III: Risk Determination: The third phase is to analyze the identified hazards for their potential impacts to safety. Some hazards may have the potential to result in minor injury (e.g., a ramp employee slipping on ice), while others might have catastrophic results (such as the collision of two aircraft at a runway intersection). In addition, some hazards may have the potential to result in safety impacts more often than others. The combination of a hazards potential to cause safety impacts on both a frequency basis and a severity basis determines the risk of that hazard to the overall safety of the airport system.

Phase IV: Risk Assessment: The fourth phase of SRM is to apply the likelihood and severity of any system failure attributed to a given hazard to the FAA's predictive risk matrix illustrated in **Fig. 6-6**.

The matrix illustrated in Fig. 6-6 defines three levels of risk, based on the composite of a hazards likelihood of occurrence and the severity of the worst credible outcome of the occurrence. As illustrated, hazards that have no safety effect are considered low risk, regardless of the likelihood that the particular hazard causes some sort of operational failure. Conversely, those hazards that may result in catastrophic outcomes are nearly always considered high risk, regardless of their frequency of occurrence.

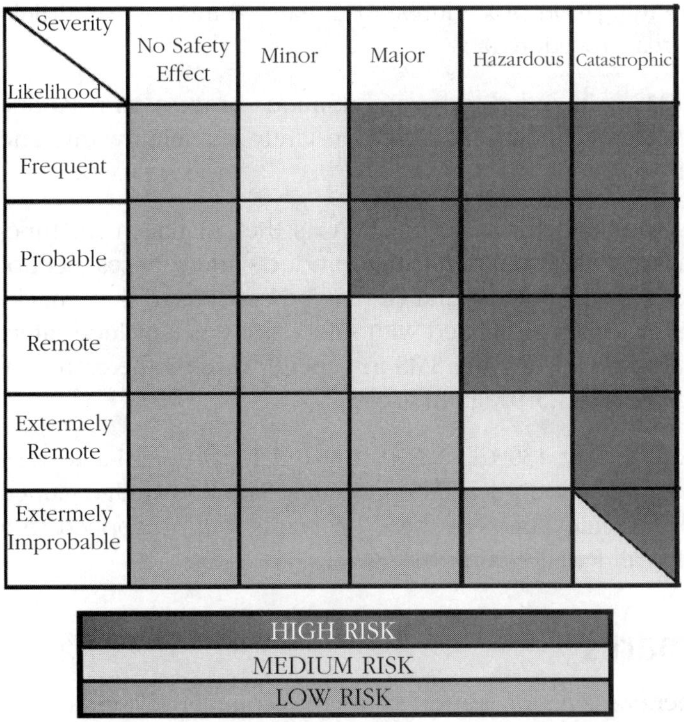

Severity / Likelihood	No Safety Effect	Minor	Major	Hazardous	Catastrophic
Frequent					
Probable					
Remote					
Extermely Remote					
Extermely Improbable					

HIGH RISK
MEDIUM RISK
LOW RISK

Figure 6-6 *SMS Predictive Risk Matrix.*

The FAA considers **low risk** as the target level of risk. At such risk levels, hazards attributed to low risk should receive a lower level of active management. These hazards should still be under continuous observation to prevent any occurrences where the risk level associated with these hazards may rise.

The FAA considers **medium risk** as an "acceptable level of risk." While these hazards and any associated mitigation policies meet the minimum standards for acceptable risk, efforts should be made to actively manage these hazards.

The FAA considers **high risk** as an "unacceptable level of risk." SMS programs are to give the highest priority for finding ways of reducing either the likelihood or the severity of any failures attributed to these hazards.

Phase V: Risk Mitigation: In the fifth SRM phase, airport management addresses each of the risks associated with each identified hazard, prioritizes these risks in accordance with the predictive risk matrix, and then establishes mitigation procedures to reduce the risks associated with each hazard as necessary.

Examples of risk mitigation include *avoidance,* such as creating a rule to disallow any operations that cause a certain level of risk (e.g., prohibiting airfield maintenance vehicles to perform work on certain areas of the airfield under low visibility conditions), or *control*, creating control measures to mitigate risk (e.g., targeted wildlife mitigation procedures, or enhanced training for airfield users on movement area procedures).

The SRM portion of SMS is intended to be a continuous process, as risks due to hazards, as well as hazards themselves, are constantly changing within and around the airfield environment.

Safety Assurance is further defined by the FAA as the "management functions that systematically provide confidence that products and services meet or exceed safety requirements." Safety assurance may be considered the "checks and balances" of SMS. Practices associated with safety assurance include safety audits, data analysis, and reviews of the SMS in general to see if there are any areas of improvement in the SMS program itself.

As of 2010, more than 20 Part 139 Class I, II, III, and IV airports have been working with the FAA to develop pilot SMS programs. The knowledge gained from the design and implementation of these programs will be applied to a future potential rule for all Part 139 airports.

Concluding remarks

Whether at a small general aviation airport or a large commercial service airport, the proper management of operations on the airfield is essential to the safety and efficiency of aircraft operations. For airports serving most air carrier operations, a written plan of operations management, addressing specific areas of operations and certain mandated specifications, is required by the Federal Aviation Administration as written in 14 CFR Part 139. For all airports, however, it is suggested that the areas of operations described in 14 CFR Part 139 be addressed, because the hazards that accompany wildlife, climate, and the potential for accidents resulting from aircraft operations have the potential of occurring regardless of the presence of commercial air carrier service.

Key terms

ACM (Airport Certification Manual)

AOC (Airport operating certificate)

ARFF (aircraft rescue and fire fighting)

flexible pavement, asphalt

rigid pavement, concrete

subgrade

spalling

pavement maintenance

pavement rehabilitation

pavement reconstruction

NDT (nondestructive testing)

dynamic/viscous hydroplaning

AFFF (aqueous film-forming foam)

FOD (foreign object debris)/damage)

SMS (Safety management systems)

SICP (Snow and Ice Control Plan)

Questions for review and discussion

1. To what airports does 14 CFR Part 139 apply?
2. What are the differences between rigid pavements and flexible pavements?
3. What is meant by vibratory or dynamic testing?
4. What are some of the symptoms of potential pavement failures?
5. What are the typical useful life spans of airfield pavements?
6. What is the difference between pavement maintenance and pavement rehabilitation?

7. How is an airport's ARFF index determined?

8. What are the ARFF requirements of an airport based on its index?

9. Why is a snow and ice control plan so important at some airports?

10. What are the most common methods for snow and ice removal from airfields?

11. What are some of the goals of an airport's safety inspection program?

12. What are some of the more common procedures associated with an airport's safety inspection program?

13. What are some of the hazards associated with birds and wildlife in the vicinity of an airfield?

14. What are some of the control techniques associated with bird and wildlife hazard management?

Suggested readings

FAA Advisory Circular 150/5200-18C—*Airport Safety Self-Inspection*.

FAA Advisory Circular 150/5200-30C—*Airport Winter Safety and Operations*.

FAA, *Wildlife* Strikes to Civil Aircraft in the United States 1990–2001, June 2002.

FAR Part 121—*Operating Requirements: Domestic, Flag, and Supplemental Operations*.

FAR Part 139—*Certification* of *Airports*.

FAA Advisory Circular 150/5200-37—Introduction to Safety Management Systems (SMS) at Airports.

FAA Airports Cooperative Research Program (ACRP) Report 1: *Safety Management Systems for Airports*, Volume 1 & 2, 2009.

7

Airport terminals and ground access

Outline

- Introduction
- The historical development of airport terminals
  - Unit terminal concepts
  - Linear terminal concepts
  - Pier finger terminals
  - Pier satellite and remote satellite terminals
  - The mobile lounge or transporter concept
  - Hybrid terminal geometries
  - The airside-landside concept
  - Off-airport terminals
  - Present-day airport terminals
- Components of the airport terminal
  - The apron and gate system
  - Aircraft gate management
    - Gantt charts
  - The passenger handling system
    - Passengers and their required processing facilities
    - Passenger check-in
    - Security screening
    - At-gate processing
    - Customs and border patrol facilities
    - Ancillary passenger terminal facilities
    - Vertical distribution of flow
    - Baggage handling
    - Baggage claim

- Airport ground access
  - Access from the CBD and suburban areas to the airport boundary
  - Access modes
  - Factors influencing demand for ground access
  - Coordination and planning of ground access infrastructure
  - Access from the airport boundary to parking areas and passenger unloading curbs at the terminal building
  - Vehicle parking facilities
    - Off-airport parking
    - Employee parking
    - Car rental parking
    - Terminal curbs
  - Technologies to improve ground access to airports

Objectives

The objectives of this section are to educate the reader with information to:

- Understand the development of airport terminals from the early days of commercial aviation to present-day terminal design concepts.
- Identify the facilities within an airport terminal that facilitate the transfer of passengers and baggage to and from aircraft.
- Describe the essential and ancillary processing facilities, including terminal concessions, located within airport terminals.
- Be familiar with the various modes of transportation that comprise airport ground access systems.
- Describe various technologies that are being implemented to improve ground access to airports.

Introduction

The airport terminal area, comprised of passenger and cargo terminal buildings, aircraft parking, loading, unloading, and service areas such as passenger service facilities, automobile parking, and public transit stations, is a vital component to the airport system. The primary goal of an airport is to provide passengers and cargo access to air transportation, and thus the terminal area achieves the goal of the airport by providing the vital link between the airside of the airport and the landside. The terminal area provides the facilities, procedures, and processes to efficiently move crew, passengers, and cargo on, and off, commercial and general aviation aircraft.

The term *terminal* is in fact somewhat of a misnomer. Terminal implies ending. Although aircraft itineraries begin and end at an airport's terminal area, the itineraries of passengers and baggage do not. It is vitally important to understand that the airport terminal is not an end point, but an area of transfer along the way. As will be discussed in this section, the building configurations, facilities, and processes that comprise an airport terminal area require careful planning and management to ensure the efficient transfer of passengers and cargo through the airport and aviation system.

The historical development of airport terminals

Just as there were no runways or other airfield facilities during the very earliest days of aviation, there certainly were no terminals, at least the way they are recognized today. The first facilities that could be remotely considered airport terminal areas evolved in the early 1920s with the introduction of airmail service. Airmail operations required small depots in order to load and unload mail, fuel aircraft, and perform any required maintenance. Little in the way of formal passenger or cargo processing was required, and hence, airport terminal facilities were little more than single-room structures with the most basic of infrastructure.

The introduction of commercial passenger air service in the late 1920s resulted in the need to develop certain basic passenger processing policies. The earliest passenger processing strategies evolved from the major intercity transportation mode of the day, the railroads. Tickets and boarding passes were issued for passengers, and similar to policies set for rail transport, cargo rates were also charged, typically by the weight of the cargo being transported. (Sometimes passengers were weighed as well, primarily to ensure that the aircraft did not exceed its maximum takeoff weight!) (Fig. 7-1). The facilities required for performing basic ticketing and weighing functions, as well as for aircraft boarding and alighting the relatively few passengers and little cargo that used civil air transportation could be, and were often, incorporated into one-room facilities, strikingly similar to the facilities that served the railroads.

Unit terminal concepts

These first terminals were the earliest **centralized facilities,** centralized meaning that all passenger processing facilities at the airport are housed in one building. These first centralized facilities became known as the earliest **simple-unit terminals,** because they contained all required passenger processing facilities for a given air carrier in a single-unit building. In addition to passenger processing facilities, the airport's administrative offices,

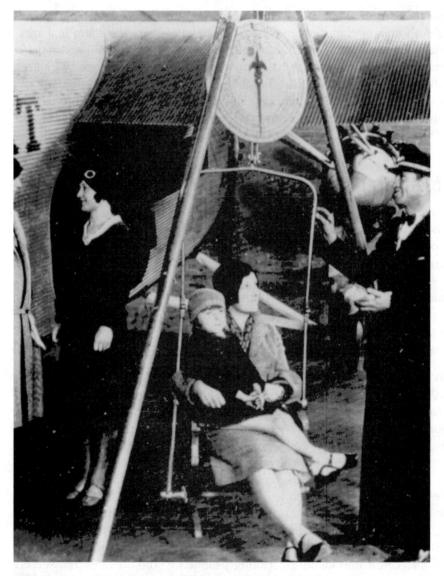

Figure 7-1. *Passengers weigh in prior to departure from Chicago's Midway Airport in 1927.*

and even air traffic control facilities, were located within the unit terminal building (Fig. 7-2).

As air service became more popular, particularly in the 1940s and 1950s, airport terminals expanded to accommodate increasing volumes of aircraft, passengers, and cargo. As multiple airlines began to serve single communities, airport terminals expanded in two ways. In smaller communities, two or

Figure 7-2. *Allegheny County, Pennsylvania, historical unit terminal concept.* (Photo courtesy Allegheny County Airport)

more airlines would share a common building, slightly larger than a simple-unit terminal, but have separate passenger and baggage processing facilities. This configuration became known as the **combined-unit terminal.** In larger metropolitan areas, separate buildings were constructed for each airline, each building behaving as its own unit terminal. This terminal area configuration became known as the **multiple-unit terminal** concept (Fig. 7-3). Even though the multiple-unit terminal area consists of separate facilities for each airline, thereby considered by some as a decentralized terminal environment, each individual unit terminal in the multiple-unit terminal concept is still considered an individual *centralized* facility because all passenger and cargo processing required for any given passenger or piece of cargo to board any given flight still exists in one facility.

The early centralized terminals, including the simple-unit, combined-unit, and multiple-unit terminals, employed the **gate arrival concept.** The gate arrival concept is a centralized layout that is aimed at reducing the overall size of terminal areas by bringing automobile parking as close as possible to aircraft parking. The simple-unit terminal represents the most fundamental type of gate arrival facility, consisting of a single common waiting and ticketing area with exits onto a small aircraft parking apron. Even today, the gate arrival concept is adaptable to airports with low airline activity and is particularly

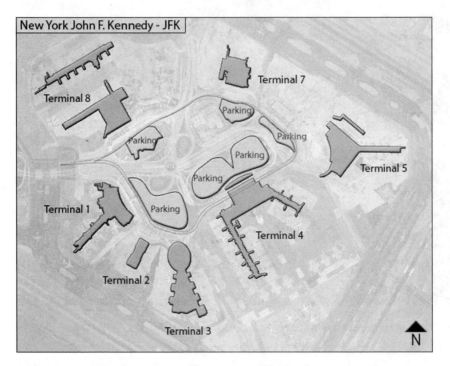

Figure 7-3. *JFK International Airport in New York City provides an example of a multiple-unit terminal concept.* (Figure Courtesy ifly.com)

applicable to general aviation operations whether a smaller general aviation terminal is located separately from a larger terminal for commercial air carriers or is the operational center for an airport used exclusively for general aviation.

Where the terminal serves airline operations, close-in parking is usually available for three to six commercial aircraft. Where the simple-unit terminal serves general aviation only, the facility is within convenient walking distance of aircraft parking areas and adjacent to an aircraft service apron. The simple-unit terminal facility normally consists of a single-level structure where access to aircraft is afforded by a walk across the aircraft parking apron (Fig. 7-4).

Linear terminal concepts

As airports expanded to meet the growing needs of the public, as well as the growing wingspans of aircraft, simple-unit terminals expanded outward in a rectangular or *linear* manner, with the goal of maintaining short distances between the vehicle curb and aircraft parking that existed with unit terminals. Within linear terminals, ticket counters serving individual airlines were introduced and loading bridges were deployed at aircraft gates to allow passengers

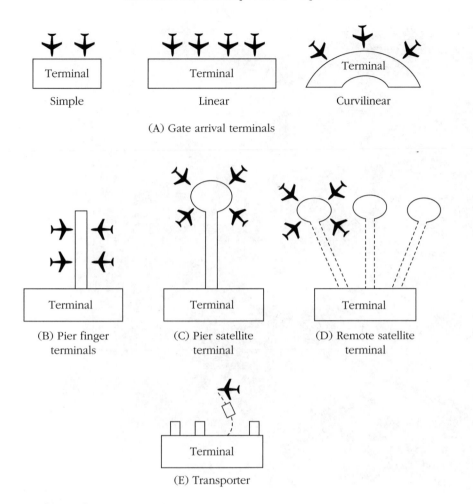

Figure 7-4. *Terminal design concepts.* (Source: FAA)

to board aircraft without having to be outside on the apron, thereby improving convenience and safety for passengers.

In some instances airports were extended in a **curvilinear** fashion, allowing even more aircraft to park "nose-in" to the terminal building while maintaining short walking distances from the airport entrance to the aircraft gate (Fig. 7-5).

In many respects, the linear and curvilinear terminal concepts are mere extensions of the simple-unit terminal concept. More sophisticated linear terminals, particularly those that serve high volumes of passengers, often feature two-level structures where enplaning passengers are processed on one level and deplaning passengers on the other level. Passenger walking distances from the "curb to the gate" are typically short, on the order of 100 feet. The linear

Figure 7-5. *DFW Airport, whose terminal area employs a multiple-unit curvilinear terminal concept, now accommodates a large percentage of transfer passengers.* (Photo courtesy FAA)

configuration also lends itself to the development of automobile parking that is close to the terminal building, and provides extended curb frontage for loading and unloading of ground transportation vehicles.

One of the main disadvantages of linear terminals becomes evident as the length of the terminal building increases. Walking distances between facilities, particularly distantly separated gates, become excessive for the passenger whose itinerary requires a change in aircraft at the airport. Prior to airline deregulation the percentage of these transfer passengers was insignificant. After 1978, however, this percentage increased dramatically and the issue of long walking distances between gates became a major issue, particularly at the hub airports.

Pier finger terminals

The **pier finger terminal** concept evolved in the 1950s when gate *concourses* were added to simple-unit terminal buildings. Concourses, known as *piers* or *fingers,* offered the opportunity to maximize the number of aircraft parking spaces with less infrastructure. Aircraft parking was assigned to both sides of a pier extending from the original unit terminal structure. The pier finger terminal is the first of what are known as **decentralized facilities,** with some of the required processing performed in common-use main terminal areas, and other processes performed in and around individual concourses.

Many airports today have pier finger terminals in use. Since the earliest pier finger designs, very sophisticated and often convoluted forms of the concept have been developed with the addition of hold rooms at gates, loading bridges, and vertical separation of enplaning and deplaning passengers in the main-unit terminal area.

As pier finger terminals expanded, concourse lengths at many terminal buildings became excessive, averaging 400 feet or more from the main terminal to the concourse end. In addition, as terminals expanded by adding additional piers, distances between gates and other facilities became not only excessive in distance, but also confusing in direction. Moreover, often the main-unit terminal facility and corridors connecting the individual fingers were not expanded along with the construction of additional concourses, leading to passenger crowding in these areas (Fig. 7-6).

Another of the disadvantages of pier finger terminals is that expansion of terminals by adding or lengthening concourses may significantly reduce the amount of apron space for aircraft parking and movement. Also, the addition of concourses to the terminal tends to put constraints on the mobility of aircraft, particularly those that are parked closer to the main terminal building.

Pier satellite and remote satellite terminals

Similar to pier finger terminals, **pier satellite terminals** formed as concourses extended from main-unit terminal buildings with aircraft parked at the end of the concourse around a round atrium or *satellite* area. Satellite gates are usually served by a common passenger holding area.

Satellite terminal concepts, developed in the 1960s and 1970s, took advantage of the ability to create either underground corridors or **Automated Passenger Movement Systems (APMs)** to connect main terminal buildings with concourses (Fig. 7-7). Such terminals are said to be built on the **remote satellite concept.**

The main advantage of the remote satellite concept is that one or more satellite facilities may be constructed and expanded when necessary while providing

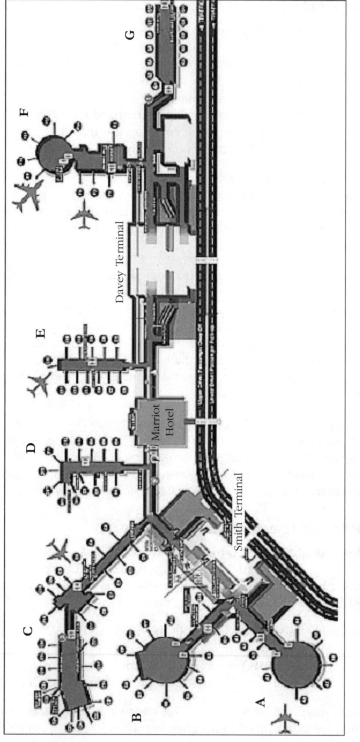

Figure 7-6. *The old pier finger terminal complex at Detroit's Metropolitan Airport.* (Figure courtesy Detroit Metropolitan Airport)

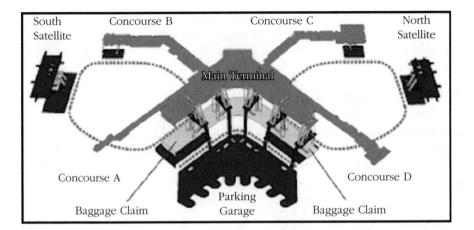

Figure 7-7. *Terminal configuration at Seattle–Tacoma International Airport, one of the first airports to employ APMs to reach remote satellite terminals.* (Figure courtesy Seattle–Tacoma International Airport)

sufficient space for aircraft taxi operations between the main terminal building and satellites. In addition, although distances from the main terminal to a satellite may be quite large, APMs or other people-mover systems such as moving walkways or shuttle buses are provided to reduce walking distances.

Another of the advantages of the satellite concept is that it lends itself to a relatively compact central terminal with common areas for processing passengers, because aircraft with large wingspans, which for all intents and purposes dictate the size of terminal gate areas and thus concourses and satellite, are parked at remote satellites rather than at the central facility.

As with the pier finger concept, the expansion of pier satellite and remote satellite concept terminals tend to result in terminal facilities that not only have large distances between key points within the terminal, but also often become confusing for passengers in their attempts to find their way to their respective gates, baggage claim areas, or other desired facilities.

The mobile lounge or transporter concept

In 1962 the opening of Dulles International Airport west of Washington, D.C., designed as the first airport specifically for the new jet aircraft of the day, introduced the **mobile lounge** or *transporter concept* of airport terminals. Sometimes known also as the *remote aircraft parking concept,* the Washington Dulles terminal area attempted to maximize the number of aircraft that may be parked and maximize the number of passengers that may be processed, with minimal concourse infrastructure. In this concept, aircraft are parked at remote parking locations away from the main-unit terminal building. To travel between aircraft and the terminal building, passengers would board transporters, known as mobile lounges, that would roam the airfield among ground vehicles and taxiing aircraft (Fig. 7-8).

With the mobile lounge concept, walking distances were held to a minimum because the main, relatively compact, terminal building contains common passenger processing facilities, with automobile curbs and parking located in close proximity to the terminal building entrances. Theoretically, expansion to accommodate additional aircraft is facilitated by the fact that there is no need to physically expand concourses, piers, or satellites, just merely add additional mobile lounges, if necessary.

Despite its theoretical advantages, the mobile lounge concept did not on the whole win approval from passengers. Mobile lounge boarding areas in the main terminal often became excessively congested as passengers with carry-on baggage would crowd the area, often arriving early so as not to miss their assigned mobile lounge boarding time. Moreover, the relatively small mobile lounges offered far less room for passengers than the aircraft from or to which they are transitioning, especially in comparison to large "wide-body" aircraft

Figure 7-8. *Washington Dulles International Airport terminal.*
(Photo courtesy Metropolitan Washington Airports Authority)

introduced in the late 1960s, leaving passengers crowded and often uncomfortable while on the mobile lounge. In addition, mobile lounges require constant maintenance, which over time becomes an excessive cost element of operations (Fig. 7-9).

In the mid-1990s Dulles in effect abandoned the mobile lounge concept by constructing satellite or *midfield* concourses on the airfield. In 2010, mobile lounges at Dulles were phasing out of service upon the completion of an underground APM system that connects the original unit terminal with newly built remote concourses.

In the United States, no other airports have relied entirely on the mobile lounge concept for their terminal areas, with the exception of providing shuttle bus services to aircraft that must be parked in remote parking spots because of lack of available gate space at the terminal building or concourses.

Hybrid terminal geometries

With the volatile changes in the amount and behavior of civil aviation activity in the 1970s, with increasing numbers of large aircraft (with high seating capacities and large wingspans), volumes of passengers, and changes in route structures,

Figure 7-9. *Mobile lounge attached to aircraft at Washington's Dulles International Airport, circa 1970.*
(Photo courtesy Metropolitan Washington Airports Authority)

particularly after airline deregulation in 1978, airport management had to expand and modify terminal areas to accommodate almost constantly changing environments. As a result, many airport terminal geometries expanded in an ad hoc manner, leading to *hybrid terminal geometries* incorporating features of two or more of the basic configurations (Fig. 7-10). In addition, for airports that accommodate an airline's hub, airport terminal planning became necessary to accommodate up to 100 or more aircraft at one time and efficiently handle record volumes of passengers, particularly those passengers transferring between aircraft.

It is no coincidence that in the 1970s and 1980s public sentiment for the planning and management of many airport terminals in the United States declined considerably. Issues including congestion, long walking distances, confusing directions, as well as limited amenities and passenger services became popular issues of criticism. As a result, airport planners began to redevelop terminal area designs, focusing on strategic planning and design of terminals that can accommodate requirements of accessing ground vehicles, passengers, and aircraft, with sufficient flexibility to adapt to ever-changing levels of growth and system behavior.

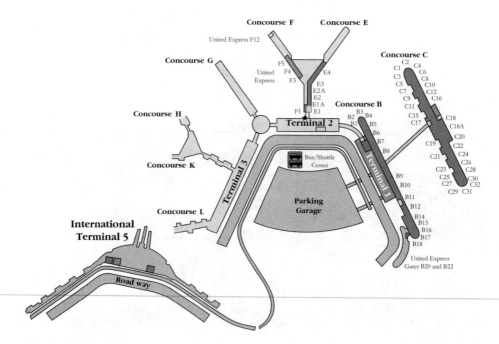

Figure 7-10. *Chicago O'Hare International Airport combining unit, linear, pier, and satellite terminal concepts.* (Figure courtesy United Airlines)

The airside-landside concept

The most significant terminal area concept to emerge involved a more physical separation between facilities that handle passengers and ground vehicles and those that deal primarily with aircraft handling. The **airside-landside concept** emerged with the opening of the Tampa International Airport in 1972, and has proliferated throughout the United States at airports such as Pittsburgh International Airport and Orlando International Airport (Fig. 7-11).

The airside-landside concept relies heavily on automated pedestrian movement systems to quickly and efficiently shuttle passengers to and from two separate facilities. In the landside facility, all passenger and baggage processing can be performed without being physically close to an aircraft. In addition, sufficient ancillary facilities, such as concessions, atriums, and the like, are located in landside facilities to provide amenities to facilitate a pleasurable experience for the passenger. Airside facilities, which have been built in various shapes and sizes, from X shapes to long concourses, focus on the efficient servicing of aircraft, including fueling, loading, and unloading. Separating each of the two processes allows greater flexibility in adapting to changes in either environment, be it new aircraft or changes in passenger processing policies.

Off-airport terminals

In the 1980s the airside-landside concept formed the basis for a series of experimental concepts known as **off-airport terminals.** With the notion that certain

Figure 7-11. *Example of airside-landside concept, Tampa International Airport.* (Figure courtesy Hillsborough County Airport Authority)

passenger processes, such as ticketing and baggage check-in, and certainly automobile parking, did not need to be within any proximity of aircraft, such processes weren't necessarily required to be performed on airport property. As a result, facilities located miles away from the airport itself were introduced whereby passengers could park their personal vehicles, check themselves and their baggage in for their flights, and then take a shuttle bus to the airport. With the use of these off-airport terminals, passengers would avoid the often significantly more crowded passenger processing facilities at the main terminal. Also the passenger would not be required to find parking at the often more crowded and expensive parking facilities at the main terminal.

Off-airport terminals serving the San Francisco Bay Area, Los Angeles, and Las Vegas were met with positive response, with increased passenger convenience being the prime characteristic of the systems. Because of increased security measures following the attacks of September 11, 2001, however, off-airport terminals have had to discontinue any passenger or baggage check-in processes, and are now primarily used merely as off-airport parking facilities. However, the off-airport terminal concept set the precedent for implementing the idea of passenger processing at sites away from the main airport terminal, setting the stage for the potential future of airport terminal planning.

Present-day airport terminals

With more than 1 billion passengers traveling around the world through airports each year, each with different agendas, itineraries, needs, and desires, airport terminals have become complex systems in their own rights. Modern-day airport terminals incorporate necessary passenger and baggage processing services, as well as a full spectrum of customer service, retail shopping, food and beverage, and other facilities, to make the passengers' transition between the airside and landside components of the airport system as pleasant as possible.

It is clear that no single airport terminal configuration is best for all airports. The airfield, schedules of airlines, types of aircraft, volumes of passengers, and local considerations, such as local architecture, aesthetics, and civic pride, dictate different choices from airport to airport and from one time to another. The airport terminal planner has the dubious task of anticipating conditions up to 10 years in the future in an environment that seems to change by the day. To ensure that present-day airport terminal plans will be effective in the future, the airport planner must rely on the fundamental requirements of airport terminals and behaviors of passengers, and also must plan with the idea of flexibility in mind, such as considering facilities that can be expanded modularly or can provide the opportunity for relatively low-cost, simple modifications that future circumstances might demand. In addition, the demands of the twenty-first century have required airport terminals to be both technologically adaptable and environmentally sustainable.

For airport management, airport terminal areas, when properly planned and managed, have provided significant sources of revenue from airline leases to retail concessions. Airport terminals have also become a sense of pride for communities in general, as they are typically the first impression that visitors get of their destination city and the last experience they get before leaving. Many airport terminals today appear more to be shopping malls than passenger processing facilities, and other airport terminals are fully equipped with hotels and conference centers. These facilities have actually encouraged visitors to use the facilities at the airport without ever intending to board an aircraft.

The size and shape of airport terminal configurations has both an uncertain yet exciting future. Security regulations imposed by the Transportation Security Administration have established the need to expand airport security facilities, whereas advances in information technologies have suggested the ability to reduce the size of other passenger processing facilities such as staffed ticket counters. No matter how policies, regulations, technologies, and behaviors change, however, the basic function of the airport terminal area, that of efficiently linking passengers and cargo to the airside and landside components of the civil aviation system, should always be understood by airport managers and planners alike.

Components of the airport terminal

The airport terminal area is in the unique position of accommodating the needs of both aircraft and the passengers that board them. As such, the component systems of the airport terminal area may be thought of as falling into two primary categories: the **apron and gate system,** which is planned and managed according to the characteristics of aircraft, and the *passenger and baggage handling systems,* which are planned and managed to accommodate the needs of passengers and their baggage in their transition to or from the aircraft.

The apron and gate system

The apron and gates are the locations at which aircraft park to allow the loading and unloading of passengers and cargo, as well as for aircraft servicing and preflight preparation prior to entering the airfield and airspace.

The size of aircraft, particularly their lengths and wingspans, is perhaps the single greatest determinant of the area required for individual gates and apron parking spaces. In fact, the grand size of airport terminals is a direct result of large numbers of gates designed to accommodate aircraft of wingspans reaching 200 feet in length. The size of any given aircraft parking area is also determined by the orientation in which the aircraft will park, known as the *aircraft parking type*. Aircraft may be positioned at various angles with respect to the terminal building, may be attached to loading bridges or *Jetways,* or may be freestanding and adjoined with *air stairs* for passenger boarding and deplaning. Some aircraft parking types require aircraft to be maneuvered either in or out of their parking spaces by the use of *aircraft tugs,* whereas other parking types allow the movement of aircraft in and out under their own power. The five major **aircraft parking** types are *nose-in parking, angled nose-in, angled nose-out, parallel parking,* and *remote parking* (Fig. 7-12).

Most large jet aircraft at commercial service airports park **nose-in** to gates at the terminal and connect directly to the terminal building by loading bridges. Aircraft are able to enter nose-in parking spaces under their own power, and tend to be pushed out by an aircraft tug and oriented so that they may move forward on the apron without coming into contact with any other structures. The primary advantage to nose-in parking is that it requires less physical space for aircraft than any other aircraft parking type. The majority of commercial service airports, particularly those with large volumes of jet aircraft operations, have primarily nose-in parking. With nose-in parking, only the front-entry door on the aircraft is typically used for boarding, because the rear doors are typically too far from the terminal building to extend a loading bridge (although some airports have used air stairs and a supervised pathway on the ramp to allow passengers to deplane from the rear of the aircraft at nose-in parking gates). This has some, but not an entirely significant, impact on the efficiency of passenger boarding and deplaning (Fig. 7-13).

Angled nose-in parking brings aircraft as close to the terminal building as possible while maintaining enough maneuvering room so that aircraft may exit the parking space under its own power. Angled nose-in parking is typically used by smaller aircraft, such as turboprops or small regional jets. Air stairs are typically used to board and deplane passengers, removing the necessity

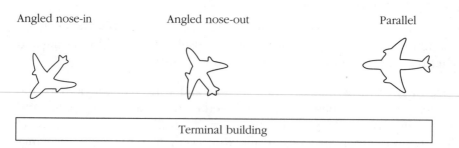

Figure 7-12. *Aircraft parking positions.*

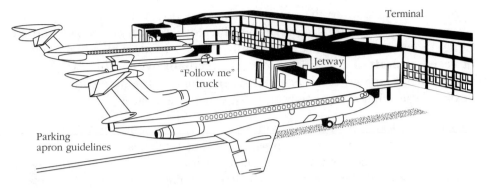

Figure 7-13. *Nose-in parking.*

for loading bridges. Angled nose-in parking requires slightly more parking area over nose-in parking for aircraft of similar size. However, because smaller aircraft tend to use angled nose-in parking, the difference in sizes of the two parking areas is not significantly different.

Angled nose-out parking brings aircraft slightly farther from the terminal building than nose-in and angled nose-in parking, because the blast from jets or large propellers has the potential of causing damage to terminal buildings if too close to the facility. Angled nose-out parking is typically used by larger general aviation aircraft and at facilities with relatively low levels of activity.

Parallel parking is said to be the easiest to achieve from an aircraft maneuvering standpoint, although each space tends to require the largest amount of physical space for a given size of aircraft. In this configuration, both front and aft doors of the aircraft on a given side may be used for passenger boarding by loading bridges. Typically, however, parallel parking is employed only by smaller general aviation aircraft with relatively large amounts of parking space near the terminal building. In addition, cargo aircraft may parallel park at their respective cargo terminals to facilitate the loading and unloading of their respective loads.

Remote parking may be employed when there is limited parking area available at the terminal building itself or when aircraft parked may be stationed there overnight or for longer durations. Remote parking areas are typically comprised of a series of rows of parking spaces, sized to accommodate varying sizes of aircraft. Smaller commercial and general aviation aircraft may be boarded and deplaned from the remote parking areas with the use of shuttle buses or vans. Larger commercial aircraft are typically taxied to a close-in parking space prior to passenger loading.

Most airports have more than one aircraft parking type to accommodate the various types of aircraft that serve the different terminal geometries and air carrier or general aviation activities. Furthermore, airports with a high number of **based aircraft** or air carrier aircraft that **remain overnight (RON)** at the airport, must take into consideration higher volumes of remote parking that is flexible to accommodate aircraft of various shapes and sizes.

Taxilanes are found on airport aprons to direct aircraft taxiing between airfield taxiways and aircraft parking areas on the apron. Taxilanes exist as *single-lane taxiways,* where there exists sufficient room for one aircraft, and *dual-lane taxiways,* with sufficient room for two aircraft taxiing in opposite directions to move simultaneously. Dual-lane taxilanes are typically found at the busiest of airports serving larger aircraft.

Aircraft gate management

One of the most important and sometimes most challenging aspects of planning and managing the apron concerns the number of aircraft parking areas, or gates, that are required for efficient operations. The number of commercial aircraft gates required at an airport, for example, over any given operating day is dependent on a series of factors, including: the number and type of aircraft scheduled to use a gate, each aircraft's scheduled *turnaround time* (also known as *gate occupancy time*), and the type of *gate-usage agreement* that each air carrier has with the airport.

The number and type of each aircraft scheduled to use a gate is of course vital to the planning of gate facilities. For each type of aircraft that uses the airport, there should be at least one aircraft parking area that can accommodate the aircraft. For smaller airports that are frequented by larger aircraft on a sporadic basis, a remote parking facility with sufficient space may be appropriate, whereas aircraft that operate more often should be considered for their size when constructing permanent gate facilities. At many airports, gates for larger aircraft are planned for the ends of linear terminals or satellite configurations, where aircraft wingspans are accommodated with minimal sacrifice of space for additional aircraft, and gates for smaller aircraft tend to be located nearer the center of the terminal.

The *turnaround time* of each aircraft directly affects the number of aircraft that can use a gate over the course of a day. Turnaround times of aircraft vary widely, based in part on the size of aircraft, the itinerary of the aircraft, the number of passengers, the volume of cargo to be loaded and unloaded, and the schedules of the air carrier. Turnaround times of smaller commercial service aircraft flying relatively short routes, carrying less than 50 passengers, for a regional airline, for example, may be as low as 15 minutes, whereas wide-body aircraft flying on international routes may require 2 or more hours turnaround time. As such, a gate serving small regional air carrier aircraft gates may serve 30 or more aircraft in an operating day, and gates serving international flights may accommodate only three to five aircraft per day.

The *gate-usage agreement* that each air carrier has with airport management also plays a significant role in the total number of required gates at the airport terminal. The three most common types of gate-usage agreements are *exclusive-use, shared-use,* and *preferential-use* agreements.

As the name implies, under an **exclusive-use agreement,** an air carrier retains sole authority to use a particular gate or set of gates at an airport terminal. This agreement gives the air carrier flexibility when adjusting flight schedules, assuring the carrier that gates will always be available when needed. Operationally, however, this type of agreement leads to inefficiencies in overall gate use, because when the air carrier is not currently using its gates, the gate sits idle, despite the fact that another air carrier may desire a gate parking space at that time. Air carriers signing exclusive-use agreements usually do so for a premium, and for a relatively long contract period, and thus are identified typically as *signatory carriers* at the airport. Signatory carriers tend to have the majority of operations at the airport, thus warranting exclusive-use agreements.

Under **shared-use agreements,** air carriers and other aircraft schedule use of gates in coordination with airport management and other air carriers serving the airport. Thus individual gates may be shared by multiple air carriers. Shared-use agreements are usually arranged by air carriers that have relatively few operations scheduled at the airport. For example, international air carriers tend to arrange shared-use agreements with United States airports, because they each have perhaps only a few operations per day at any given airport. For air carriers that have many operations at an airport, shared-use agreements reduce the flexibility in schedule planning. From an airport management perspective, however, shared-use agreements are operationally efficient, maximizing the number of aircraft that may use gates over the course of a schedule day.

Preferential-use agreements are hybrids of the exclusive-use and shared-use agreements. Under a preferential-use agreement, one air carrier has preferential use of the gate. However, should that air carrier not be using the gate during some period of the day, other air carriers subscribing to the agreement may use the gate, as long as its use does not interfere with upcoming operations from the preferential carrier. Preferential-use agreements are typically signed by one carrier that has moderate levels of service at the airport, and one or more carriers or charter aircraft that have relatively few operations. From an operational perspective, the overall number of aircraft utilizing gates under shared-use agreements depends primarily on the number of operations served by, as well as the typical turnaround time of, the preferential carrier. The greater number of operations and greater turnaround time of, the preferential carrier tends to lead to fewer numbers of aircraft using the gates over the course of an operating day.

Gantt charts

The management and planning of gate utilization at airport terminals can be a challenging venture, particularly when high volumes of operations occur during busy or *peak* periods. One tool used to assist with the scheduling and management of gate operations is a variation of a graphical scheduling management tool developed by Henry Gantt in 1917. A *Gantt chart* (also known as a ramp chart or gate utilization chart) is a graphical representation of the utilization of aircraft gates over a given period of time.

On the basis of each aircraft's operating schedule and scheduled turnaround time, and on the basis of each gate's gate-usage agreement, aircraft are allocated gate space, represented by rows on the Gantt chart, during their projected gate utilization periods, represented by columns on the chart. From plotting each aircraft's operation on the Gantt chart, terminal planners and gate managers can visually identify inefficiencies in gate utilization and potential conflicts, particularly during irregular operations, such as when an aircraft must stay at the gate past its scheduled *push back* time because of unforeseen circumstances, or when an aircraft arrives early to the airport.

Figure 7-14 represents a Gantt chart example for a given set of flight schedules, with gates 1 and 2 operating under shared-use agreements and gate 3 operating under an exclusive-use agreement.

The passenger handling system

The commercial airport terminal's **passenger handling system** is a series of links and processes that facilitate the transfer of passengers between an aircraft and one of the modes of the local ground transportation system. These processes include the *flight interface, passenger processing,* and *access/processing interface.*

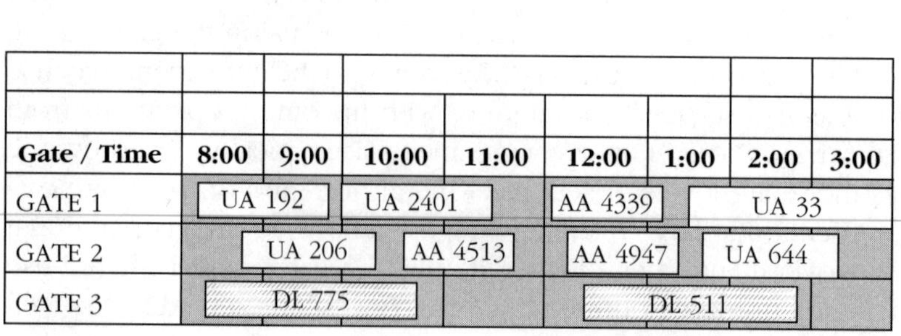

Gate / Time	8:00	9:00	10:00	11:00	12:00	1:00	2:00	3:00
GATE 1	UA 192		UA 2401		AA 4339		UA 33	
GATE 2		UA 206		AA 4513		AA 4947	UA 644	
GATE 3		DL 775				DL 511		

Figure 7-14. *Sample Gantt gate utilization chart.*

The **flight interface** provides the link between the aircraft gates and passenger processing facilities. The flight interface includes gate lounges and service counters, moving sidewalks, buses, and mobile lounges; loading facilities such as loading bridges and air stairs; and facilities for transferring between flights, including corridors, waiting areas, and mobile conveyance facilities (Fig. 7-15).

Passenger processing facilities accomplish the major processing activities required to prepare departing passengers for use of air transportation and arriving passengers to leave the airport for ground transportation to their ultimate destinations. Primary activities include ticketing, baggage check, security, passport check, baggage claim, customs, and immigration. Facilities include ticketing and baggage check-in counters, baggage and passenger security stations, information kiosks, baggage claim carousels, customs facilities, and rental car and other ground transportation desks.

The **access/processing interface** makes up the facilities that coordinate the transfer of passengers between ground transportation and the terminal building, where passenger processing facilities are typically located. Activities at the access/processing interface include loading and unloading of passengers and baggage from vehicles at the curb and transit stations, and pedestrian circulation from vehicle parking facilities. The access/processing interface includes the vehicular

Figure 7-15. *Loading bridges are part of the flight interface.*
(Picture courtesy Dallas Regional Chamber)

drive and terminal curb, sidewalks, shuttle buses, automated conveyance systems to and from parking facilities, and bus stops, taxi stands, and rail stations.

In addition, the **access/egress interface** facilitates the movement of passengers and ground vehicles between origins and destinations in the community and the airport property. The access/egress interface is a component of the airport's ground access system.

Passengers and their required processing facilities

One of the greatest challenges of managing airport terminal operations is the challenge of accommodating the necessary and desired processing needs of a wide spectrum of passengers. It is staggering to think that nearly every one of the more than 1 billion passengers who travel around the world annually on commercial air carriers has a unique itinerary, and unique needs, that must be accommodated. **Passengers** may be categorized in several manners, some of which include a passengers' segment of itinerary, trip purpose, group size, type of baggage carried, and type of ticket, and whether the passenger is an international or domestic traveler. Each passenger, by nature of the various categories that passenger may fall into, requires certain facilities, known as *essential processing* facilities within the airport terminal area. The understanding of each of these facilities on an individual basis, as well as an understanding of how each facility interacts with the other facilities, is itself essential for terminal operations to be successful.

Passenger processing requirements and other needs vary widely on the basis of the **segment of itinerary** the passenger is on while at the airport. The three primary itinerary segments are *departing, arriving,* and *transferring.* **Departing passengers** are those passengers who are entering the terminal from the ground access system through the access/processing interface. **Arriving passengers** are those passengers who have just deplaned an aircraft and entered the terminal from the flight interface with the intentions of leaving the airport terminal for their final destinations through the access/egress interface. **Transfer passengers** are entering the terminal from the flight interface with the intention of boarding other flights for their ultimate destinations within a relatively short period of time, again through the flight interface.

Passengers traveling within the United States (or within the confines of any country, for that matter) are considered *domestic passengers.* In the United States, even those passengers that are not United States citizens are considered domestic passengers if their itinerary is within the confines of the United States. In other countries, noncitizens may be considered international passengers, even when traveling within the confines of the country. Passengers traveling to or from the United States are considered international passengers, regardless of their citizenship, and are processed accordingly.

The **trip purpose** of a passenger has traditionally been an indicator of the passenger's individual needs. The two most common trip purposes identified in the industry are **traveling on business,** or **traveling for leisure,** although it is understood that many travelers' itineraries combine both business and leisure activities.

The group size of passengers plays a significant role in determining the most efficient manner for passenger processing, particularly through the access/processing interfaces and processing system. Group sizes of passengers tend to be categorized as either *traveling individually* (or *in small groups*), or *traveling in large groups* (typically of 20 or more passengers in the same group).

The type of baggage carried by passengers may determine not only the processing required by such passengers but also the design and planning of **baggage handling** facilities. Passengers are said to be carrying either no baggage, carry-on baggage, baggage to be checked in, and/or oversized or oddly shaped baggage (such as golf clubs or skis).

Most recently, the type of ticket that a passenger purchases from the air carrier has contributed to determining the type of processing required. Since the early 1990s, passengers have been able to purchase either traditional *paper tickets* or *electronic tickets*. Electronic ticketing facilitates the processing of departing passengers by removing the necessity of carrying a paper ticket to a ticket counter for initial processing. Furthermore, beginning in the early 2000s the ability of passengers to check-in and printout airline boarding passes from tickets purchased over the Internet or receive digital boarding passes on their mobile devices has further removed the need for passengers to check-in at a ticket counter at the airport.

The true challenge of airport terminal planning and management is to accommodate the needs of all passengers, as well as their friends and families who meet them or see them off (commonly known as *meeters/greeters*), airport employees, airline employees, concession workers, and government staff, while minimizing the conflict between any individuals or groups.

Although every airport terminal is different in the number, type, and arrangement of passenger processing facilities, there are a series of **essential processing facilities** that must be present to ensure appropriate processing for passengers traveling on each itinerary segment.

For all departing passengers, these facilities include *passenger check-in* (traditionally known as *ticketing*) and *passenger security screening*. For those passengers traveling with baggage to be checked in, *baggage explosive detection screening* processing is required. Finally, departing passengers require some form of processing just prior to boarding at the gate.

Passenger check-in

The **passenger check-in** process has come a long way since the early days of passenger processing at airport terminals, although some characteristics dating back to the original ticketing policies, including the term *ticketing,* remain. Traditional check-in counters are facilities staffed by air carrier personnel. As with gates, check-in counters may be configured for exclusive use or common use.

Exclusive-use check-in counters are typically configured with information systems, computers, and other equipment specific to one air carrier. The number of positions at the counter is typically determined by the airline on the basis of the estimated number of departing passengers over the course of the operating day, particularly at busy, or *peak,* times. Most scheduled air carriers with consistent volumes of scheduled operations, tend to have exclusive-use check-in facilities at commercial service airports.

Common-use check-in facilities counters are typically configured for use by multiple air carriers. Many common-use ticketing facilities are equipped with **common-use terminal equipment (CUTE),** a computer-based system that can accommodate the operating systems of any air carrier that shares the check-in facility (Fig. 7-16). A growing number of airport terminals serving air carriers that have infrequent service to the airport, charter carriers, and

Figure 7-16. *CUTE (common-use terminal equipment) with variable signage.*

international carriers have implemented common-use check-in facilities, which provide the ability to serve more air carriers and passengers with less physical space than their exclusive-use counterparts.

The traditional processing that occurs at an airline check-in counter includes the purchasing of airline tickets either for trips on the day of purchase or for future travel, the assignment of seats, and the issuance of boarding passes. For passengers checking in baggage, the ticket counter has traditionally served as the location where bags would be checked and entered into the baggage handling system.

For the first 60 years of commercial aviation, much of the functions performed at the ticket counter were done manually. In recent years, the implementation of computer technology, information sharing, and automation have allowed much of the traditional processes to be distributed among other locations, many of which are not located at the airport terminal itself. The purchasing of airline tickets through travel agents, over the telephone, and most recently through the Internet or via mobile devices comprise the vast majority of airline ticketing transactions. Furthermore, the ability to acquire seating assignments, and receive boarding passes, through automated systems renders the airport terminal's check-in process an unnecessary part of many departing passengers' travels through the terminal.

Moreover, nearly all commercial air carriers have implemented automated kiosks, located near traditional check-in counters, and these kiosks perform many of the essential services of the traditional ticket counter. In addition, some airports have employed **common-use self-service (CUSS) kiosks, which offer check-in for multiple air carriers** (Fig. 7-17), while an increasing number of air carriers around the world are facilitating the complete removal of the check-in process at the airport by allowing passengers to receive digital boarding passes on their mobile devices (Fig. 7-18).

Despite the vast changes in technology and policies over time, the traditional check-in counter may never become obsolete. During periods of irregularity, such as when flights are delayed or canceled, or when passengers need special assistance with their itineraries, the check-in counter often becomes the first location that passengers go to in order to find an airline representative for assistance.

Security screening

The processing of passengers and baggage for the purpose of ensuring the security of the civil aviation system has undergone a virtual overhaul following the terrorist attacks on the United States on September 11, 2001. Since 2003, passenger and baggage security screening has been managed and operated

Figure 7-17. *Common use self-service (CUSS) kiosk.*

by the Transportation Security Administration (TSA). Although the TSA has ultimate authority over the facilities and procedures that comprise the security screening processes, airport managers and planners should be keenly aware of the security screening process, because the process has presented the most significant impacts on airport terminal planning and operations in recent years. A detailed description of security screening processes may be found in Chap. 8 of this text. From the perspective of overall airport terminal operations, passenger and baggage screening processes are perhaps the most challenging of all processing at the airport, as all passengers and baggage must be screened at the airport (as opposed to another, remote location). In many airports, security screening has become the single largest passenger-processing component within the terminal.

Figure 7-18. *Digital boarding pass received via a mobile device.* (Photo courtesy Cathay Pacific Airlines)

At-gate processing

The remaining processing to be performed on a passenger prior to boarding an aircraft typically occurs at the gate area. Each air carrier has its own method of boarding passengers onto aircraft. Some air carriers board in order of fare class, first class first, coach class next. Others board passengers in order by the row number of their assigned aircraft seats (rear to front), or most recently by predetermined boarding groups, known as *zones*, as identified on passengers' boarding passes.

At times, gate processing has also incorporated security screening policies. Early policies employed by the Transportation Security Administration called for randomly selecting boarding passengers for additional passenger and carry-on baggage screening. This policy was phased out in the early months of 2003, but has returned from time-to-time based on TSA threat assessments.

In addition to boarding, passenger processing within the gate area also includes administrative issues regarding a passenger's ticket, including seat assignment changes, requests to stand by for a flight, and any irregular issues that may arise.

Customs and border patrol facilities

Passengers arriving on international flights must generally undergo customs and immigration formalities at the airport of their initial landing in the United States. **Federal Inspection Services (FIS)** conducts these formalities, which include passport inspection, inspection of baggage, and collection of duties on certain imported items, and sometimes inspection for agricultural materials, illegal drugs, or other restricted items. FIS is operated by the United States Customs and Border Patrol (CBP), which is administered under the Department of Homeland Security.

In recent years, introduction of streamlined procedures for returning U.S. citizens, enhanced procedures for clearing non-U.S. citizens, and computerized access to records at inspection stations have substantially sped the flow of passengers at many airports. Flights from some Canadian and Caribbean airports are precleared at the originating airport, so arrival formalities are substantially reduced or eliminated. Similar procedures exist at many airports throughout the world, as well.

Ancillary passenger terminal facilities

Although not technically required for passengers, ancillary, or nonessential facilities, are provided in airports to improve the overall travel experience. Nonessential facilities include food and beverage services, retail shops, common waiting areas, information kiosks, post offices, places of worship, hotels, conference centers, bars, and smoking lounges. These facilities, known as *concessions,* when properly managed, not only offer benefits to passengers, but also may generate significant levels of revenue to support the operations of the airport (Fig. 7-19).

The management of concessions within airport terminals continues to evolve. At many large commercial service airports, where large volumes of passengers provide significant market potential for retail products and services, airport terminals have established concessions programs that offer brand name products and services, ranging from fast-food, to specialty items.

Many airports include concessions that promote and support the local economy. These programs may include the presence of shops that offer locally made products, or products associated with the area. In addition, many airports have DBE (Disadvantaged Business Enterprise) programs that offer minority- and woman-owned businesses to set up shop in the airport, at reduced lease rates, at part of its concessions program.

By locating passenger processing facilities, both essential and nonessential, in convenient locations and in a logical order, terminal planners aim to keep

Figure 7-19. *The atrium at Orlando International Airport is surrounded by concessions, including a food court, sit-down restaurant retail products, and a hotel and conference center.* (Photo courtesy Greater Orlando Airport Authority)

passengers moving through airports with a minimal amount of confusion and congestion. To fully understand the behavior of passengers within a terminal, terminal passenger flow diagrams are constructed.

Passenger flow diagrams illustrate the direction and volume of passengers traveling from one processing facility in a terminal to another. On the basis of this information, airport terminal facilities may be appropriately sized and managed to maintain efficient operations (Fig. 7-20).

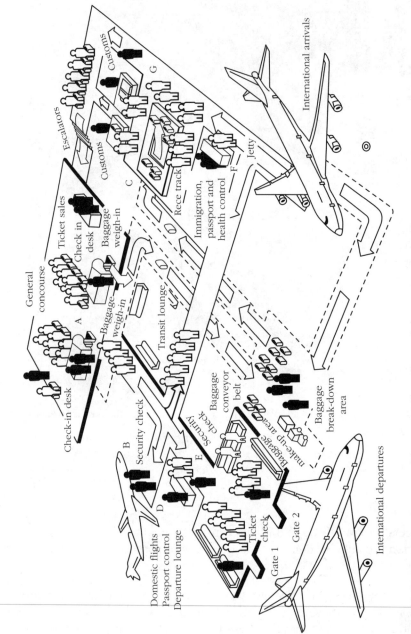

Figure 7-20. *Passenger and baggage flow through airport terminals.*

Vertical distribution of flow

Many of the larger airports distribute the passenger flow over several levels within the airport terminal. The primary purpose of distributing passenger processing activities over several levels is to separate the flow of arriving and departing passengers. The question of how many levels a terminal building should have depends primarily on the volume of passengers. It is also influenced by the type of passengers: domestic, international, and transfer. Figure 7-21 shows a cross section of the major functional areas in a multilevel passenger terminal. Departing passengers park their vehicles (1) and proceed via the bridge level (3) into the terminal or are dropped off at the vehicular circular drive (enplane drive) (5). Lobby (6), concourse (11), and gate area (14) are all on the first level. Arriving passengers proceed from the gate area (14) through the concourse (11) to the baggage claim area (7). After claiming their baggage, they proceed to the parking facility (1) via the bridge level (3) or are picked up at the ground level (deplane drive—4). Notice that the airport offices (10); mechanical, storage, and maintenance facilities (8); and service vehicle drive (2) are located above or below the passenger flow. Transit shuttles (9) and satellite transit tunnel (13) leading to a satellite terminal (normally for long-haul domestic or international flights) are located on the lower level. Variations in this basic design might occur when traffic volumes or types of traffic require. For example, at large airports where transportation between terminals operate, a special level might be needed to provide access to these systems. In addition, some airports use special levels to accommodate high-occupancy vehicles, such as shuttle vans or coach buses.

Baggage handling

Baggage handling services include a number of activities involving the collection, sorting, and distribution of baggage. An efficient flow of baggage through the terminal is an important element in the passenger handling system.

Departing passengers normally check their baggage at one of a number of sites including curbside check-in, the check-in counter in the terminal building, or at a TSA-designated baggage drop-off point, typically located near check-in counters in the terminal lobby. The bags are then screened by the TSA; sent to a central sorting area, where they are sorted according to flights; and then sent to the appropriate gate to be loaded aboard the departing aircraft. Arriving baggage is unloaded from the aircraft and sent to the central sorting area. Sorted bags are sent to a transferring flight, or to the baggage claim areas (Fig. 7-22).

At most airports, baggage handling is the responsibility of the individual air carriers. Some airports operate a consolidated baggage service, either with airport personnel or contracted out to a private third party aircraft service operator.

One of the simplest and most widely applied methods to expedite baggage handling is curbside check-in. This separates baggage handling from other

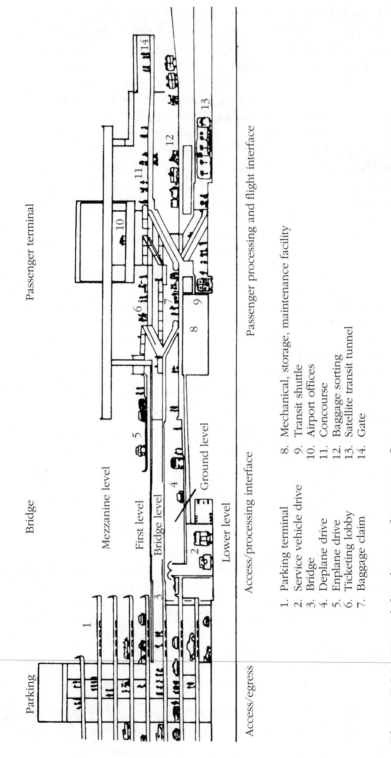

Parking Bridge Passenger terminal

Mezzanine level

First level

Bridge level

Ground level

Lower level

Access/egress Access/processing interface Passenger processing and flight interface

1. Parking terminal
2. Service vehicle drive
3. Bridge
4. Deplane drive
5. Enplane drive
6. Ticketing lobby
7. Baggage claim

8. Mechanical, storage, maintenance facility
9. Transit shuttle
10. Airport offices
11. Concourse
12. Baggage sorting
13. Satellite transit tunnel
14. Gate

Figure 7-21. *Vertical distribution of passenger flow.*

Figure 7-22. *Baggage being loaded onto palettes for loading onto aircraft.*

check-in, thereby disencumbering those locations and allowing baggage to be consolidated and moved to aircraft more directly.

Sorting baggage, moving it to and from the apron, and aircraft loading and unloading are time-critical and labor-intensive operations. Technologies to improve this process include high-speed conveyors to transport baggage between the terminal and the flight line, often used in conjunction with pallets or containers that can be put on and taken off aircraft with labor-saving equipment. Computerized baggage-sorting equipment, capable of distributing bags with machine-readable tags, has been installed at several airports.

Baggage claim

For passengers who checked baggage at the airport prior to departure, facilities for claiming their baggage must exist at the airport as well. Baggage claim facilities are typically located in an area conveniently positioned near facilities that accommodate ground transportation from the airport, including parking lots, shuttle vans, taxi cabs, and rental car counters.

Baggage is typically presented to arriving passengers in the baggage claim area by use of a baggage claim carousel, configured in such a way as to provide sufficient carousel frontage to accommodate all passengers desiring access to their baggage, while minimizing the total amount of space required for the claim area (Fig. 7-23).

Figure 7-23. *Baggage claim facilities at*
Las Vegas McCarran International Airport.
(Figure courtesy Clark County, Las Vegas McCarran International Airport)

Carousels are typically shared between air carriers in a given terminal. This is feasible because limited infrastructure is required specifically for one particular carrier in these areas. Typically, however, each air carrier will have its own administrative area, primarily to handle cases of lost, unclaimed, or damaged baggage.

Airport ground access

Access to the airport from the surrounding community is an integral part of the overall passenger and baggage processing system. The *access/egress link* of an airport's passenger handling system includes all of the ground transportation facilities, vehicles, and other modal transfer facilities required to move the passenger to and from the airport. Included in the access/egress link are highways, intercity and metropolitan rail service, autos, taxicabs, buses, shuttles, limousines, and transfer stations, including off- and on-airport parking sites and rail stations.

Airport access is usually divided into two major segments:

- Access from the **CBD (central business district)** and suburban areas via highway and rapid transit systems to the airport boundary
- Access from the airport boundary to parking areas and passenger unloading curbs at the terminal building

Access from the CBD and suburban areas to the airport boundary

The segment connecting the airport with the surrounding metropolitan area is a part of the overall regional or urban transportation system and serves general and airport traffic. State and local highway departments and local transit authorities bear the major responsibility for the administration, design, and construction of this segment. Airport management, however, is responsible for developing the requirements of airport traffic that must be served within this segment. They are also responsible for promoting the development of facilities to serve that demand. Regional, state, and local planning bodies, commonly known as **metropolitan planning organizations (MPOs),** are relied upon to bring together the general needs of urban transportation and the specialized needs of airports by the development of comprehensive transportation plans for metropolitan or regional areas as a whole. At the federal level, the Department of Transportation and the Department of Housing and Urban Development provide national inputs through programs such as the Federal Highway Grants-in-Aid Program, and urban transportation planning funds. With this diversification of responsibility, careful coordination is required if the first segment of the airport access problem is to be effectively resolved.

Access modes

Unless the ultimate destination of any travel itinerary is the airport itself, every trip on a commercial aircraft and nearly every trip on a general aviation aircraft includes an additional mode of transportation. A *mode* of transportation is defined as a type of vehicle used to travel from one point to another.

The Transportation Research Board defines the most common modes of airport access as:

- *Private vehicles:* Vehicles used to transport airline passengers or visitors (e.g., family members, employees, friends, or clients), without payment of a fare by the passenger, which are privately owned and privately operated.
- *Rental cars:* Vehicles used to transport airline passengers or visitors, which are leased by the passenger or visitor from an agency doing

business at or near the airport and rented for the duration of the passenger's or visitor's trip. Vehicles rented under a long-term lease (i.e., greater than 3 months) are considered private vehicles, not rental cars.

- *Courtesy vehicles:* Door-to-door, shared-ride transportation provided for customers of hotels, motels, rental car agencies, parking lots (both those privately operated and airport operated), and other services. Typically, no fare is charged because the transportation service is considered part of (or incidental to) the primary service being provided. Service is provided using a variety of vehicles, including full-size buses, minibuses, vans, and station wagons.

- *Airline crew vehicles:* Shared-ride transportation between airports and hotels provided at no charge for airline crew members by the employer. Service is provided using a variety of vehicles, including full-size buses, minibuses, vans, and station wagons.

- *Taxicabs:* Privately operated door-to-door, on-demand, exclusive transportation (i.e., for a single party, typically up to five persons). Fares are typically calculated according to trip length and travel time using a taximeter and according to rates established by a city or county licensing agency (e.g., a taxicab commission or public services commission), but fares may be zone fares, flat fares (predetermined fares between certain points, such as the airport and downtown), or negotiated fares. Typically, the fare is for use of the entire vehicle, although some communities allow extra fares per passenger or piece of baggage.

- *Town cars (on-demand limousines):* Privately operated door-to-door, on-demand ground transportation services that typically charge premium fares calculated on a per-mile and per-hour basis, available at the curbsides of some airports. These exclusive transportation services are typically provided using luxury town cars or limousines.

- *Prearranged limousines:* Door-to-door services that provide exclusive transportation and require reservations. Fares may be flat, calculated on a per-hour basis, or negotiated, regardless of the number of persons transported, according to rates approved by local or state licensing agencies. Such agencies sometimes also specify the geographic area that can be served and the tariff (or maximum fee) that can be charged. Prearranged limousine services are typically provided using luxury vehicles and include private car services (black cars), luxury limousine services, and suburban taxicabs (i.e., prearranged taxicab service provided by an operator not licensed to provide on-demand service at the airport). These services typically require prior reservations but may also be dispatched by radio requests. Prearranged limousines are not permitted to respond to hails or on-demand requests for transportation. Privately owned and privately operated luxury limousines are considered private vehicles, as are those operated or leased by a corporation. However, most surveys do not distinguish between privately owned and other types of limousines.

- *Chartered buses and vans:* Exclusive, door-to-door transportation services requiring reservations or prior arrangements. Fares are typically calculated on a per-hour basis regardless of the number of persons transported, according to tariffs approved by local or state licensing agencies. Chartered bus and van services are provided using buses, minibuses, and vans (seating eight or more passengers) and include tour buses, cruise ship buses, and other prearranged transportation for more than five passengers.

- *Shared-ride, door-to-door vans:* Shared-ride, door-to-door transportation services, which charge customers a predetermined flat fare per passenger or zone. Typically, transportation from the airport is on-demand, but transportation to the airport requires prior reservations. Vehicles may be licensed as shared-ride vans, airport transfer vans, or, in some communities, as taxicabs or prearranged/chartered vans. In most communities, the service is operated using radio-dispatched, eight-passenger vans, but station wagons, limousines, and sedans are also used.

- *Scheduled buses:* Scheduled service operating to established stops or terminals, typically on a scheduled basis, along a fixed route, which charges a predetermined flat fare per passenger or zone. In many communities, there are two classes of bus service:

  - *Express (including semiexpress) transportation* between the airport and major destinations in the region, often provided by a private operator licensed by state or regional agencies but in some communities are provided by a public operator. Sometimes referred to as "airporters."

  - *Multistop transportation* between the airport and the region, typically operated by a public agency (i.e., traditional bus service).

- *Rail service:* Fixed-route rail service operating to established stops or terminals on a scheduled basis. Customers are charged a predetermined flat fare per passenger or zone. Types of trains used to provide this service include light rail, commuter rail, and rapid transit.

In the United States, the vast majority of all ground access trips to and from airports are served by private automobile, rental car, taxicab, or shared-ride van, all of which are modes that carry relatively few passengers per vehicle, and rely directly on an infrastructure of public roads and highways to connect the airport with ultimate origins and destinations, and curb frontage and parking facilities for vehicle loading, unloading, and parking at the airport itself.

Many larger commercial service airports serving high-population metropolitan areas are served by modes of public transportation, including rail service and scheduled buses. The *mode share* for public transportation, that is, the percentage of passengers using public transportation to access the airport, of these airports range from less than 2 percent to slightly less than 20 percent. Most smaller commercial service airports and virtually all general aviation airports have public transportation mode share near zero, reflecting the fact that virtually all access to the airport is made through private or privately hired vehicle (Fig. 7-24).

Several airports around the world, however, have much higher public transportation mode shares, in some cases nearing 60 percent. These airports reflect concerted efforts to provide public transportation ground access by the metropolitan planning organizations along with regional environments whose populations are less dependent on transportation by automobile than those in the United States (Fig. 7-25).

Factors influencing demand for ground access

Demand for ground access, that is, the volume of people that wish to have access between the airport and their respective origins and destinations at commercial service airports, is primarily generated by the number of enplaning and deplaning passengers using the airport. These volumes are generated in part by the provision of air service by the air carriers that serve the airport. Characteristics of this air service include destinations served, the type of aircraft used, and the daily departure and arrival schedules of the air carriers.

In addition to passengers themselves, airports are accessed by those people seeing off or meeting passengers at the airport. These people are known as

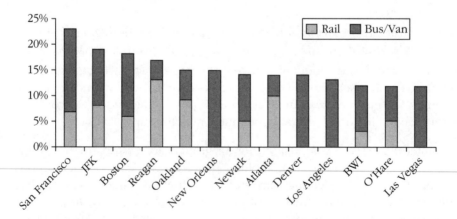

Figure 7-24. *Public transit mode share from selected U.S. airports.*

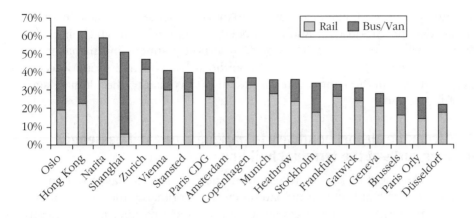

Figure 7-25. *Public transportation mode share of selected airports outside the United States.* (Source: Transportation Research Board, National Academy of Sciences)

meeters/greeters. The demand for airport access by meeters/greeters is dependent on similar characteristics as that of passengers themselves.

A significant proportion of trips made to and from airports are generated by the workforce in place at each airport, including airport, airline, and government employees, as well as employees of the many private companies that do business at the airport, including concessionaires, contractors, and suppliers. These trips are less dependent on available flight service. They are more associated with the travels that occur during any business day, including morning and evening commutes and trips associated with business delivery. In addition, as many functions in the airport operate as much as 24 hours per day, there are a number of trips to the airport that occur outside normal business hours.

Coordination and planning of ground access infrastructure

To effectively develop ground access requirements to the airport from the CBD and suburban areas, it is important to gain an understanding of the geographic region from which passengers access the airport. This region is known as an airport's *capture area.* For commercial service airports, the geographic size of a capture area varies greatly, depending primarily on the population density in the region and the availability and cost of air carrier service from the airport, as well as from other airports within the region. General aviation airports typically serve more local areas, such as one CBD, suburban area, or outlying community. Many communities fall into multiple airport capture areas illustrating the fact that passengers in fact choose to access different airports from the same region on the basis of the characteristics of each airport, offered air service, and the ground access system.

Although not the most significant determinant of passenger volumes, the ability to access one airport over another indeed has an effect on which airport a passenger will choose to use. The ability of airport planners and managers to identify the airport's capture area and coordinate an effective ground access system from within the capture area to the airport is vital to the ultimate success of the airport.

Access from the airport boundary to parking areas and passenger unloading curbs at the terminal building

The second segment of airport access, from the airport boundary to the parking area and terminal building unloading curbs, is primarily the responsibility of airport management. This segment includes vehicle parking facilities, curb frontage at the terminal, intra-airport public transit systems such as shuttle buses or light rail systems, and vehicle roads that connect facilities existing on airport property.

Vehicle parking facilities

Parking facilities at or near the airport must be provided for passengers, visitors accompanying passengers, people employed at the airport, car rentals and limousines, and those doing business with airport tenants.

Public parking facilities are provided for airline passengers, meeters/greeters, and other members of the public doing business at the airport. Most commercial service airports have separate parking facilities for short-term and long-term parking. Surveys at a number of major airports indicate that a large number (75 percent or more) park 3 hours or less and a much smaller group parks from 12 hours to several days or longer; however, short-term parkers, due to their relatively short parking durations, typically represent only about 20 percent of the total maximum vehicle accumulation. Consequently, many airports designate relatively few parking spaces to short-term parkers, typically the most convenient (closest area) spaces. Parking rates for short-term parking are typically higher than that for longer-term parking. This rate strategy achieves two goals. First, it provides incentive for those intending to park their vehicles for a relatively long period of time to use long-term parking facilities, thereby leaving spaces available for short-term parkers in the closer, more convenient, short-term parking area. Second, it tends to maximize the amount of total revenue generated by the parking system to the airport (Fig. 7-26).

The number of parking spaces required to provide adequate service levels is normally greater than total parking demand. This is because at a large parking facility in which many areas cannot be seen simultaneously—for example, in a multilevel garage or extensive open lot—it is more difficult to find the last

Figure 7-26. *New multilevel parking garage at Washington Dulles International Airport.* (Figure courtesy Metropolitan Washington Airports Authority)

empty spaces. Thus a large parking facility may be considered full when 85 to 95 percent of the spaces are occupied, depending on its use by long- or short-term parking, size, and configuration (Fig. 7-27).

Off-airport parking

Surrounding many airports often exist public airport parking facilities located off-airport property and operated by independent private operators. These facilities typically offer parking at lower rates than airport-operated facilities. Although they tend to be located farther away from the airport terminal, frequent shuttle service between the parking facility and the terminal often offsets the extra distance. In addition, some off-airport parking facilities offer extra amenities ranging from free coffee and newspapers for customers, to automobile washes and valet service. The success of off-airport parking facilities can have a direct, significant effect on airport revenues, because these facilities do not pay any portion of their revenue to the airport.

Employee parking

Separate parking facilities are normally provided for employees working at the airport. Employee parking lots may be located as far as several miles from the terminal area. In these cases, employees are bused to the airport from the outlying facility.

Car rental parking

The car rental parking areas are often located in various locations on airport property. The strategies for location of car rental parking, as well as automobile pickup and drop-off, have varied greatly among airports in recent years, depending on the size of airports, the volume and type of passengers renting vehicles, and the number and strategies of the private car rental companies serving the airport.

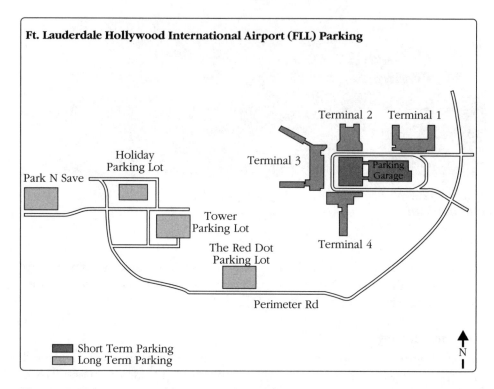

Figure 7-27. *Location of airport parking facilities at Fort Lauderdale Hollywood International Airport.* (Courtesy ifly.com)

Traditionally, car rental pickup and drop-off facilities had been located close to the terminal building in order to minimize passenger walking distances from the terminal to their vehicles. Rental cars in inventory are often parked in a special area away from the terminal building and driven to the car rental area upon request. At larger airports with limited space close to the terminal buildings, individual car rental companies will locate all but reservation counters at off-airport property. These companies provide customer access between the terminal and their vehicles by using shuttle vans or buses.

Recent trends in car rental facilities include the construction of on-airport consolidated rental car facilities (CRCF). These facilities combine the operations of several individual rental car companies, providing central locations for reservations, automobile pickup and drop-off, and vehicle inventory. Access to consolidated car rental facilities is provided by the airport in the form of buses or automated people-mover systems. These facilities have been met with varying degrees of criticism and praise. Although the existence of a single facility to handle car rental operations seems more agreeable to passengers than individual dispersed facilities, the fact that these facilities are often located a considerable distance from terminal facilities, and the fact that high volumes of travelers and their luggage must travel together to a consolidated facility, result in the consideration that consolidated rental car facilities create more inconvenience than they do enhancement to the car rental process (Fig. 7-28). The operation of CRCFs has begun to mature since their inception in the late 1990s. Higher-frequency bus service or the implementation of rail links connecting terminal buildings and CRCFs has facilitated the movement of passengers between the two. As a result, the proliferation of CRCFs around the United States continues.

Terminal curbs

The **terminal curb** front provides temporary vehicle storage during passengers' transition between the terminal and the landside, and it is at the curbside that all passengers, except those using nearby parking or transit facilities, either enter or leave some form of ground transportation. A variety of pedestrians, private automobiles, taxis, buses, commercial delivery trucks, and shuttle vans use the terminal curb area. At the terminal curb, passengers might be carrying luggage to or from the terminal building, checking luggage at curbside facilities, and waiting for access to taxis or other vehicles. At some airports, passengers must cross frontage roads to reach parking areas from the terminal curb.

Figure 7-28. *Buses take passengers to and from the consolidated rental car facility at Houston's George Bush Intercontinental Airport.* (Courtesy Houston Airport System)

The primary determinants in the amount of curb frontage space required at a terminal are the number of vehicles that arrive to the curb over a given period of time, the types of vehicles that use the curb, and the length of time that vehicles stop for loading and unloading, referred to as the dwell time. Dwell times of vehicles range from a little as 1 minute for private vehicles and taxi-cabs dropping off passengers, to greater than 5 minutes for shuttle vans and buses waiting for arriving passengers to transport to their ultimate destinations on the landside.

Because private automobiles are the dominant ground access mode at most airports, they are the principal source of terminal curb frontage demand. Such demand can be reduced at some airports by increasing availability of convenient parking, which typically raises the proportion of motorists who enter or exit parking areas directly without stopping at the curb frontage, or by encouraging passengers to use off-airport check-in facilities if these are available.

Demand for curb frontage is also determined by flight schedules and particularly by the arrival pattern of originating passengers (how far in advance of the scheduled departure time they arrive at the airport) and the route through the terminal of terminating passengers (how long it takes them to travel from an arriving flight to the curb). Type of flight and trip purpose also influence terminal curb demand. For example, originating passengers on international flights are requested to arrive at the airport earlier than those aboard domestic flights. Terminating international passengers also typically take more time than domestic passengers to reach the curb frontage because of required customs and immigration procedures.

Passengers on business trips tend to arrive at the airport closer to their departure times than leisure passengers. Deplaning business passengers, who might carry all their baggage aboard an aircraft and thus not need to stop at the baggage claim, tend to reach the curb frontage in less time than those deplaning passengers who have checked bags. Transfer passengers at some airports use buses operating on frontage roads and thus also contribute to the demand on terminal curb facilities.

The curb frontage demand resulting from shuttle buses and courtesy vans might be related to the number of trips per hour they make to the terminal and not directly to the number of passengers. The operators of these vehicles, seeking to ensure that all passengers are picked up promptly and reliably, may provide frequent service operated on specific headways and allow some vehicles to be underutilized in order to reduce waiting time for their patrons. Vehicle dwell time varies with type of vehicle, number in the vehicle, and baggage loads of passengers.

The most common forms of physical improvement at terminal curbs are additional curb frontage, bypass lanes, multiple entry and exit points in the terminal building, remote park and ride facilities, and pedestrian overpasses or underpasses. These improvements are intended to increase the utilization of curb frontage by vehicular traffic or, in the case of park and ride, to reduce demand on the curb front by diverting passengers from private cars to high-volume vehicles. Walkways to segregate foot and vehicular traffic promote pedestrian safety and facilitate roadway traffic by eliminating conflicts between pedestrians and vehicles.

Some terminal curbs are designed initially, or are retrofitted, with two levels, an enplaning area on one level and a deplaning area on the other level. In some cases, procedural changes—either alone or in conjunction with low-cost physical modifications such as signage or lane dividers—are an effective alternative to expensive construction or remodeling of the curb front. For example, parking restrictions combined with strict enforcement will reduce curbside congestion and dwell time in discharging and boarding passengers. Similarly, separation of private cars from taxis, buses, and limousines can diminish conflicts among these kinds of traffic and improve the flow to and from the curb front.

An effective approach at some airports has been provision of bus service from remote parking to the terminal and regulations to discourage bringing private automobiles to the terminal building. None of these measures is a substitute for adequate curbside capacity, but they can lead to more efficient use of the facilities available and perhaps compensate for deficiencies in terminal and curb frontage design.

Technologies to improve ground access to airports

A variety of technologies are in development and implementation to improve both segments of airport ground access, including advanced traveler information systems (ATIS); emerging bus, rail, and automated people-mover technologies; as well as alternative strategies for off-site airport check-in. ATIS systems throughout the United States have adopted the "511" moniker, representing the telephone information number and Internet and mobile device access addresses.

Advanced traveler information systems allow travelers to better estimate the travel time to the airport and in some cases offer the passenger alternative routes or modes that may offer reduced travel time or monetary cost of travel. Much of this information is gathered from real-time monitoring of traffic volumes on major access roads, and operational status of public transit systems (Figs. 7-29 and 7-30).

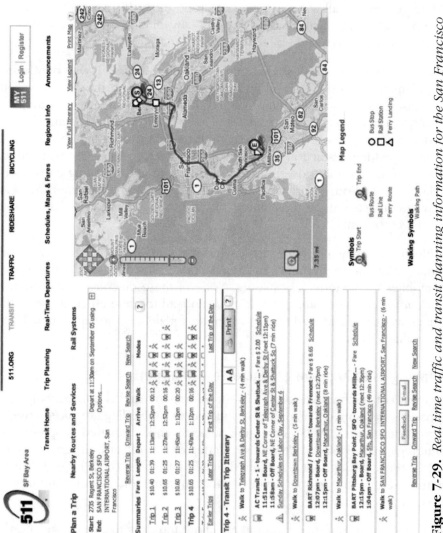

Figure 7-29. *Real time traffic and transit planning information for the San Francisco Bay Area.* (Figure courtesy 511.org)

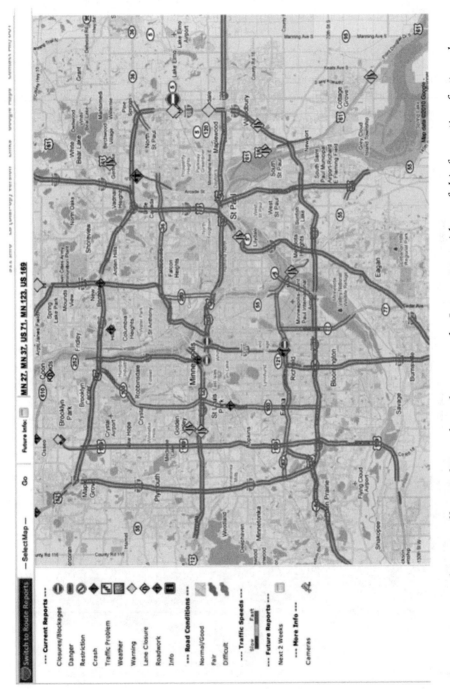

Figure 7-30. *Real-time traffic conditions broadcast over the Internet provide useful information for travelers accessing airports.* (Figure courtesy Minnesota Department of Transportation)

Advanced traveler information systems may also be used to improve the performance of public parking facilities by providing information to travelers regarding specific locations where spaces within parking facilities are available (Fig. 7-31).

The implementation of latest-generation public transportation systems connecting airports with regional transportation centers seeks to improve ground access to airports by providing convenient access to airports using the existing public transportation infrastructure and reducing the demand of private and private-hire automobile traffic on the surrounding road systems (Fig. 7-32).

Concluding remarks

An airport's terminal serves as a uniquely fundamental component of the airport system, requiring planning and management to accommodate a wide variety of aircraft and passenger types. Although fundamental operational and planning concepts apply to every airport terminal facility, there are no two airport terminals in the world that are exactly alike. As a result, specific understanding of the operations of a particular airport terminal facility is necessary to operate and plan for the goal of accommodating both passengers and aircraft in the most efficient and high-quality manner.

Equally important to the terminal itself is the ability for passengers to access the terminal and other airport facilities from the surrounding area. Airport management has the responsibility to manage ground access systems within the airport boundary and to promote efforts to facilitate ground access throughout the airport's capture area by coordinating with the area's local governments

Figure 7-31. *Smart parking facilities provide specific parking space availability.*

Figure 7-32. *The APM system at Newark Liberty International Airport connects the airport terminal with parking and rental car facilities, as well as the regional rail transportation centers.* (Picture courtesy Port Authority of New York/New Jersey)

and metropolitan planning organizations. Airport ground access is vital to the airport not only for the passengers a working system brings to the airport, but also by generating revenue for the airport.

Airport terminals and ground access systems are set to benefit from new technologies that will make operations of these systems more efficient. Airport planners and managers that apply these technologies, combined with an understanding of the fundamentals of terminal and ground access operations, have the potential of developing future facilities that will conveniently and efficiently handle future volumes of airport users.

Key terms

airport terminal concepts
    centralized facilities
        simple-unit terminal
        combined-unit terminal
        multiple-unit terminal
        gate arrival concept
        linear terminal

curvilinear terminal
decentralized facilities
  pier finger terminal
  pier satellite Terminal
  APM (Automated Passenger Movement Systems)
  terminal concourses
  remote satellite terminal
  mobile lounge concept
  airside-landside concept
off-airport terminals
Apron/gate system
aircraft parking
  nose-in
  angled nose-in
angled nose-out
  parallel
  remote
RON (remain overnight)
turnaround time
gate-usage agreements
  exclusive-use
  shared-use
  preferential-use
  Gantt chart
passenger handling system
  flight interface
  passenger processing
  access/processing interface
  access/egress interface
passenger types
  segment of itinerary
    departing passengers
    arriving passengers
    transferring passengers
  trip purpose

business travel

leisure travel

baggage handling

essential processing facilities

ticketing

exclusive-use counters

CUTE (common-use terminal equipment)

CUSS (common-use self-service) kiosks

security screening

at-gate processing

FIS (Federal Inspection Services)

ancillary processing facilities

concessions

baggage handling

baggage claim

carousels

airport ground access

CBD (central business district)

MPO (metropolitan planning organization)

vehicle parking

terminal curb

Questions for review and discussion

1. What are some of the various terminal design concepts that have existed over the history of civil-use airports?

2. What are some of the advantages and disadvantages of each type of airport terminal geometry?

3. How did the advent of APM systems affect the construction of airport terminals?

4. What is the mobile lounge concept?

5. What are off-airport terminals? What potential do they have in the future of airport terminals?

6. What are the different aircraft parking configurations that may exist at airports? When is each type of parking configuration most applicable?

7. What is a Gantt chart? How can Gantt charts help airport management?

8. What are the different types of gate-usage agreements that airports negotiated with aircraft operators? What are the advantages and disadvantages of each type of agreement?

9. What are the different processes that comprise the passenger handling system at airport terminals?

10. In what ways are passengers categorized while traveling through airport terminals?

11. What are the required passenger processing facilities that exist at the airport terminal?

12. What are CUTE systems?

13. What are FIS? What passengers typically require FIS?

14. Into what two categories is airport ground access typically divided?

15. What are MPOs? What authority do MPOs have over airport ground access?

16. What are the different modes that typically provide ground access to airports?

17. What are some factors influencing the demand for ground access?

18. What are the various parking facilities that are needed at airports?

19. What are some of the technologies that exist and are being developed to improve ground access to airports?

Suggested readings

Airport Landside Planning and Operations. Special Report 1373. Washington, D.C.: Transportation Research Board, 1992.

Airport Terminal and Landside Design and Operation. Special Report 1273. Washington, D.C.: Transportation Research Board, 1990.

Doganis, Rigas. *The Airport Business*. New York: Routledge, Chapman and Hall, Inc., 1992.

FAA, *Intermodal Ground Access to Airports: A Planning Guide,* FAA Report No. DOT/FAAIPP/96-3, Washington, D.C., 1996.

Hart, Walter. *The Airport Passenger Terminal*. Malabar, Fla.: Krieger Publishing, 1991.

Horonjeff, Robert, Francis X. McKelvey, William J. Sproule, and Seth B. Young. *Planning and Design of Airports*. New York: McGraw-Hill, 5th edition, 2010.

Measuring Airport Landside Capacity. Special Report 215. Washington, D.C.: Transportation Research Board, 1987.

Improving Public Transportation Access at Large Airports. TCRP Special Report 62. National Academy Press, 2000, Washington, D.C.

Ground Access to Major Airports by Public Transportation, ACRP Report 4. National Academies Transportation Research Board, 2008, Washington, D.C.

Airport Passenger Terminal Planning and Design, Volume 1: Guidebook, ACRP Report 25. National Academies Transportation Research Board, 2010, Washington, D.C.

Reference Guide on Understanding Common Use at Airports, ACRP Report 30. National Academies Transportation Research Board, 2010, Washington, D.C.

Innovations for Airport Terminal Facilities, ACRP Report 10. National Academies Transportation Research Board, 2008, Washington, D.C.

8

Airport security

Outline

- Introduction
- History of airport security
- The Transportation Security Administration
- Security at commercial service airports
  - Law Enforcement, Contingencies, and Incident Response
  - Passenger screening
  - Checked-baggage screening
  - Employee identification
  - Controlled access
  - Biometrics
  - Perimeter security
- Security at general aviation airports
  - The twelve-five and private charter programs
- The future of airport security

Objectives

The objectives of this section are to educate the reader with information to:

- Be familiar with the history of airport security threats and associated legislative action.
- Describe the organizational structure of the Transportation Security Administration.
- Define the various security sensitive areas around airports.
- Describe the facilities located at airports that are part of the post-September 11, 2001, security environment.

- Understand the differences in security procedures between commercial service and general aviation airports.
- Be familiar with the various technologies that are being developed to enhance airport security.

Introduction

One of the most significant issues facing airports in the early twenty-first century is that of airport security. Most users of commercial service airports are subjected to security infrastructure, policies, and procedures within the airport terminal area. Airport security is not limited to the terminal area, however. Airport security concerns all areas and all users of the airport. In the history of commercial aviation, there have been more than 600 aircraft hijackings and more than 100 aircraft bombings attributed to terrorism throughout the world. While many rules, policies, and operating procedures have been put in place over the years to mitigate the threats to aviation security, terrorism and other activities still continues throughout the world well into the twenty-first century. A number of these events have resulted in major legislative and policy changes in how civil aviation operates, and how airport managers and planners design and operate modern-day airports.

Airport security procedures are designed to deter, prevent, and respond to criminal acts that may affect the safety and security of the traveling public. Criminal activity includes the hijacking of aircraft, known as **air piracy,** damaging or destroying aircraft with explosives, and other acts of **terrorism,** defined as the systematic use of terror or unpredictable violence against governments, publics, or individuals to attain a political objective. Criminal activity also includes acts of assault, theft, and vandalism against passengers and their property, aircraft, and all airport facilities.

History of airport security

In the earliest days of civil aviation, when the greatest concerns were simply the safety of flight, there was little concern over airport security, or aviation security in general. Aviation security first became an issue in 1930, when Peruvian revolutionaries seized a Pan American mail plane with the aim of dropping propaganda leaflets over Lima. Between 1930 and 1958, a total of 23 hijackings were reported, mostly committed by eastern Europeans seeking political asylum. The world's first fatal aircraft hijacking took place in July 1947 when three Romanians killed an aircrew member.

The first major act of criminal violence against a U.S. air carrier occurred on November 1, 1955, when a civilian by the name of Jack Graham placed a bomb in luggage belonging to his mother. The bomb exploded in flight, killing all 33 people on board. Graham had hoped to cash in on his mother's life insurance policy, but instead was found guilty of sabotaging an aircraft and sentenced to death. A second such act occurred in January 1960, when a heavily insured suicide bomber killed all aboard a National Airlines aircraft. As a result of these two incidents, demands for luggage inspection at airports serving air carrier aircraft surfaced.

While the majority of hijackings up through the 1960s were either criminals attempting to escape from the United States or to hold people for ransom, the rise of Fidel Castro in Cuba in 1959 came a significant increase in the number of aircraft hijackings, at first by those wishing to escape from Cuba, then by those hijacking U.S. aircraft to Cuba. In May 1961 the federal government began using armed guards on select air carrier aircraft to prevent hijackings.

In August 1969, Arab terrorists carried out the first hijacking of a U.S. aircraft flying outside the Western Hemisphere when they diverted an Israel-bound TWA aircraft to Syria. Another incident that October involved a U.S. Marine who sent a TWA plane on a 17-hour circuitous journey to Rome. This was the first time that FBI agents attempted to thwart a hijacking in progress and that shots were fired by the hijacker of a U.S. plane. In March 1970, a copilot was killed and the pilot and hijacker seriously hurt during a hijacking. The first passenger death in a U.S. hijacking occurred in June 1971.

Following the hijacking of eight airliners to Cuba in January 1969, the Federal Aviation Administration created the Task Force on the Deterrence of Air Piracy. The task force developed a hijacker "profile" that could be used along with metal detectors (magnetometers) in screening passengers. In October, Eastern Air Lines began using the system, and four more airlines followed in 1970. Although the system seemed effective, a hijacking by Arab terrorists in September 1970, during which four airliners were blown up, convinced the White House that stronger steps were needed. On September 11, 1970, President Richard Nixon announced a comprehensive antihijacking program that included a federal air marshal program.

Between 1968 and 1972, hijacking of U.S. and international aircraft was at its peak. During the 5-year period, the U.S. Department of Transportation recorded 364 hijackings worldwide. As a result, security issues had become a significant concern for the traveling public, and created the need for congressional action.

On March 18, 1972, the first airport security regulations were made effective, later formalized within the FAA as Federal Aviation Regulations Part 107— Airport Security. Under this regulation, airport operators were required to prepare and submit to the FAA a security program, in writing, containing the following elements:

- A listing of each **air operations area (AOA),** that is, those areas used or intended to be used for landing, takeoff, or surface maneuvering of aircraft
- Identification of those areas with little or no protection against unauthorized access because of a lack of adequate fencing, gates, doors with locking means, or vehicular pedestrian controls
- A plan to upgrade the security of air operations with a time schedule for each improvement project

Under FAR Part 107, airport operators were required to implement an *airport security program (ASP)* in the time frame approved by the FAA. In addition, airports were required to have all persons and vehicles allowed in the AOA suitably identified. Airport employees allowed in the AOA were subject to background checks prior to receiving proper identification and permission to enter into air operations areas.

FAR Part 107 was limited to security "as it affects or could affect safety in flight," reflecting the focus of the FAA to protect air carrier aircraft, and not other areas of the airport environment. FAR Part 107 did not extend to security in automobile parking lots or terminal areas distant from the air operations area.

In October 1972, four hijackers bound for Cuba killed a ticket agent. The next month, three criminals seriously wounded the copilot of a Southern Airways flight and forced the plane to take off even after an FBI agent shot out its tires. These violent hijackings triggered a landmark change in aviation security. In December, the FAA issued an emergency rule making inspection of carry-on baggage and scanning of all passengers by airlines mandatory at the start of 1973. An antihijacking bill signed in August 1974 sanctioned the universal screening. The FAA incorporated these regulations as FAR Part 108—Airplane Operator Security, in 1981. Prior to 1981, security programs were required for airlines as defined in FAR Part 121.

These stringent measures paid off, and the number of U.S. hijackings never returned to the worst levels before 1973. No scheduled airliners were hijacked in the United States until September 1976, when Croatian nationalists commandeered a jetliner. Two fatal bombings did occur, though: a bomb exploded in September 1974 on a U.S. plane bound from Tel Aviv to New York, killing all 88 persons aboard, and a bomb exploded in a locker at New York's LaGuardia Airport in December 1975, killing 11. That bombing caused airports to locate lockers where they could be monitored.

In June 1985, Lebanese terrorists diverted TWA Flight 847 leaving Athens for Beirut. One passenger was murdered during the 2-week ordeal; the remaining 155 were released (Fig. 8-1). This hijacking, as well as an upsurge in Middle East terrorism, resulted in several U.S. actions, among them the International Security and Development Cooperation Act of 1985 that made federal air marshals a permanent part of the FAA workforce.

On December 21, 1988, a bomb destroyed Pan American flight 103 over Lockerbie, Scotland (Fig. 8-2). All 259 people aboard the London-to-New York flight, as well as 11 on the ground were killed. Investigators found that a bomb concealed in a radio-cassette player had been loaded on the plane in Frankfort, Germany. This tragedy followed an FAA bulletin issued in mid-November that warned of such a device and one on December 7 of a possible bomb to be placed on a Pan Am plane in Frankfort. Early in 2001, a panel of Scottish judges convicted a Libyan intelligence officer for his role in the crime. Security measures that went into effect for U.S. carriers at European and Middle Eastern airports after the Lockerbie bombing included requirements to x-ray or search all checked baggage and to reconcile boarded passengers with their checked-in baggage, known as **positive passenger baggage matching (PPBM)**. PPBM had actually been implemented regionally throughout the world in 1985, after twin bombings attributed to a Singh separatist group who placed bombs on aircraft at the Vancouver International Airport on June 23 that the same year and subsequently killed 329 people on Air India Flight 182 and killed 11 at the Tokyo Narita International Airport. PPBM had not been applied to Pan Am 103, which was considered one of the reasons why that bombing occurred.

Figure 8-1. *Hijackers hold the pilot of TWA flight 847 hostage, June 1985.* (Source: www.abcnews.com)

Figure 8-2. *Pan Am 103 lies in ruins near Lockerbie, Scotland, after a bomb stowed on board exploded in flight, December 1988.*
(Source: www.terrorvictims.org)

In response to the Lockerbie bombing, President George Bush established the President's Commission on Aviation Security and Terrorism to review and evaluate policy options in connection with aviation security. As a result of the workings of the commission, President Bush signed the Aviation Security Improvement Act, which, in part, called for increased focus on developing technology and procedures for detecting explosives and weapons intended to be stowed on commercial air carrier aircraft.

Throughout the 1990s and into the twenty-first century, the FAA sponsored research on new equipment to detect bombs and weapons and made incremental improvements to aviation security that included efforts to upgrade the effectiveness of screening personnel at airports. In 1996, two accidental airline crashes resulting from in-flight explosions, TWA flight 800 and ValuJet flight 592, focused attention on the danger of explosives aboard aircraft, including those caused by hazardous cargo. The FAA's response, based on results of a commission led by Vice President Al Gore, included banning certain hazardous materials from passenger airplanes. The Gore Commission resulted in the Aviation Security and Anti-Terrorism Act of 1996, which mandated 10-year employment background checks for all airport employees and a fingerprint-based criminal history record check for those who could not pass the employment check (known as an Access Investigation). The Act also authorized the deployment of additional canine explosive sniffing teams, deployed explosive detection systems for limited use in secondary screening, and created the CAPPS—computer assisted passenger pre-screening system. The 1997 federal appropriation to the FAA provided funds for more airport security personnel and for new security equipment.

In the late 1990s and into 2000, airport security procedures were sometimes faulted by the media and by the Department of Transportation's Office of the Inspector General (OIG), an independent government office that assesses federal programs and operations and makes recommendations. In 1999, for example, a report issued by the OIG criticized the FAA for being slow to limit unauthorized access to secured areas in airports, stating that its investigators were able to penetrate these areas repeatedly. In 2000, it also faulted the agency for issuing airport identification used to access security-sensitive airport areas without sufficient checks. But for the 10 years following February 1991, there were no airline hijackings in the United States.

During this period of time, airport security issues began to focus on other acts of criminal activity. Efforts to reduce the amount of theft of passenger property and efforts to reduce smuggling of contraband on commercial aircraft were increased. Also, increases in acts of minor passenger violence, known as *air rage,* thought to be a result of the increases in congestion and delays and decreases in the customer service quality of commercial air carriers, were addressed.

The 10-year lull from airline security tragedies ended with the historical events of September 11, 2001. The worst international terrorist attack in history, involving four separate but coordinated aircraft hijackings, occurred in the United States on September 11, 2001, by a total of 19 alleged operatives of the Al-Qaida terrorist network. Details of the events of September 11, 2001 are described in Chap. 3 of this text.

The Transportation Security Administration

As a result of the events of September 11, 2001, and the subsequent signing of the **Aviation and Transportation Security Act (ATSA),** the practice of airport security began to undergo radical changes, beginning with the creation of the Transportation Security Administration.

With the signing of the ATSA, the **Transportation Security Administration (TSA)** was incorporated into the organizational structure of the U.S. Department of Transportation, to be operated in close coordination with all other transportation administrations, including the FAA, and headed by an undersecretary of transportation security. On December 10, 2001, Secretary of Transportation

Norman Mineta announced the appointment of then Chief of the Bureau of Alcohol, Tobacco, and Firearms, and former Secret Service agent, John Magaw, as the TSA's first undersecretary of transportation.

In May 2002, Undersecretary Magaw resigned his post, amid feelings by airport operators that the TSA was not sympathetic to the transportation-related needs of airport management that was necessary to create an efficient security system. Admiral James Loy, former administrator of the United States Coast Guard, was appointed temporary undersecretary of the TSA.

In March of 2003, the TSA, along with the Coast Guard, Customs Service, and Immigration and Naturalization Service, was formally moved into the newly formed United States Department of Homeland Security, led by Secretary Tom Ridge. At the same time, James Loy was appointed the first administrator of the TSA.

By 2003, the TSA employed a workforce of over 55,200 passenger and baggage screeners at 429 commercial service airports in the United States, supervised by a team of 155 **federal security directors (FSDs)** each assigned to one or more airports, along with an administrative staff of over 600 regional and national managers. As of 2010, congress has capped the number of TSA screeners to 45,000 and has allocated funding for 43,000 TSA screening staff.

The mission of the TSA is to protect all of the nation's transportation systems to ensure freedom of movement for people and commerce. Since its inception in 2001, the TSA has concentrated the vast majority of its efforts on securing the transportation of passengers on commercial air carriers traveling through the nation's airports through the implementation of passenger and baggage screening requirements set forth in the ATSA.

Regulations regarding the security of airport and other civil aviation operations have been moved to the TSA. They are published under Title 49 of the Code of Federal Regulations (49 CFR—Transportation) and have traditionally been known as TSRs. TSRs are enforced by the TSA. A listing of TSRs of specific relevance to airport security may be found in Chap. 3 of this text.

The Transportation Security Regulations define specific areas of the airport that are subject to various security measures. These areas are defined as air operations areas, secured areas, sterile areas, SIDA areas, and exclusive areas. Under the Transportation Security Regulations, each airport operating under Federal Aviation Regulations Part 139—*Certification of Airports*, must have an **airport security program (ASP)** which, in part, defines the following areas on its property.

The air operations area (AOA) is defined as a portion of an airport, specified in the airport security program, in which security measures are carried out.

This area includes aircraft movement areas, aircraft parking areas, loading ramps, safety areas for use by aircraft, and any adjacent areas (such as general aviation areas) that are not separated by adequate security systems, measures, or procedures. This area does not include the secured area. The AOA is required to be secured via an Access Control System of some sort. This includes perimeter security and controlled access measures, and positive identification procedures.

The **secured area** is defined as a portion of an airport, specified in the airport security program, in which certain security measures specified in 49 CFR Part 1542—*Airport Security* are carried out. This area is where aircraft operators and foreign air carriers that have a security program under 49 CFR Part 1544—*Aircraft Operator Security: Air Carriers and Commercial Operators* or 49 CFR Part 1546—*Foreign Air Carrier Security* enplane and deplane passengers and sort and load baggage and any adjacent areas that are not separated by adequate security measures. Specifically, the secured area is the area at the airport where commercial air carriers conduct the loading and unloading of passengers and baggage between their aircraft and the terminal building. Each commercial service airport must designate at least one Secured Area. Access control systems in the Secured Area are required to have the ability to allow access to authorized personnel, immediately deny access to unauthorized personnel, and distinguish access for personnel within the secured area. This is usually accomplished through a computerized Access Control and Monitoring System (ACAMS).

The **sterile area** is defined as a portion of an airport defined in the airport security program that provides passengers access to boarding aircraft and to which the access generally is controlled by TSA, or by an aircraft operator under 49 CFR Part 1544 or a foreign air carrier under 49 CFR Part 1546 through the screening of persons and property. Specifically, the sterile area is that part of the airport to which passenger access must be gained through TSA passenger screening checkpoints.

The **security identification display area (SIDA)** is defined as a portion of an airport, specified in the airport security program, in which security measures specified in the TSRs are carried out. This area includes the secured area and may include other areas of the airport such as the AOA, airport and tenant administrative areas, fuel farms, and navigational aid facilities. Within the SIDA, all persons must display proper identification or be accompanied by an authorized escort.

An **exclusive area** is defined as any portion of a secured area, AOA, or SIDA, including individual access points, for which an aircraft operator or foreign air carrier that has a security program under 49 CFR Part 1544 or 49 CFR Part 1546 has assumed responsibility for the security of its area. Examples of exclusive areas include air carrier aircraft storage and maintenance hangars. Other tenants of the airfield are eligible for Airport Tenant Security Programs, which allow for a supervised self-security program. Such tenants include fixed-base operators (FBOs) serving general aviation and charter aircraft.

Areas that do not fall within the above definitions are considered public areas, and are not directly subject to TSA security regulations concerning restricted access. These areas include portions of airport terminal lobbies, parking lots, curb frontage.

Security at commercial service airports

The events of September 11, 2001, the associated legislative action of the ATSA, and the formation of the TSA have all contributed to the changing rules, regulations, policies, and procedures associated with airport security. In addition, state and local governments, along with organizations representing members of the aviation industry, from the Air Line Pilots Association, to the American Association of Airport Executives, to the Aircraft Owners and Pilots Association, have made major contributions to the potential future security for the users of the nation's commercial service and general aviation airports.

Security at commercial service airports must be in accordance with regulations **49 CFR 1542—Airport Security,** as well as with elements of **49 CFR 1544—Aircraft Operator Security: Air Carriers and Commercial Operators**. Similar to the requirements of 14 CFR Part 139, each commercial airport is required to have an approved **airport security program (ASP)**. Within the ASP, the following are included:

1. The name and contact information of an appointed **Airport Security Coordinator (ASC)**, who will be assigned primary responsibility for security at the airport

2. A narrative and graphical description of the AOA, SIDA, and all sterile and secured areas, and measures for controlling access and movement within these areas

3. A narrative and graphical description of all passenger and baggage screening procedures as found in each aircraft operator's **Air Carrier Security Operations Program (ASOP)**

4. Procedures used to comply with the required employee background checks and identification requirements

5. A description of escort and challenge procedures

6. A description of all security-related training programs

7. Incident management procedures

8. A description of all administrative matters relating to security, including record keeping, audit, and general management functions

Law Enforcement, Contingencies, and Incident Response

In addition, elements within the Airport Security Program require the airport operator to have a minimum level of law enforcement personnel to respond to incidents, to support the ASP, and to respond to increases in the Homeland Security Advisory System (HSAS).

Commercial service airports are required to have enough law enforcement personnel to respond to incidents that have occurred or are occurring on inbound aircraft, to respond to alarms and issues that arise at screening checkpoints, and to support the requirements of the Airport Security Program. The number of law enforcement officers (LEOs) is different for each airport as airports vary in size and levels of passenger service. While unarmed security personnel are often used at airports to staff access gates and patrol the airfield, the Part 1542 requires that a minimum number of LEOs who are certified peace officers, able to carry a weapon, must wear a uniform and have arrest authority.

In addition, each Airport Security Program must include the specific measures the airport operator will take when the HSAS is raised. These are known as contingency measures. The ASP must also include incident response procedures for bomb threats, suspicious devices, hijackings, and other threats to civil aviation. The difference between the contingency plans and incident response plans is that contingency plans are responses to the HSAS being raised or lowered, while incident management plans are what the airport implements for an actual incident.

Other elements of the Airport Security Program are Security Directives (SDs). SDs are issued by TSA and contain information on how the airport operators are required to make adjustments and changes to their security programs in response to an identified threat.

An airport's ASP must be approved by TSA.

At commercial service airports, areas of airport security are commonly categorized as passenger screening, baggage screening, employee identification, and controlled access and perimeter security.

Passenger screening

The processing of passengers and baggage for the purpose of ensuring the security of the civil aviation system has undergone a virtual overhaul following the terrorist attacks on the United States on September 11, 2001. Since 2002, passenger and baggage security screening has been managed and operated by the TSA. Even though the TSA has ultimate authority of the facilities and procedures that comprise the security screening processes, airport managers and planners should be keenly aware of the security screening process, because the process has presented the most significant impacts on airport terminal planning and operations in recent years.

Since 2001, only ticketed passengers have been allowed to pass through security checkpoints. With the exception of a short period of time following the first Gulf War in 1990, prior to 2001, the general public was allowed to enter the sterile area provided they go through security passenger screening. This allowed friends and family to meet or see off passengers at the gate. With the implementation of the new rules, additional security measures were put in place to ensure that those entering security screening were indeed ticketed passengers on an upcoming flight.

Identification control procedures now include the requirement that each passenger present a government-issued photo identification and his/her boarding pass as a first step in the security screening process. Since 2007, the identification control function has been performed by TSA security officers.

Passenger screening facilities include an automated screening process, conducted by a **magnetometer**, or **walk-through metal detector (WTMD)**, that attempts to screen for weapons potentially carried on by a passenger. As a passenger walks through a WTMD, the presence of metal on the passenger is detected. If a sufficient amount of metal is detected, based on the sensitivity setting on the WTMD, an alarm is triggered. Passengers who trigger the WTMD are then subject to a manual search by a TSA security officer. Manual searches range from a further check of metal on the passenger's person with the use of a handheld wand, to a manual pat down (Fig. 8-3). Unfortunately, magnetometers do not detect explosives. As a result, the industry has been deploying Advance Imaging Technology (AIT), also known as whole body imagers. AITs are able to spot certain weapons and explosives that conventional magnetometers cannot.

The importance of the AIT systems was highlighted in December 2009, when an individual attempted to blow up a Northwest Airlines flight by concealing a bomb in his underwear. Previously, in 2004, two Chechen female suicide bombers destroyed two Russian airliners by concealing explosives on their bodies.

An example of AIT systems being deployed at airports as of 2010 includes full body scanners, as illustrated in Fig. 8-3b.

Carry-on baggage screening facilities are located at security screening stations to examine the contents of passengers' carry-on baggage for prohibited items such as firearms, sharp objects that may be used as weapons, volatile liquids, or plastic or chemical-based *trace explosives*. All carry-on baggage is first

Figure 8-3. *Passenger screening checkpoint.* (Courtesy Denver International Airport)

inspected through the use of an x-ray machine. Bags selected because of suspicions as a result of the x-ray examination, or selected on a random basis, are further inspected through the use of **explosive trace detection (ETD) equipment** (Fig. 8-4) and/or by manual search. In addition, personal electronic items such as laptop computers or cellular phones are frequently inspected by being turned on and briefly operated to check for authenticity.

Since x-ray technologies at screening checkpoints remained largely the same throughout the past 20 years, TSA has slowly been upgrading x-ray machines to provide better resolution and multiple viewing angles. The new machines are known as Advanced Technology or **AT x-rays**.

Prior to September 11, 2001, passenger and carry-on baggage screening fell under the responsibility of the commercial air carriers whose aircraft provided passenger service at any given airport, as dictated by FAR Part 108—Aircraft Operator Security, Air Carriers and Commercial Operators. Under this regulation, air carriers typically subcontracted security responsibilities to private firms. Studies of these firms conducted through 2001 revealed a work environment

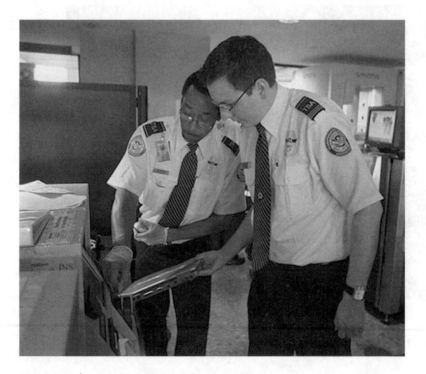

Figure 8-4. *Carry-on baggage screened using ETD.* (Photo courtesy USA Today)

characterized by low, almost minimum, wages, high turnover rates of 100 to 400 percent annually, low levels of training, and low performance quality, illustrated by independent audits that illustrated the ability to bring prohibited items, such as firearms and other weapons, through checkpoints.

Since November 2002, passenger screening at nearly all commercial service airports has been performed by the TSA-employed screener workforce under 49 CFR Part 1544. The TSA workforce is provided higher wages than their pre-September 11 private force counterparts, receives higher levels of training, including 44 hours of classroom and 60 hours of on-the-job training, and by some measures, exhibits higher performance quality. TSA passenger screening procedures also call for more strict screening standards, including a wider range of prohibited items, more thorough hand searches, removal of passenger shoes for inspection, implementation of new security screening technologies, and identification checks.

Approximately 20 commercial service airports do have a private security screening workforce, contracted to the TSA, as part of the TSA's Screening Partnership Program (SPP). The SPP was established in accordance with the ATSA to provide the ability for airports to "opt-out" of a federal screening workforce. The SPP was based on a pilot program in which five airports located in San Francisco, CA; Kansas City, MO; Rochester, NY; Jackson Hole, WY; and Tupelo, MS contracted private sector firms to perform passenger screening in 2002.

The initial months following the implementation of TSA passenger screening were also characterized by significantly higher levels of passenger delay at screening checkpoints. In addition, those critical of security processing at airports noted an increase in a newly defined "hassle factor." These negative impacts were a result of the increased amount of time and the increased amount of physical interaction required to process passengers. Over time, the negative impact issues have somewhat decreased and because the TSA added screening stations and staff, processes were made more efficient, and the traveling public became more accustomed to the new environment (Fig. 8-5).

With the implementation of TSA passenger and carry-on screening policies came a mandate of "no tolerance." This mandate effectively gave the TSA the authority to fully evacuate all or part of an airport upon the occurrence of a security breach of any magnitude. As a result, dozens of airport evacuations, affecting hundreds of air carrier operations, and tens of thousands of passengers, have occurred. As of 2010, the frequency of these events have declined significantly as both the TSA and the traveling public have adapted to the new security environment.

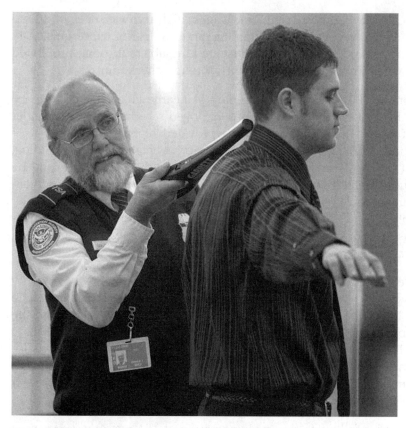

Figure 8-5. *Passenger screening performed by TSA personnel.*
(Photo courtesy USA Today)

Checked-baggage screening

Facilities to conduct screening of checked baggage for explosives have been placed at airports to adhere to the requirement implemented by the TSA on January 1, 2003, to have every piece of checked baggage screened by certified explosive detection equipment prior to being loaded onto air carrier aircraft (known as the 100 percent EDS rule). The primary piece of equipment used to perform checked-baggage screening, the **explosive detection system (EDS),** uses computed tomography technology, similar to the technology found in medical CT scan machines, to detect and identify metal and trace explosives that may be hidden in baggage (Fig. 8-6).

Because of the size, expense, and production rates of this system, it took several years for most airports to install sufficient EDS equipment to handle the volume of checked luggage. In addition, some oversized or unusually shaped baggage cannot fit inside the EDS. In these instances, checked baggage is

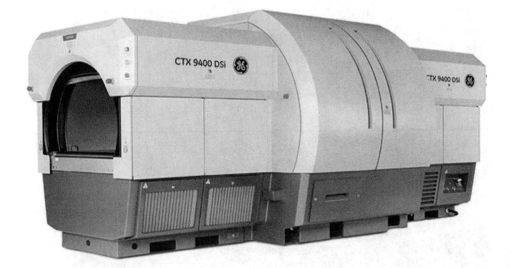

Figure 8-6. *EDS—explosive detection system.* (Source: GE Inc.)

screened by the use of electronic trace detection (ETD) systems, or manually by TSA baggage screeners.

Because the configuration of each airport terminal and the volume and behavior of each terminal's passengers are unique, and because the very short time line between the mandate of 100 percent checked-baggage screening in November 2001 and its implementation in January 2003, the location of checked-baggage screening has varied significantly from airport to airport. Checked-baggage screening locations have ranged from terminal lobbies, to facilities next to ticket counters, to curbside locations, to back rooms where baggage sorting is performed prior to being loaded on aircraft (Fig. 8-7). As of 2010, many of the nation's commercial service airports have built their infrastructure to have formal "in-line" EDS systems located near baggage sorting systems, or long-term lobby-based EDS systems in place.

Employee identification

TSA regulations require any person who wishes to access any portion of an airport's SIDA must display appropriate identification. This identification, known typically as a *SIDA badge,* is usually in the form of a laminated credit card–sized identification badge with a photograph and name of the badge holder. Persons typically requiring a SIDA badge include airport employees, air carrier employees, concessionaires, contractors, and government employees such as air traffic controllers and airport security staff.

Figure 8-7. *EDS machines located "in line" in baggage sorting areas.* (Photo courtesy Boeing Corp.)

In many instances the SIDA badge is color coded or otherwise marked to identify the areas within the airport the badge holder may access. In addition, many identification badges are equipped with magnetic strips, bar codes, or other formats readable by electronic means that carry detailed data regarding access authority of the badge holder, including any associated personal identification numbers needed to enter through certain access points, areas of authorization, as well as an electronic badge expiration date.

Prior to obtaining an identification badge, persons must complete an application, and undergo a fingerprint-based criminal history records check, and a TSA security threat assessment (STA). Any of the following criminal histories result in the disqualification for obtaining an SIDA badge:

1. Forgery of certificates, false marking of aircraft, and other aircraft registration violation
2. Interference with air navigation
3. Improper transportation of a hazardous material
4. Aircraft piracy
5. Interference with flight crew members or flight attendants
6. Commission of certain crimes aboard aircraft in flight
7. Carrying a weapon or explosive aboard aircraft
8. Conveying false information and threats
9. Aircraft piracy outside the special aircraft jurisdiction of the United States
10. Lighting violations involving transporting controlled substances

11. Unlawful entry into an aircraft or airport area that serves air carriers or foreign air carriers contrary to established security requirements
12. Destruction of an aircraft or aircraft facility
13. Murder
14. Assault with intent to murder
15. Espionage
16. Sedition
17. Kidnapping or hostage taking
18. Treason
19. Rape or aggravated sexual abuse
20. Unlawful possession, use, sale, distribution, or manufacture of an explosive or weapon
21. Extortion
22. Armed or felony unarmed robbery
23. Distribution of, or intent to distribute, a controlled substance
24. Felony arson
25. Felony involving a threat
26. Felony involving:
    i. Willful destruction of property
    ii. Importation or manufacture of a controlled substance
    iii. Burglary
    iv. Theft
    v. Dishonesty, fraud, or misrepresentation
    vi. Possession or distribution of stolen property
    vii. Aggravated assault
    viii. Bribery
    ix. Illegal possession of a controlled substance punishable by a maximum term of imprisonment of more than 1 year.
27. Violence at international airports
28. Conspiracy or attempt to commit any of the criminal acts listed above

Upon approval, a SIDA badge is issued to the applicant. Upon issuance of the badge, the person must display the SIDA badge at all times while in any portion of the SIDA. Typical policies within an airport security program require the badge to be displayed right side up, above the waist, on the outermost garment, in clear view, by the badge holder.

To enforce the use of proper identification, airports employ *challenge programs* designed to encourage persons within the SIDA to ask to see proper identification of those persons whose SIDA badges are not clearly displayed. In addition, airports often impose penalties to those not displaying proper identification, ranging from temporary confiscation of the person's SIDA badge, to termination of employment. Lack of proper identification within a SIDA area may also be considered a federal criminal offense.

Controlled access

A variety of measures is used around airports to prevent, or more appropriately, control the movement of persons and vehicles to and from security-sensitive areas of the airport property.

At most commercial service airports, **controlled access** through doors that provide access to the AOA, secured areas, sterile areas, and other areas within the SIDA, as well as many employee-only restricted areas, is enforced by the use of control systems. These systems range from simple key locks to smart-access technologies, such as keypad entry systems requiring proper pass code. In many cases, pass codes are calibrated with a person's SIDA badge, requiring both a presentation of the person's badge and proper pass code entry to gain access.

One weakness associated with door entry to security-sensitive areas, regardless of their access control measures, is the ability to allow unauthorized persons to enter through the door after an authorized person has opened the door. This situation is known as *piggybacking,* and is almost always a violation of security policies.

In some instances, revolving turnstiles with a one-rotation limit per access, rather than typical door systems, have been used to restrict the number of persons achieving access through these areas.

Biometrics

Advanced identification verification technologies, including those that employ biometrics, are continuously being developed to enhance access control at airports. **Biometrics** refers to technologies that measure and analyze human body characteristics such as fingerprints, eye retinas and irises, voice patterns, facial patterns, and hand measurements, especially for identification authentication purposes.

Biometric devices typically consist of a reader or scanning device, software that converts the scanned information into digital form, and a database that stores the biometric data for comparison.

For the most part, biometric technologies have initially been found to be most applicable when controlling the access of those with SIDA badges at the airport. Controlling the access of the general public using biometrics proves more difficult, because previously recorded data are required to authenticate the identification of the person. If anything, however, biometrics provides another technology to prevent unauthorized access to security-sensitive areas (Fig. 8-8).

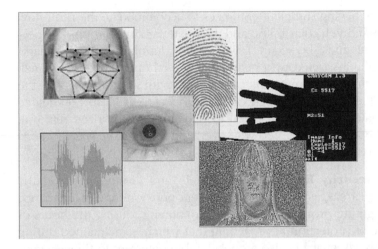

Figure 8-8. *Various biometric technologies are being tested for enhanced access control at airports.*

Perimeter security

An important part of an airport's security program is its strategy for protecting the areas that serve as the border between secured and unsecured areas of the airport, known as the airport perimeter. Four of the most common methods for securing the airport's perimeter, known as **barriers**, are physical barriers such as buildings and fences, natural barriers such as water or heavily wooded areas, electronic technology such as radio-frequency identification and airport surface detection radar, and access points such as gates and doors connected to an access control and monitoring system.

Controlled access gates provide a way for persons and especially vehicles to enter the secured area of the airport through the airport perimeter. Similar to controlled access doors, controlled access gates typically use some form of controlled access mechanism, ranging from simple key entry or combination locks, to advanced identification authentication machines, involving either the entry of a personal access code or verification through biometric technology. In addition, some controlled access gates are manned by guard personnel, further enhancing the security of the perimeter.

It is required that the number of access gates surrounding an airport's perimeter be limited to the minimum required for the safe and efficient operation of the airport. Active perimeter entrances of manned sites should be designated in order to enable guard force personnel the opportunity to maintain full control without unnecessary delay in traffic or reduction of operational efficiency. This largely is a matter of having sufficient entrances to accommodate the peak flow of both pedestrian and vehicular traffic and adequate lighting for rapid inspection. Unmanned gates must be secured, illuminated during the hours of darkness, and periodically inspected by a guard or assigned operational personnel. Gates must be constructed of materials of equal strength and durability as the fence and must either slide fully open or swing open to at least a 90-degree angle. Hinges of gates should be installed to preclude unauthorized removal. Gates should be topped with a barbed wire overhang meeting the specifications for the fence.

At most airports, **security lighting** is located in and around heavy traffic areas, aircraft service areas, as well as well as other operations and maintenance areas. Protective lighting provides a means of continuing, during the hours of darkness, a degree of protection approaching that which is maintained during daylight hours. This safeguard is also a considerable deterrent to thieves, vandals, and potential terrorists.

Various lighting systems include:

- *Continuous lighting*. This is the most common protective lighting system. It consists of a series of fixed lights arranged to flood a given area with overlapping cores on a continuous basis during the hours of darkness.
- *Standby lighting*. Lights in this system are either automatically or manually turned on when an interruption of power occurs or when suspicious activity is detected.
- *Movable lighting*. This type of lighting consists of manually operated movable floodlights.
- *Emergency lighting*. This system may duplicate any one of the aforementioned systems. Its use is limited to periods of power failure or other emergencies and is dependent upon an alternate power source.

Patrolling by airport operations staff, as well as local law enforcement, often contributes to enhancing airport perimeter security. Patrols of the airport perimeter, for the most part, are performed on a routine basis. In addition, air traffic control towers, responsible for the movement of aircraft and vehicles on the movement areas of an airport's airfield, are able to keep a consistent watch over activities within the airport perimeter. Because of the nature of the task, most air traffic control towers are situated so that they have an optimal view of the entire airfield. This facilitates the ability for air traffic controllers to spot potential security threats. Coordination between air traffic controllers, airport operations staff, and local law enforcement further enriches the security of the airport perimeter.

Security at general aviation airports

Historically, the Federal Aviation Administration focused virtually all of its aviation security programs toward the commercial aviation sector of the industry. The FAA's justification for this strategy was that nearly 100 percent of all passenger air travel takes place at commercial airports using the airlines or other large aircraft.

The majority of general aviation activity, on the other hand, is performed by private pilots, using their own aircraft for the purposes of personal travel or recreation. In addition, the majority of general aviation aircraft have dramatically less mass than commercial airliners and cargo aircraft, making them relatively less suited for use as kinetic energy weapons or "guided missiles." This in turn has led local law enforcement officials to historically labeling GA airports as "low security threats." In addition, because most general aviation airports are relatively small and used by relatively few, frequent users, the people using the airport are usually known by one another. However, two recent notable incidents have focused attention on the security of general aviation airports, specifically the 2002 intentional crash of stolen Cessna 172 by a teenager into the Bank of America building in Tampa, Florida, and the intentional 2010 crash of a Piper Cherokee into the IRS building in Austin, Texas. Also, part of an earlier plot in 1995 involving Ramzi Yousef (convicted of bombing the World Trade Center in 1993) included flying a small, explosives-laden airplane into CIA headquarters or the U.S. Capitol.

General aviation airports have a number of characteristics that make them prone to potential security risks. In many cases, aircraft owners, pilots, and passengers have access to the airfield with relatively little outside supervision. What supervision is preferred is done so by the users of the airport themselves, including aircraft owners, fixed-base operators, and airport employees. Most other security measures in place at general aviation airports, such as fencing or controlled access gates, are designed more for deterrence rather than security measures.

The biggest threat for general aviation facilities, however, is the fact that the load-carrying capability of aircraft of general aviation airports, even if limited, enables the delivery of explosives, compensating for their relative lack of kinetic energy or fuel. Another potential risk is that general aviation aircraft could be used to strike ground-based targets. Given the ubiquity of general aviation aircraft and airports, such aircraft are never far from major urban centers, critical infrastructures, and other targets. Another important issue is the fact that most localities do not have the staff to assign someone to patrol the airport; therefore, patrolling is usually done on an infrequent basis which can range anywhere from once every few hours to weeks. An equally serious factor that reduces the security at general aviation airports is the fact that of much of the equipment (i.e., fencing, gates, etc.) surrounding the airport is antiquated or undermaintained for supporting unauthorized access. In fact, the main purpose of gates and fencing at a number of general aviation airports is to keep out animals and/or deter people from accidentally walking/driving onto the airport.

Similar to commercial service airports, common security attributes that general aviation airports may be equipped with are:

- Personnel and vehicle identification procedures
- Perimeter fencing
- Controlled access gates
- Security lighting
- Locks and key control
- Patrolling

Contrary to commercial service airports and other airports operating under TSA regulations, however, implementation of a security plan, and enforcement of security procedures is largely assumed by airport management, as well as the users of the airport, including fixed-base operators, aircraft owners, and members of the public who frequent the airport.

The regulations administered by the TSA are applicable to airports regularly serving aircraft operations for scheduled passenger service, public charter passenger operations, and private charter passenger operations operating aircraft with a maximum certificated takeoff weight of 12,500 pounds or more. Although all commercial service airports fall under these regulations, general aviation airports in the United States do not. Nevertheless, security at general aviation airports requires consideration, not only for management of these airports, but also for the civil aviation system in general.

In the wake of the events of September 11, 2001, general aviation airports were among the last to be reopened for use, amid concerns over the lack of current security regulations at general aviation airports, the sheer volume of general aviation aircraft, and the proximity of many general aviation airports to potential terrorist targets. Eventually, most general aviation airports reopened, first to aircraft filing under IFR flight plans only, then to most VFR operations. General aviation airports near urban areas were placed under strict airspace classifications, and many temporary flight restrictions, known as TFRs, were placed around security-sensitive sites. As of 2010, three general aviation airports continue to fall under permanent TFR, College Park Municipal Airport, Washington Executive/Hyde Field Airport and Potomac Airfield, due to their close proximity to Washington D.C., while others around the United States are often restricted by temporary TFRs. Such TFRs occur during periods of special events or when heads of state are in the vicinity.

Professional organizations in support of general aviation have pressed hard since the events of September 11, 2001, to ensure that general aviation remains a safe, secure, efficient, and fully available method of transportation in the wake of new security restrictions. In particular, the Aircraft Owners and Pilots Association (AOPA) has acted as the leader in preventing further restrictions on general aviation airports and the activity they serve. In addition, AOPA has been proactive in emphasizing the fact that general aviation airports are largely self-enforcing in their security practices, and have been very successful in contributing to the very limited amount of criminal activity that occurs within general aviation.

AOPA, in coalition with the Experimental Aircraft Association (EAA), the General Aviation Manufacturers Association (GAMA), the Helicopter Association International (HAI) and the National Business Aviation Association (NBAA), delivered to the TSA a series of recommendations to enhance general aviation security. These recommendations included suggestions for improving general aviation security for passengers, aircraft, and for airports. Those suggestions for airports included:

- Outdoor signage should be prominently displayed near areas of public access warning against tampering with aircraft or unauthorized use of aircraft. In addition, signage indicating the phone number for reporting suspicious activity should be placed in areas where pilots and/or ramp personnel gather.
- Pilots should be advised to be on the lookout for suspicious activity on or near airports, including:
  - Aircraft with unusual or unauthorized modifications
  - Persons loitering for extended periods in the vicinity of parked aircraft or in air operations areas
  - Pilots who appear to be under the control of other persons
  - Persons wishing to obtain aircraft without presenting proper credentials or persons who present apparently valid credentials but do not have a corresponding level of aviation knowledge
  - Anything that doesn't "look right" (i.e., events or circumstances that do not fit the pattern of lawful normal activity at an airport)

In addition, AOPA has partnered with the TSA to develop a nationwide Airport Watch Program that uses general aviation users as eyes and ears for observing and reporting suspicious activity. **AOPA Airport Watch** is supported by a centralized government-provided toll-free hotline (1-866-GA-SECURE) and system for reporting and acting on information provided by general aviation pilots (Fig. 8-9). AOPA's airport watch program has been adopted by TSA to a broader "transportation-wide" program, known as "See Something, Say Something."

Figure 8-9. *AOPA's Airport Watch Program encourages self-enforcement of security procedures at general aviation airports.* (Source: AOPA)

The twelve-five and private charter programs

Transportation Security Regulations do require larger general aviation aircraft to apply certain security requirements. Specifically, 49 CFR Part 1550.7 states that any aircraft with a maximum certified takeoff weight of 12,500 pounds or more must be thoroughly searched before departure and all passengers, crew members, and other persons and their accessible property, such as carry-on items, must be screened before boarding the aircraft, as part of what TSA defines as the **twelve-five program.** 49 CFR Part 1544.101 states that all aircraft used for private charter operations with maximum certified takeoff weight of 45,000 kilograms (100,309.3 pounds) or with a passenger seating configuration of 61 or more must ensure that all passengers and their carry-on baggage are screened prior to aircraft boarding. This is known as the **private charter program.** General aviation airports that serve these types of aircraft operations should provide adequate space to allow for security compliance under these regulations.

The future of airport security

Since the first criminal threats to civil aviation, reactive policies to prevent further occurrences of current threats have been implemented. This reactive paradigm has resulted in two consequences: (1) the reduction in the number of attacks from a current type of threat and (2) the creation of new threats against civil aviation that the system has not been prepared to mitigate. This has been evidenced by the historical development of different threats, from nonviolent hijackings, to violent hijackings using firearms, the placing of unattended explosives on aircraft, suicide hijackings, attempted suicide bombings, and most recently, attempts to down aircraft using shoulder-fired missiles near airports where aircraft are at relatively low altitudes and speeds.

As a result, thoughts regarding the future of airport security suggest a shift of policy, from a reactive approach to screening for the placement of weapons or explosives on aircraft, to a proactive approach to protecting against violent or other criminal acts by persons in and around the entire airport environment. This proactive approach requires technological and human expertise to screen persons for suspicious activity, rather than simply screening them for unauthorized possessions. As such, there has been a new emphasis on behavioral screening, whereby TSA screeners and other personnel attempt to focus on suspicious persons based in part on their mannerisms and overall psychological appearance.

It is also clear that there will be a continued push toward implementing advanced technologies to increase the probability of detecting otherwise hard to distinguish items placed on and in all parts of a would-be terrorist's body, carry-on luggage, and checked baggage.

Finally, the future of airport security lies in continuing to examine all points of access to the aviation system. This includes further screening of employees, vendors, and other non-passengers with access to commercial aircraft. In addition, airport security policies continue to expand into the cargo and general aviation sectors of the industry.

Concluding remarks

The events of September 11, 2001 were certainly most tragic, and as a result future concerns regarding the security of airports, and the aviation system in general, may prove to be addressed in a much more proactive manner. Prioritizing airport security has resulted in rapid developments in security technology and significantly increased security funding, and has led to addressing issues long considered a concern by many members of the traveling public.

Protecting against unknown future threats is an imperfect science, and as such, the future of airport security will always be an unknown entity. Concerns for the safe, secure, and efficient travel of passengers and cargo domestically and internationally will always be a top priority for the civil aviation system, and it can be assured that efforts to make the system as secure as possible will continue to be held in top priority, by all levels of government, as well as airport management, for the foreseeable future.

Key terms

air piracy

terrorism

air operations area (AOA)

positive passenger baggage matching (PPBM)

Aviation and Transportation Security Act of 2001 (ATSA)

Transportation Security Administration (TSA)

federal security director (FSD)

airport security program (ASP)

secured area

sterile area

security identification display area (SIDA)

Air carrier security operations plan (ASOP)

Airport Security Coordinator (ASC)

Walk-Through Metal Detector (WTMD)

exclusive area

passenger screening

magnetometer

explosive trace detection (ETD)

explosive detection system (EDS)

controlled access

biometrics

perimeter fencing

security lighting

AOPA Airport Watch

twelve-five program

private charter program

Questions for review and discussion

1. How have threats to aviation security evolved since the beginning of civil aviation?

2. How has airport security traditionally adapted to civil aviation threats?

3. What are some of the most significant changes in airport security as a result of the events of September 11, 2001?

4. What is the organizational structure of the Transportation Security Administration?

5. What are the various security-sensitive areas found at airports, as defined by Transportation Security Regulations?

6. What is required for an applicant to receive an SIDA badge?

7. What can cause an applicant to be refused an SIDA badge?

8. What are the procedures that exist as part of passenger screening?

9. What technologies are used to perform the screening of carry-on baggage at airports?

10. What technologies are used to perform the screening of checked luggage at airports?

11. What is piggybacking?

12. What are some of the technologies that are used to control access to sensitive security areas at airports?

13. What is biometrics? What are some of the technologies that are considered to apply biometrics to the airport security environment?

14. How does airport security differ between commercial service airports and general aviation airports?

15. How might airports better prepare themselves for future threats to civil aviation security?

Suggested readings

Advisory Circular 107-1—*Aviation Security, Airports,* Washington, D.C.: Federal Aviation Administration, 1972.

Airport Access Control. Office of Inspector General Audit Report. Federal Aviation Administration. Report No. AV-2000-017.

Aviation Security: Long-Standing Problems Impair Airport Screeners' Performance. GAO Report RCED-00-75, June 28, 2000.

Aviation Security: Office of Inspector General Report AV-1998-134. Washington, D.C.: United States Department of Transportation, May 27, 1998.

Aviation Security: Registered Traveler Program Policy and Implementation Issues. Report GAO-03-253, United States General Accounting Office, November 2002.

Aviation Security: Terrorist Acts Illustrate Severe Weaknesses in Aviation Security. Report GAO-01-1166T, September 20, 2001.

Aviation Security: Transportation Security Administration Faces Immediate and Long-Term Challenges. Report GAO-02-971T, September 25, 2001.

Aviation and Transportation Security Act. Public Law 107-1, 107th Congress, Washington, D.C, November, 19, 2001.

Controls over Airport Identification Media. Office of Inspector General Audit Report. Federal Aviation Administration. Report No.: AV-2001-010, December 7, 2000.

Kent, Jr., Richard J. *Safe, Separated, and Soaring: A History of Federal Civil Aviation Policy, 1961–1972.* Washington, D.C.: U.S. Department of Transportation, Federal Aviation Administration, 1980.

Preston, Edmund. *Troubled Passage: The Federal Aviation Administration during the Nixon-Ford Term, 1973–1977.* Washington, D.C.: U.S. Department of Transportation, Federal Aviation Administration, 1987.

Remarks by Secretary of State Colin L. Powell with Lockerbie Family Members, February 8, 2001, Washington, D.C.

Rumerman, Judith, *Aviation Security.* U.S. Centennial of Flight Commission, Federal Aviation Administration, 2003, Washington, D.C.

St. John, Peter. *Air Piracy, Airport Security, and International Terrorism: Winning the War against Hijackers.* New York: Quorum Books, 1991.

Transportation Security: Post-September 11th Initiatives and Long-Term Challenges. Testimony GAO-03-616T, United States General Accounting Office, April 1, 2003.

U.S. Department of Transportation, Federal Aviation Administration, Office of Civil Aviation Security. *U.S. and Foreign Registered Aircraft Hijackings, 1931–1986.* Washington, D.C.: Federal Aviation Administration, 1986.

49 CFR 1542—Airport Security, 2002, Washington D.C.

Part III

Airport administrative management

9

Airport financial management

Outline

- Introduction
- Airport financial accounting
  - Operating expenses
- Liability insurance
  - Airport liability coverage
  - Operating revenues
- Planning and administering an operating budget
- Revenue strategies at commercial airports
  - The residual cost approach
  - The compensatory cost approach
  - Comparing residual and compensatory approaches
  - Net income
  - Majority-in-interest (MII) clauses
  - Term of use agreements
- Pricing of airport facilities and services
  - Pricing on the airfield area
  - Terminal area concessions
  - Landside and ground transportation facilities
  - Airline leased areas
  - Other leased areas
- Variation in the sources of operating revenues
- Rise in airport financial burdens
- Airport funding
- Grant programs
  - Airport Improvement Program (AIP)

- Passenger facility charges (PFCs)
- Other federal funding sources
- Facilities and equipment program
- Federal letters of intent
- State grant programs
- Grant assurances
- Airport financing
  - General obligation bonds
  - General airport revenue bonds
  - Special facilities bonds
  - Financial and operational factors
  - Airline rates and charges
  - Community economic base
  - Current financial status and debt level
  - Airport management
  - Bond ratings
  - Interest costs
  - Defaults
- Private investment
  - Build, operate, and transfer (BOT) contracts
  - Lease, build, and operate (LBO) agreements
  - Full privatization

Objectives

The objectives of this section are to educate the reader with information to:

- Understand the difference between operation and maintenance (O&M) and capital improvement expenses.
- Be familiar with the process of airport financial accounting.
- Explain the need for liability insurance at airports.
- Describe the various operating and nonoperating revenues at airports.
- Be familiar with planning and operating budgets.
- Recognize the differences between the various forms of airport-airline financial agreements.
- Describe the concept of a majority-in-interest clause.
- Describe the different types of funding programs available to airports.

- Distinguish between the different types of financial bonds available to airports.
- Identify the different levels of privatization that may exist at airports.

Introduction

The vast amount of property, infrastructure, and labor that is required to operate, maintain, and improve airports requires significant levels of financial resources. Such resources are realized through a number of strategies available to airport management. With each source of funding available to airports, however, come rules and policies that determine which strategy airport management may employ to cover a portion of the airport's cost burdens.

Airport expenses may be described as falling into two types: **capital improvement expenses** and **operation and maintenance (O&M)** costs. Operation and maintenance costs consist of those expenses that occur on a regular basis and are required to maintain the current operations at the airport. Such expenses typically include wages and salary of airport employees, costs of utilities such as power, water, and telecommunications, and a broad spectrum of regularly needed supplies, from individual airfield lights to office supplies.

Capital improvement expenses, on the other hand, are very large, periodic expenses that contribute to significant airport infrastructure improvement or expansion. Capital improvement expenses include the costs of major construction projects such as airfield and terminal expansion, the acquisition of major utilities such as air rescue and fire fighting vehicles, and the purchase of land for future expansion.

In general, revenues from the operation of the airport are used to cover the airport's O&M expenses. To manage the balance of operating revenues and expenses, financial accounting is typically employed.

Airport financial accounting

The nature of airport expenses depends upon a number of factors including the airport's geographic location, organizational setup, and financial structure. Airports in warmer climates, for example, do not experience the sizable snow removal and other cold weather–related expenses that airports in colder climates must face. Some municipalities, counties, or local authorities absorb the costs of certain staff functions, such as accounting, legal, planning, and public relations. Certain operating functions such as emergency service, policing, and traffic control might also be provided by local fire departments and local law enforcement agencies at some airports. In addition, the ever-changing demand characteristics of passengers, service characteristics of air carriers and other aircraft operators, as well as aircraft, navigation, communication, and information technologies affect the need to invest in projects involving airport capital improvements.

Airport accounting involves the accumulation, communication, and interpretation of economic data relating to the financial position of an airport and the results of its operations for decision-making purposes. It differs from accounting procedures found in business firms because airports vary considerably in terms of goals, size, and operational characteristics. As such, it is very difficult to derive a unified accounting system that can be used by all airports. A system tailored to the needs of a large commercial airport might be impractical for a small GA airport or vice versa. Many airports have different definitions of what elements constitute operating and nonoperating revenues and expenses and sources of funds for airport development. A good accounting system is needed for a number of reasons:

- Financial statements are needed to inform governmental authorities and the local community regarding details of the airport's operations.
- A good accounting system can assist airport management in allocating resources, reducing costs, and improving control.
- Negotiating charges for use of airport facilities can be facilitated.
- Financial statements can influence the decisions of voters and legislators.

Operating expenses can be divided into four major groupings: airfield; terminal; hangars, cargo, other buildings, and grounds; and general and administrative expenses.

Operating expenses

Operating and maintenance expenses associated with the airfield area include:

- Runways, taxiways, apron areas, aircraft parking areas, and airfield lighting systems maintenance
- Service on airport equipment
- Other expenses in this area, such as maintenance on fire equipment and airport service roads
- Utilities (electricity) for the airfield

Operating and maintenance expenses associated with the terminal include:

- Buildings and grounds—maintenance and custodial services
- Improvements to the land and landscaping
- Loading bridges and gates—maintenance and custodial services
- Concession facilities and services

- Observation facilities—maintenance and custodial services
- Passenger, employee, and tenant parking facilities
- Utilities (electricity, air-conditioning and heating, and water)
- Waste disposal (plumbing)—maintenance
- Equipment (air-conditioning, heating, baggage handling)—maintenance

Operating and maintenance expenses associated with hangars, cargo facilities, other buildings, and grounds include:

- Buildings and grounds—maintenance and custodial services
- Improvements to the land and landscaping
- Employee parking—maintenance
- Access roadways—maintenance
- Utilities (electricity, air-conditioning and heating, and water)
- Waste disposal (plumbing)—maintenance

General and administrative expenses include all payroll expenses for the maintenance, operations, and administrative staff of the airport. Other operating expenses for materials and supplies are included under general and administrative expenses.

Airports also often incur nonoperating expenses including the payment of interest on outstanding debt (bonds, notes, loans, etc.), contributions to governmental bodies, and other miscellaneous expenses. In addition, some airports compute depreciation on the full value of facilities including federal and other aid, whereas other airports limit depreciation to only their share of the construction costs.

Liability insurance

An increasingly large percentage of airport expenses are derived from required insurance to cover various areas of liability. Airports and their tenants have the same general type and degree of liability exposure as the operator of most public premises. People sustain injuries and damage their clothing when they fall over obstructions or trip over concealed obstacles, and their automobiles are damaged when struck by airport service vehicles on the airport premises. Claims from such accidents can be for large amounts, but claims stemming from aircraft accidents have even greater catastrophe potential. The occupants of aircraft might be killed or severely injured and expensive aircraft damaged or destroyed, not to mention injury to other persons or other types of property at or near an airport. Liability in such instances can stem from a defect in the surface of the runway, from the failure of airport management to mark obstructions properly, or failure to send out the necessary warnings and to close the airport when it is not in usable condition.

Airports and their tenants are liable for all damage caused by their failure to exercise reasonable care. The principal areas in which litigation arises can be summarized under three main headings:

- *Aircraft operations.* Liability to tenants and the general public arising out of aircraft accidents, fueling, maintenance and servicing, and rescue efforts.
- *Premises operations.* Liability to tenants and the general public arising out of automobile and other vehicle accidents, elevators and escalators, police and security enforcement, tripping and falling, contractual obligations, airport construction, work performed by independent contractors, and special events such as airshows.
- *Sale of products.* Liability to tenants and the general public arising out of maintenance and servicing, fueling, and food and beverage services.

Airport operators require that all tenants purchase their own insurance as appropriate for their particular circumstances and with certain minimum limits of liability. Generally, the airport operator is included as an additional insured under the tenant's insurance coverage; however, this does not relieve the airport operator from securing its own liability protection under a separate policy. The comprehensive coverage and limits of liability needed by most major airports far exceed what is required by the average tenant.

Airport liability coverage

The basic airport premises liability policy is designed to protect the airport operator for losses arising out of legal liability for all activities carried on at the airport. Coverage can be written for bodily injury and property damage. A number of exclusions apply to the basic policy, and consequently the insuring agreements must be amended to add certain exposures. By endorsement, the basic contract can be extended to pick up any contractual liability the airport might assume under various agreements with fuel suppliers, railroads, and so forth. Elevator liability and liability arising out of construction work performed by independent contractors might also be covered. The basic policy can also be extended to provide coverage for the airport that sponsors an airshow or some other special event.

For those airports engaged in the sale of products or services, the premises liability policy can provide coverage for the airport's products liability exposure. Aircraft accidents arising out of contaminated fuel originally stored in airport fuel storage tanks or even food poisoning from an airport restaurant would be examples. Aircraft damaged while in the care, custody, or control of the airport for storage or safekeeping can be covered by extending the premise's liability policy to provide hangar keepers coverage.

The growth of aviation and airports during the past 30 years has increased the industry's exposure to liability claims. Airports invest thousands of dollars in purchasing adequate insurance coverage and limits of liability to protect their multimillion-dollar assets. The courts have consistently held airport operators responsible for the safety of aircraft and the public as well as for the issuance of proper warning of hazards. In many cases, municipalities have not been immune, with courts determining that the operation of an airport is a proprietary or corporate function rather than a government responsibility.

Operating revenues

Similar to operating expenses, airport operating revenues can be divided into five major groupings: airfield area, terminal area concessions, airline-leased areas, other leased areas, and other operating revenue.

The *airfield* or airside of the airport produces revenues from sources that are directly related to the operation of aircraft:

- Landing fees for scheduled and unscheduled airlines, itinerant aircraft, military or governmental aircraft
- Aircraft parking charges in hangars and on paved and unpaved areas
- Fuel flowage fees from FBOs and other fuel suppliers

Terminal concessions include all of the nonairline users of the terminal area that generate revenues by offering the following products and services:

- Food and beverage concessions (includes restaurants, snack bars, and lounges)
- Specialty stores and shops (includes boutiques, newsstands, banks, gift shops, clothing stores, duty-free shops, and souvenir shops)
- Personal services (includes beauty and barber shops, valet shops, shoeshine stands)
- Business services (includes office suites, conference rooms, and WiFi services)
- Display advertising

Landside and ground transportation facilities located on airport property outside of the terminal building:

- Automobile parking facilities and services
- Car rental facilities
- Off-terminal hotels and other properties

Airline leased areas include revenue derived from the air carriers for ground equipment rentals, cargo terminals, office rentals, ticket counters, hangars, operations, and maintenance facilities.

All of the remaining leased areas at the airport that produce revenue are brought together under other leased areas. Freight forwarders, fixed-base operators, governmental units, and businesses in the airport industrial area would be included under this category. All revenue derived from nonairline cargo terminals and ground equipment rentals to nonairline users would also be included.

Other operating revenue includes revenues from the operation of distribution systems for public utilities, such as electricity, and contract work performed for tenants. Other miscellaneous service fees are also included under this category.

Airports also generate *nonoperating revenues,* including interest earned on investments in governmental securities, local taxes, subsidies or grants-in-aid, and selling or leasing of properties owned by the airport but not related to airport operations. The magnitude of nonoperating income can vary considerably between airports.

Planning and administering an operating budget

Planning an operating budget is an integral part of airport financial management. Every airport must make short-term decisions about the allocation and scheduling of its limited resources over many competing uses; it must make long-term decisions about rates of expansion of capital improvements and funding sources. Both short-term and long-term decisions require planning. Planning is important because it:

- Encourages coordinated thinking. No one department can act independently. A policy decision in a particular department affects the airport as a whole.
- Helps develop standards for future performance. Without plans, the airport's measure of financial performance can be based only on historical standards. Although past operating statements help to set these standards for the future, they should not necessarily serve as standards themselves.
- Assists management in controlling the actions of subordinates. By planning, employees are provided a goal or standard to achieve.
- Might help reveal potential problems for which remedial measures can be taken earlier.
- Promotes smoother-running operations. For example, new equipment can be ordered in advance of its anticipated usage. With smooth, uninterrupted operations, the overall efficiency of the airport can be increased.

Once the airport has decided upon a plan of action for the future, these plans are incorporated into a written financial budget. Budgets are simply the planned dollar amounts needed to operate and maintain the airport during a definite period of time such as a year. Budgets are established for major capital expenditures such as runway resurfacing, taxiway construction, and new snow removal equipment as well as for operating expenses during the planning period.

In an airport maintenance department, there are labor expenses and a variety of other expenses for supplies, minor equipment purchasing and repair, and mechanical systems maintenance. The real expenses incurred during the year are a measure of the actual performance. The difference between actual expenses and the budgeted amount is called a variance. The variance measures the efficiency of the department.

Airports generally operate under one of three different forms of budget appropriation: lump sum appropriations, appropriation by activity, and line-item budgeting.

A **lump sum appropriation** is the simplest form of budget and generally only utilized by small GA airports. There are no specific restrictions as to how the money should be spent. Only the total expenditure for the period is stipulated. This is the most flexible form of budgeting.

Under an **appropriation by activity** form of budget, appropriated expenses are planned according to major work area or activity with no further detailed breakdown. Appropriation by activity enables management to establish capital and operating expense budgets for particular areas such as airside facilities, terminal building area, and so forth. It also permits flexibility in responding to changing conditions.

The **line-item budget** is the most detailed form of budgeting, used quite extensively at the large commercial airports. Numerical codes are established for each operating and capital expense item. Budgets are established for each item and often adjusted to take into consideration changes in volume of activity. For example, as the number of passenger enplanements changes, budgets for the terminal building maintenance can be adjusted accordingly.

A very popular approach to budgeting at many airports is the zero-based budget. The **zero-based budget** derives from the idea that each program or departmental budget should be prepared from the ground up, or base zero for each budget cycle. This is in contrast to the normal budgeting practice, which builds on the base of a previous period. By calculating the budget from a zero base, all costs are newly developed and reviewed entirely to determine their necessity. Various programs are reviewed and costed thoroughly and then ranked in degree of importance to the airport. Managers are presumably forced to look at a program in its entirety rather than as an expense add-on to an existing budget.

In drawing up a budget, the first step normally involves an estimate of revenues from all sources for the coming year. The next step is to establish budgets for the various areas of responsibility. When budgets are being investigated, pre-determined, and integrated, the department managers who must live within the budgets are consulted about the amount of money available and help draw up budgets for their departments for the coming period. A manager who has some say about the budget and expenses is more inclined to make an added effort to keep down the actual expenses of the department. Actual expenses are then checked against budgeted expenses frequently during the period that the budgets are in effect. Managers are supplied with figures of actual expenses so that they can compare them with budgeted expenses and investigate variances.

Revenue strategies at commercial airports

At most commercial airports, the financial and operational relationship between the airport operator and the air carriers serving the airport is defined in legally binding agreements that specify how the risks and responsibilities of running the airport are to be shared. These contracts, commonly termed airport use agreements, establish the terms and conditions governing the air carriers' use of the airport. The term **airport use agreement** is used generically to include both legal contracts for the air carriers' use of airfield facilities and leases for use of terminal facilities. At many airports, both are combined in a single document. A few commercial airports do not negotiate airport use agreements with the air carriers, but instead charge rates and fees set by local ordinance. The airport use agreements also specify the methods for calculating rates air carriers must pay for use of airport facilities and services; and they identify the air carriers' rights and privileges, sometimes including the right to approve or disapprove any major proposed airport capital development projects.

Although financial management practices differ greatly among commercial airports, the airport-airline relationship at major airports typically takes one of two very different forms, with important implications for airport pricing and investment:

- The **residual cost approach,** under which one or more air carriers collectively assume significant financial risk by agreeing to pay any costs of running the airport that are not allocated to other users or covered by all other sources of revenue.

- The **compensatory cost approach,** under which the airport operator assumes the major financial risk of running the airport and charges the air carriers fees and rental rates set so as to recover the actual costs of the facilities and services that they use.

The residual cost approach

In the 1950s the city of Chicago entered into a precedent-setting agreement with United Airlines regarding its financial obligations for operating out of the O'Hare Field. Citing the unique relationship the air carrier had with the airport, United Airlines entered into a 50-year agreement which stated that, although the airport should generate as much revenue from other sources as possible, United would cover all expenses by the airport that exceed revenues. That is, United would pay the difference, or *residual,* between revenues and expenses. This agreement is known as an **O'Hare Agreement** or **United contract.** This contract represented the first residual cost contract.

In general, under this approach, the airlines that enter into such a contract assume significant financial risk. They agree to keep the airport financially self-sustaining by making up any deficit—the residual cost—remaining after the costs identified for all airport users have been offset by other sources of revenue (automobile parking and terminal concessions such as restaurants, newsstands, snack bars, and the like), as well as revenue from other, *nonsignatory,* air carriers. The residual between costs and revenues provides the basis for calculating the rates charged the air carriers for their use of facilities. Any surplus revenues would be credited to the airlines and any deficit charged to them in calculating airline landing fees or other rates for the following year.

The residual cost approach had become the standard agreement between airports and air carriers in the years before airline deregulation, and still exists in the postderegulation era particularly at airports where the air carrier dominates in market share.

The compensatory cost approach

Contrary to the residual cost approach, a compensatory agreement between an airport and a serving air carrier requires the air carrier to pay rates and charges equal to the costs of the facilities the air carrier uses, as determined by cost accounting. Under a compensatory agreement, the airport operator assumes the financial risk of airport operations. Furthermore, in contrast to the situation at airports operating with residual cost agreements, the air carriers operating under a compensatory agreement provide no guarantee that fees and rents will be sufficient to allow the airport to meet its annual operating and debt service requirements.

Under a compensatory approach, air carriers are not explicitly charged for public space within airport terminals, such as terminal lobbies. Rather, air carriers, as well as all other tenants of the airport pay rent for space and use of facilities in proportion to their percentage of activity hosted at the airport. Unlike the residual cost approach, a compensatory contract does not offer airlines any reduced charges as a result of the airport generating greater revenues from nonaeronautical uses. As such, the airport provides itself the opportunity to generate revenues in surplus of its overall expenses.

Comparing residual and compensatory approaches

Residual cost and compensatory approaches to financial management of major commercial airports have significantly different implications for pricing and investment practices. In particular, they help determine:

- An airport's potential for accumulating net income for capital development.
- The nature and extent of the air carriers' role in making airport capital investment decisions, which can be formally defined in majority-in-interest clauses included in airport use agreements with the airlines.
- The length of term of the use agreement between the airlines and the airport operator.

Net income

Although large and medium commercial airports generally must rely on the issuance of debt to finance major capital development projects, the availability of substantial revenues generated in excess of expenses can strengthen the performance of an airport in the municipal bond market. It can also provide an alternative to issuing debt for the financing of some portion of capital development. A residual cost contract guarantees that an airport will always break even, thereby ensuring service without resort to supplemental local tax support, but it precludes the airport from generating earnings substantially in excess of costs.

By contrast, an airport using a compensatory approach lacks the built-in security afforded by the air carriers' guarantee that the airport will break even every year. The public operator undertakes the risk that revenues generated by airport fees and charges might not be adequate to allow the airport to meet its annual operating costs and debt service obligations. On the other hand, because total revenues are not constrained to the amount needed to break even, and because surplus revenues are not used to reduce airline rates and charges, compensatory airports may earn and retain a substantial surplus, which can later be used for capital development. Because the pricing of airport concessions and consumer services need not be limited to the recovery of actual costs, the extent of such retained earnings generally depends on the magnitude of the airport's nonaeronautical revenues.

Majority-in-interest (MII) clauses

In exchange for the guarantee of solvency, air carriers that are signatory to a residual cost-use agreement often exercise a significant measure of control over

airport investment decisions and related pricing policy. These powers are em-
bodied in **majority-in-interest (MIT) clauses,** which are a much more common feature of airport use agreements at residual cost airports than at airports
using a compensatory approach.

Majority-in-interest clauses give the airlines that represent a majority of traffic
at an airport the opportunity to review and approve or veto capital projects
that would entail significant increases in the rates and fees they pay for the
use of airport facilities. The combination of airlines that can exercise majority-
in-interest powers varies. A typical formulation would give majority-in-interest
powers to any combination of "more than 50 percent of the scheduled airlines
that landed more than 50 percent of the aggregate revenue aircraft weight during the preceding fiscal year" (standard document wording).

This arrangement provides protection for the air carriers that have assumed
financial risk under a residual cost agreement by guaranteeing payment of
all airport costs not covered by nonaeronautical sources of revenue. For
instance, without some form of majority-in-interest clause, the airlines at
a residual cost airport could be obligating themselves to pay the costs of
as-yet-undefined facilities that might be proposed in the fifteenth or twentieth year of a 30-year use agreement. Under a compensatory approach,
where the airport operator assumes the major financial risk of running the
facility, the operator is generally freer to undertake capital development
projects without consent of the signatory carrier. Even so, airport operators
rarely embark on major projects without consulting the signatory carriers
that serve the airport.

Specific provisions of majority-in-interest clauses vary considerably. At some
airports, the airlines that account for a majority of traffic can approve or disapprove all major capital development projects, for example, any project costing
more than $100,000. At other airports, projects may only be deferred for a
certain period of time (generally 6 months to 2 years). Although most airports
have at least a small discretionary fund for capital improvements that is not
subject to majority-in-interest approval, the general effect of majority-in-interest
provisions is to limit the ability of the public airport owner to proceed with any
major project opposed by the airlines. Sometimes, a group of just two or three
major carriers can exercise such control.

Term of use agreements

Residual cost airports typically have longer-term use agreements than compensatory airports. This is because residual cost agreements historically have been
drawn up to provide security for long-term airport revenue bond issues; and
the term of the use agreement, with its airline guarantee of debt service, has
generally coincided with the term of the revenue bonds. The vast majority of

residual cost airports have use agreements with terms of 20 or more years and 30 years or longer is not uncommon.

By contrast, only approximately half of the compensatory airports have use agreements running for 20 years or more. Many of the compensatory airports have no contractual agreements whatever with the airlines. At these airports, rates and charges are established by local ordinance or resolution. This arrangement gives airport operators maximum flexibility to adjust their pricing and investment practices unilaterally, without the constraints imposed by a formal agreement negotiated with the airlines, but it lacks the security provided by contractual agreements.

Pricing of airport facilities and services

Major commercial airports are diversified enterprises that provide a wide range of facilities and services for which fees, rents, or other user charges are assessed. The facilities and services provided to users generate the revenues necessary to operate the airport and to support the financing of capital development. Smaller commercial airports and GA airports typically offer a much narrower range of facilities and services, for which only minimal fees and charges often are assessed. Revenue bases shrink as airports decrease in size, and many of the smallest airports do not generate sufficient revenue to cover their operating costs, much less capital investment. Among GA airports, those that lease land or facilities for industrial use generally have a better chance of covering their costs of operation than do those providing only aviation-related services and facilities.

The combination of public management and private enterprise uniquely characteristic of the financial operation of commercial airports is reflected in the divergent pricing of airport facilities and services. The private enterprise aspects of airport operation, the services and facilities furnished for nonaeronautical use, generally are priced on a market pricing basis. On the other hand, the pricing of facilities and services for airlines and other aeronautical users is on a cost-recovery basis, either recovery of the actual costs of the facilities and services provided (the compensatory approach) or recovery of the residual costs of airport operation not covered by nonaeronautical sources of revenue. This mix of market pricing and cost-recovery pricing has important implications for airport financing, especially with regard to the structure and control of airport charges and the distribution of operating revenues.

The structure and control of fees, rents, and other charges for facilities and services are governed largely by a variety of long-term and short-term contracts, including airport use agreements with the airlines, leases, and concession and management contracts. For each of the four major groups of facilities and services outlined earlier in the chapter, the basic kinds of charges assessed at residual cost and compensatory airports can be compared in terms of the method of calculation, terms of agreements, and the frequency of contract adjustments.

Pricing on the airfield area

The major fees assessed for use of airfield facilities are landing or flight fees for commercial airlines and GA aircraft. Some airports also levy other airfield fees, such as charges for the use of aircraft parking ramps or aprons. In lieu of landing fees, many smaller airports, especially GA airports, collect fuel flowage fees, which are levied per gallon of aviation gasoline and jet fuel sold at the airport.

Under residual cost contracts, the landing fee for airlines is typically the item that balances the budget, making up the projected difference between all other anticipated revenues and the total annual costs of administration, operations and maintenance, and debt service (including coverage). Landing fees differ widely among airports, depending on the extent of the revenues derived from airline terminal rentals and concessions such as restaurants, car rental companies, and automobile parking lots. If the nonaeronautical revenues are high in a given year, the landing fee for the airlines might be quite low. At some airports, the landing fee is the budget-balancing item for the airfield cost center only. At such airports, the surplus or deficit in the terminal cost center has no influence on airline landing fees, and terminal rental rates for the airlines are set on a residual cost or compensatory basis.

The method of calculating landing fees under residual cost contracts is established in the airport use agreement and continues for the full term of the agreement. To reflect changes in operating costs or revenues, landing fees are typically adjusted at specified intervals ranging from 6 months to 3 years. At some airports, fees might be adjusted more often if revenues are significantly lower or higher than anticipated. Often, the nonsignatory airlines (those not party to the basic use agreement) pay higher landing fees than the signatory carriers. General aviation landing fees vary greatly from airport to airport, ranging from charges equal to those paid by the commercial airlines to none at all. Most landing fees are assessed on the basis of certificated gross landing weight. This practice of basing landing fees on aircraft weight tends to promote use of commercial airports by general aviation. Because most GA aircraft are relatively light (under 12,500 pounds), they pay very low landing fees at most commercial airports. The smallest GA aircraft often pay no fee. Residual cost and compensatory airports alike have landing fees for GA aircraft that are generally so small as to be negligible, either as a source of revenue to the airport or as a deterrent to use of congested facilities.

Under compensatory contracts, landing fees are based on calculation of the average actual costs of airfield facilities used by the individual air carriers. As in the case of airports operating under residual cost contracts, each airline's share of these costs is based on its share of total projected airline gross landing weights (or, in a few cases, gross takeoff weight). In addition to fees determined by this weight-based measure, some airports assess a surcharge on GA aircraft during hours of peak demand. Presently, no major airports impose such peak-hour surcharges on commercial airlines to help ease congestion problems. Some airport managers and federal authorities believe that peak-hour surcharges could reduce congestion by giving airlines and other providers of air transportation services the opportunity to save money (and lower fares) by flying during less congested periods. If peak-period demand continued to cause congestion, the increased revenue generated by the surcharges could help finance the expansion necessary to accommodate peak-hour traffic.

Landing fees at compensatory airports are established either in airport use agreements with the airlines or by local ordinance or resolution. The frequency of adjustment of the fees is comparable to that at residual cost airports.

Terminal area concessions

The structure of terminal concessions and service contract fees is similar under both compensatory and residual cost pricing approaches. Concessions contracts typically provide the airport operator with a guaranteed annual minimum payment, typically based on a rental rate of leased space, a specified percentage of the concessionaire's gross revenues, or both. Restaurants, snack bars, gift shops, newsstands, duty-free shops, hotels, and rental car operations usually have contracts of this type. Terminal concessions contracts are often bid competitively, and they range in term from month-to-month agreements to contracts of 10 to 15 years duration. (Hotel agreements generally have much longer terms, often running for 40 years or more.) Airport parking facilities might be operated as concessions; they might be run by the airport directly, or they might be managed by a contractor for either a flat fee or a percentage of revenues.

In recent years, many airports have employed a "market pricing" strategy to airport terminal area concessions. This strategy has been illustrated by the departure of the traditional airport cafeterias and gift shops and the development of multistore food courts and retail lobbies that feature brand name or specialty products priced competitively with off-airport stores. This strategy has been employed under the philosophy that concessions should operate on a competitive basis, catering to the demands of airport users, rather than as monopolies, which consider the airport user a captive market. Such strategies have resulted in significantly increased revenues from terminal area concessions for airports, thus reducing the dependency of airports for revenue from the air carriers. As a further result, airports with successful terminal area concession revenue strategies have become more inclined to offer compensatory agreements to air carriers rather than longer-term residual contracts with majority-in-interest clauses attached.

Landside and ground transportation facilities

Revenues generated from landside and ground transportation facilities make up the largest portion of most airports' nonairfield revenues. Traditionally, airport parking alone has made up the largest of an airport's nonaeronautical revenues, although with the creation of consolidated rental car facilities and associated facility charges, revenues from the operation of these CRCFs are making ever greater contributions to airport revenue budgets.

Airport Parking Revenues have traditionally been generated by charging hourly and/or daily rates to those desiring to park their car at an airport parking lot or garage. As discussed in Chap. 7, airports charge various rates depending on the duration for which the vehicle is intended to be parked. "Short-term" parking facilities, often located closest to the terminal, tend to charge higher per-hour rates, in part to discourage users from occupying these premium spaces for long periods of time, and also to maximize the revenue generated per space. "Long-term" facilities are priced to accommodate those intending to park their vehicles for longer durations, and thus have lower hourly and daily rates. These lots are typically located farther from the terminal building and transportation between these lots and the terminal building are often provided by airport shuttle vehicles or APM systems.

In recent years, airports have begun to offer premium parking services in order to increase parking revenues while offering additional services to their customers. Examples of these premium services include valet parking, reserved parking, and automobile services (such as car cleaning and minor maintenance services). Some airports have also created loyalty programs to encourage users to use airport-operated facilities, rather than off-airport parking services. Some airports have also begun to impose off-airport parking privilege fees on off-airport parking operators, in an attempt to capture otherwise lost revenue.

Car Rental Revenues are also a significant revenue source and often one of largest nonaeronautical revenue source for smaller airports. Car rental revenue sources include base rents for the use of airport facilities and charging of a percentage of the revenues gained from rental car companies at the airport, and most recently the collection of a **customer facility charge (CFC)** added on to each rental car transaction. This CFC has been used to fund the development and operation of airport CRCFs, as well as other airport expenses.

Advertising Revenues—the focus on increasing nonaeronautical revenues at airports combined with improvements in media technology has led to the maturation of airport advertising programs. Once limited to the posting of print advertising, airports have turned to LED-based multimedia changeable message signs as advertising billboards. In addition, branding of services via electronic directories and airport WiFi homepages have allowed advertising revenues to significantly contribute to airport revenues.

Airline leased areas

Under both residual cost and compensatory approaches, air carriers pay rent to the airport operator for the right to occupy various facilities (terminal space, hangars, cargo terminals, and land). Rental rates are established in the airport use agreements, in separate leases, or by local ordinance or resolution. Terminal space might be assigned on an exclusive-use basis (to a single airline), a preferential-use basis (if a certain level of activity is not maintained, the airline must share the space), or on a shared-use basis (space used in common by several airlines). Most major commercial airports use a combination of these methods. In addition, airports can charge the airlines a fee for use of any airport-controlled gate space and for the provision of federal inspection facilities required at airports serving international traffic. Some airports have long-term ground leases with individual airlines that allow the airlines to finance and construct their own passenger terminal facilities on land leased from the airport.

Among residual cost contracts, the method of calculating airline terminal rental rates varies considerably. Typically, to arrive at the airline fee total, all other revenues generated within the terminal cost center are subtracted from the total costs of the center (administration, operations and maintenance, and debt service). Each airline's share is based on the square footage it occupies, with proration of jointly used space.

Under residual cost contracts where receipts from airline landing fees alone are used to balance the airport budget, the terminal rental rates for the airlines can be set in various ways—on a compensatory basis (recovering the average actual costs of the facilities used), by an outside appraisal of the property value, or by negotiation with the air carriers. In all cases, each carrier's share of costs is based on its proportionate use of the facilities. Rental rates might be uniform for all types of space leased to the airlines, or they might differ according to the type of space provided—for example, they might be significantly higher for leases of ticket counters or office space than for rental of gate or baggage claim areas.

Under residual cost contracts, the rental term for leased areas generally coincides with the term of the airport use agreement with the airlines. The frequency of adjustment of terminal rental rates ranges considerably—annually at many airports, but up to 3 to 5 years at others.

Under compensatory contracts, the method of calculating terminal rental rates is based on recovery of the average actual costs of the space occupied. Each air carrier's share of the total cost is based on the square footage leased. Typically, rates differ according to the type of space and whether it is leased on an exclusive, preferential, or joint-use basis. The rental terms for airline-leased areas often coincides with that of the airport use agreement. (It is set by ordinance at airports that operate without agreements.) Rates are typically adjusted annually at compensatory airports.

Other leased areas

A wide variety of arrangements are employed for other leased areas at an airport, which might include agricultural land, fixed-base operations, cargo terminals, and industrial parks. The methods of calculating rental rates and the frequency of adjustment differ according to the type of facility and the nature of use. What these disparate rentals have in common is that, like terminal concessions and services, they are generally priced on a market basis, and the airport managers have considerable flexibility in setting rates and charges in the context of market constraints and their own policy objectives.

Variation in the sources of operating revenues

In general, revenue diversification enhances the financial stability of an airport. In addition, the specific mix of revenues might influence year-to-year financial performance. Some of the major sources of airport revenue (notably landing fees and terminal concessions) are affected by changes in the volume of air passenger traffic, whereas others (airline terminal rentals and ground leases) are essentially immune to fluctuations in air traffic.

The distribution of operating revenues differs widely according to factors such as passenger enplanements, the nature of the market served, and the specific objectives and features of the airport's approach to pricing and financial management. Airport size generally has a strong influence on the distribution of revenues. The larger commercial airports typically have a more diversified revenue base than smaller airports. For example, they tend to have a wider array of income-producing facilities and services in the passenger terminal complex. In general, terminal concessions can be expected to generate a greater percentage of total operating revenues as passenger enplanements increase. On average, concessions account for at least one-third of total operating revenues at large, medium, and small commercial airports, compared to about one-fifth at very small (nonhub) commercial airports and a smaller fraction still at GA airports.

Factors other than airport size also affect distribution of operating revenues. At commercial airports, for example, parking facilities generally provide one of the largest sources of nonaeronautical revenues in the terminal area. Airports that have a high proportion of transferring passengers might, however, derive a smaller percentage of their operating income from parking revenues than do so-called origin and destination airports. Other factors that can affect parking revenues include availability of space for parking, the volume of air passenger traffic, the airport pricing policy, availability and cost of alternatives to driving to the airport (mass transit and taxicab service), and the presence of private competitors providing parking facilities at nearby locations off the airport property.

The approach to financial management, because it governs the pricing of facilities and services provided to airlines, significantly affects the distribution of operating revenues. Because so many other factors play an important role in determining revenue distribution, however, the mix of operating revenues at an airport cannot be predicted on the basis of whether the airport employs a residual cost or compensatory approach. The mix of revenues varies widely among residual cost airports. With airline landing fees characteristically picking up the difference between airport costs and other revenues at residual cost airports, airfield area income differs markedly according to the extent of the airport's financial obligations, the magnitude of terminal concession income and other revenues, and the volume of air traffic.

Rise in airport financial burdens

Despite significant growth in the number of passengers during the past two decades, airport charges per passenger more than doubled during this period. The greatest percentage of increase has been in the area of rent, which approximately equaled the amount of landing fees paid in the early 1980s but is now more than double the amount of landing fees. The reasons for this shift are obvious: relatively few new runways being constructed versus many expensive terminal expansions and upgrades.

At the same time that airport costs have been increasing, airline prices (yields) have continued to decline. Most forecasters predict that airfares will continue to decline and that the industry will remain fiercely competitive. Consequently, all aspects of the airlines' cost structure will remain under pressure. It is no surprise, therefore, that steadily increasing airport costs have been a source of contention between airlines and airports.

Airport financial burdens have been driven primarily by the following factors:
- Governmental mandates, including new security, environmental, Americans with Disabilities Act, and noise-related compliance costs

- Renewal and replacement of old facilities and equipment
- Airline requirements for support facilities
- Changing airline demand patterns that require consolidation of hub facilities and reduction of activity at nonhub airports
- Additional airport security requirements

Apart from the average rate of growth of airport costs, there is also a significant disparity in cost growth by airport size. Larger airports have a greater need for infrastructure and, consequently, have experienced the greater cost increases. However, the significant increase in operating expenses at large airports is a concern, because it suggests that their expansion and modernization programs have not been accompanied by any increase in operating efficiency.

Airlines generally agree that infrastructure needs have driven a significant portion of airport cost increases, most recently with respect to the need for increased security infrastructure. The debt service associated with major airport construction projects necessary to replace aging facilities inevitably increases total costs.

Airport funding

Although the burdens of managing airport finances to cover the costs of an airport's operating and maintenance budget have increased in recent years, the significantly greater costs associated with moderate to major construction and technology improvements that define capital improvement projects have historically been far beyond that of any revenues generated. As a result, airports have relied on three alternative sources of funding to cover capital improvement costs: federal and state grant programs, bond issues, and private investment to supplement airport revenue.

Grant programs

Since the post–World War II era, the federal government has provided **grant programs** from which owners of public-use airports could acquire funds for airport development. These funds were provided without responsibility for paying any monies back to the government, and thus have been known as grant-in-aid programs. The earliest of these programs, known as the Federal-Aid Airports Program (FAAP), was established with the passage of the Federal Airport Act of 1946 and funded from the general fund of the Treasury.

A more comprehensive program was established with the passage of the Airport and Airway Development Act of 1970. This act provided for grant assistance for airport planning under the **Planning Grant Program (PGP)** and for airport development under the **Airport Development Aid Program (ADAP).** The source of funds for these programs was a new Airport and Airway Trust Fund, into which revenues were deposited from several aviation user taxes on such items as airline fares, air freight, and aviation gasoline. The act, after several amendments and a 1-year extension, expired on September 30, 1981.

Airport Improvement Program (AIP)

The successor grant program, the **Airport Improvement Program (AIP),** was established by the Airport and Airway Improvement Act of 1982. It provided assistance under a single program for airport planning and development through funding from the Airport and Airway Trust Fund. The Airport and Airway Improvement Act has been extended several times over the years, providing increasing funding authorizations through the latest extension in 2003, which subsequently expired in 2007. As of 2010, the AIP was operating under *continuing resolution*, which effectively has temporarily extended the program at the levels authorized for 2007. In 2007, under the Vision 100—Century of Aviation Reauthorization Act, approximately $3.4 billion was authorized by the congress for the program (Fig. 9-1).

AIP funds are used for four general purposes: airport planning, airport development, airport capacity enhancement, and noise compatibility programs. The trust fund relies on user fees and taxes assessed on those who benefit from the services made possible by AIP grants, such as:

A 7.5% domestic passenger ticket tax

A $3.40 domestic flight segment tax

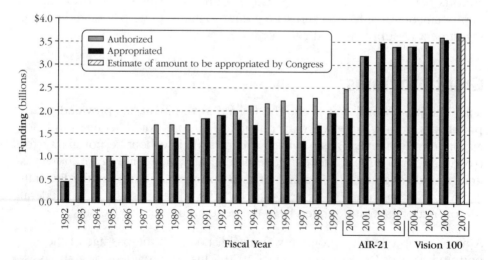

Figure 9-1. *Annual congressional AIP authorizations and appropriations.*
(Source: National Academies Transportation Research Board)

A $15.10 international departure and arrival tax

A $7.50 tax on flights between the Continental United States and Alaska or Hawaii

A 6.25% tax on domestic air cargo/mail

A fuel tax of $0.043 per gallon on fuel used for commercial aviation and $0.193 per gallon (AVGas) and $0.218 per gallon (JetA) for general aviation

An airport must be part of the National Plan of Integrated Airport Systems (NPIAS) to be eligible for AIP funding. The purpose of the plan is to identify those public-use airports that are essential to providing a safe and efficient air traffic system to support civil aviation, the military, and the U.S. Postal Service. The sponsor must also meet several legal, financial, and miscellaneous requirements. These requirements are necessary to ensure that the sponsor is capable of fulfilling the provisions stipulated in the grant obligations.

AIP funds may only be used toward specific types of projects that directly contribute to the capital improvement of airport facilities. The categories of projects approved for AIP funding are airport planning, airport development, airport capacity enhancement and preservation, and noise compatibility programs, as illustrated in Fig. 9-2.

Strategic Investments	Airport Revenues	Revenue Bonds	AIP Grants		PFCs		
			Entitlement	Discretionary	Pay-as-you-go	Bonds	Other
Land acquisition	■	■	■	■			
Runway extensions/new runways/taxiways		■	■	■		■	
New terminals/concourses		■	◇			■	
Security projects			■	■	■	■	
On-airport and access roads	■	■					
People movers		■				■	
Infrastructure for tenant/3rd party facilities	■	■					■
Public parking	■	■	●	●	●	●	
Consolidated rental car facilities		◇	●	●	●	●	■
Ongoing maintenance	■		◇	◇	●	●	
Planning and preliminary design	■						

■ Key Source   ◇ Secondary Source   ● Not Eligible/Advisable

Figure 9-2. *Allowable uses of funding sources.* (Source: National Academies Transportation Research Board)

Eligible airport planning projects can be conducted on either an areawide or individual airport basis. Areawide planning includes preparation of integrated airport system plans for states, regions, or metropolitan areas. Grants for integrated airport system planning are made to the planning agency with jurisdiction over the entire region under study. Airport system planning addresses the current and future air transportation needs of the region as a whole. Individual airport planning addresses the current and future needs of an individual airport through the airport master planning process, aviation requirements, facility requirements, and potential compatibility with environmental and community goals. Individual airport planning also includes the preparation of noise compatibility plans.

Eligible airport development projects may include the construction, improvement, or repair (excluding routine maintenance) of an airport. These projects may include land acquisition, site preparation, navigational aids, or the construction of terminal buildings, roadways, runways, and taxiways. For AIP funding purposes, airport development grants cannot be used for the construction of hangars, automobile parking areas, car rental facilities, buildings not related to the safety of persons at the airport, and art objects or decorative landscaping.

The Airport and Airway Safety and Capacity Expansion Act of 1987 allows for AIP funding of projects that significantly enhance or preserve airport capacity. Increasing airport capacity allows the national system to better accommodate its service demand and also reduces aircraft delays, particularly at the largest primary airports. Considerations for airport capacity funding include the project's cost and benefit, the project's effect on overall national air transportation system capacity, and the financial commitment of the airport sponsor to preserve or enhance airport capacity.

Federal Aviation Regulations Part 150 outlines the eligibility criteria for an airport noise compatibility program. Airports receiving noise compatibility–related grants may include the owners and operators of a public-use airport or local governments surrounding the airport.

Funds granted to airports by the AIP are provided in three different funding categories: entitlement, set-aside, and discretionary funds. Entitlement funds represent the largest funding category, making up approximately half of all AIP funding, although it is expected that entitlement funding will be greatly reduced in favor of discretionary funds in future FAA reauthorization.

Entitlement funds are based on those airports' annual enplanements. In addition, **apportionment funds** for cargo operations at these airports are based on aggregate landed weight of all cargo aircraft. **Set-aside funds** are available to any eligible airport sponsor and are allocated according to congressionally mandated requirements for a number of different set-aside subcategories. Set-aside distributions include:

- Allocations to all 50 states, the District of Columbia, and the insular areas based on land area and population
- Funds specifically for the insular areas
- Minimum funding levels for Alaska for purposes such as reliever airports, nonprimary commercial service airports, airport noise compatibility programs, integrated airport system plans, and the Military Airport Program

Discretionary funds are grants that go to projects that address goals established by the Congress, such as enhancing capacity, safety, and security or mitigating noise at all types of airports. AIP funds are typically awarded at 80 percent of the total costs of a given project. The remaining 20 percent of the costs of the project are expected to be covered by other sources, including state and local funding, bond issues, or airport revenues.

Passenger facility charges (PFCs)

In 1972 the Supreme Court ruled in *Evansville-Vanderburgh Airport v. Delta Air Lines* that tolls charged to enplaning and deplaning passengers were constitutional. This ruling prompted several airport operators to collect such tolls. However, in 1973 Congress enacted the Anti-Head Tax Act, which stated that the user-fee and tax revenues collected for the Airport and Airway Trust Fund would be sufficient to fund airport development, and banned airport tolls, or head taxes.

Some years later, critical shortages of airport capacity and the associated capital to finance airport development prompted major legislative campaigns for **passenger facility charges (PFCs).** In response to these shortages, Congress authorized domestic airports to assess PFCs on enplaning passengers as part of the Aviation Safety and Capacity Expansion Act of 1990. The act provided publicly owned commercial service airports the permission to assess a $1, $2, or $3 PFC on domestic, territorial, or international revenue passengers enplaned at the airport. The PFC must be assessed uniformly across all of an airport's passengers. A maximum of two charges may be imposed on a passenger traveling to and from an airport (either one-way, round-trip, connecting, or origin/destination). The Wendell Ford Aviation Investment and Reform Act for the 21st Century (AIR-21) allowed PFC charges to be assessed at $4 or $4.50 per passenger segment. The maximum PFC assessment is expected to increase to approximately $6.00 per segment with the next FAA funding reauthorization, as of 2010.

Revenues from PFCs may be used only to fund eligible projects that satisfy statutory goals. Projects eligible for PFC funding include those that meet one of the following three criteria:

- Preserve or enhance the capacity, safety, or security of the national air transportation system
- Reduce noise resulting from an airport
- Furnish opportunities for enhanced competition between or among air carriers

PFC revenue can finance the entire allowable cost of a project or can be used to pay debt service or related expenses for bonds issued to fund an eligible project. A PFC is considered local revenue and may be used to meet the non-federal share of projects funded under the AIP.

If a sponsor of an airport that accounts for at least 0.25 percent of total annual U.S. enplanements imposes a PFC, then that airport will lose a fraction of its AIP entitlement. This amount is equal to 50 percent of the projected PFC revenues per year. However, the reduction may not exceed 50 percent of the AIP apportionment funds (not including discretionary or set-aside funding) anticipated for that airport in that fiscal year.

PFC revenues may also be leveraged as a revenue stream to support a bond issue. The PFCs can be a fairly stable revenue stream, assuming that enplanements do not fluctuate greatly in the short run. However, several risks are associated with leveraged PFC revenue, including:

- The failure to generate the amount needed for annual debt service payments (including coverage) because enplanements, and subsequently PFC revenues, were lower than projected
- Interruption in the flow of PFC revenues if, for example, an airline that is collecting PFCs declares bankruptcy
- Expiration of authority to collect PFC revenue because of failure to obtain project approval
- Termination of PFC authority for failure to comply with necessary assurances or for a violation of federal noise regulations
- Requirement of FAA approval of amendments to an approved PFC application

A summary of the allowable uses for AIP and PFC funds, as well as for other funding sources, including bonds, which will be discussed later in this chapter, is illustrated in Fig. 9-2.

Other federal funding sources

Whereas the AIP and PFC programs are the primary forms of federal funding for airports, two additional programs are available. They are the facilities and equipment (F&E) program and federal letters of intent (LOI).

Facilities and equipment program

The **facilities and equipment (F&E)** program provides funding for airports for the installation of navigational aids and control towers, as necessary. It funds 100 percent of the costs of these requirements in the interest of navigation, air traffic control, and safety. Eligible projects under the F&E program include site preparation for navigational aids, the installation of navigational aids, and the construction of control towers.

Federal letters of intent

Federal letters of intent (LOI) represent another means of receiving government funding for airport capital improvements. In general, the Airport and Airway Improvement Act of 1982 prohibited the use of AIP funds for projects begun before an AIP grant had been formally issued. However, the Airport and Airway Safety and Capacity Expansion Act of 1987 allowed the issuance of LOI. In writing LOI, the FAA states its intent to appropriate future funds to the approved project. The FAA issues LOI for projects that will significantly enhance systemwide airport capacity.

In 1994, the FAA issued new regulations stating that it would consider LOI for primary and reliever airports only for airside development projects with significant capacity benefits. The three main criteria for determining which airports will receive LOI for certain projects are:

- The effect of the project on the overall capacity of the airport system
- Project benefit and cost
- Project sponsor financial commitment or timing

The FAA evaluates the use of LOI in terms of "aircraft delay savings," measured as the avoided cost of operating delayed flights and the value of passenger hours wasted during delays. The best project candidates are new airports, new runways, or existing runway extensions in metropolitan areas with current forecasted delays of over 20,000 hours per year. Projects are prioritized according to their function:

- Airport safety and security
- Preservation of existing infrastructure

- Aid compliance with governmental standards (e.g., noise migration)
- Upgrade of service
- Increase in airport system capacity

The use of LOI has been a subject of great interest and concern to the airport community. A number of issues limit the use of LOI as a stand-alone, tangible revenue stream. First, LOI are not a legal pledge to provide funds; the letters clearly state that the FAA is not committing funds to a proposed project. Second, LOI bear the risk that Congress may delay or even fail to grant reauthorization of AIP funding in any given year. Third, future federal budget cuts may limit the amount of AIP discretionary funds. Fourth, apportionments are based on the number of enplanements; failure to attain projected enplanement levels could result in reduced funding.

State grant programs

In addition to federal funding, many individual states in the nation offer grant programs for airport capital improvements. These sources are typically found within state Departments of Transportation, funded from the general tax base of the state, as well as state user fees on transportation-related facilities such as highway tolls, automobile and other vehicle registrations, and fuel taxes. State and local funding is offered either as supplemental funding to federal grants, or as primary funding for airports and/or airport projects not eligible for funding through AIP or PFC programs. As with federal funding programs, state grants are typically funded at some percentage (on the order of 90 percent) of the total funds required, with the airport owner obliged to pay the remaining costs.

In addition to individual state grant programs, some states have developed block grant programs. Under a block grant program, individual states apply for federal funding on behalf of their represented airports. In turn, the states may allocate the funds received from the federal government to the individual airports as they see fit.

The primary requirement for block grant eligibility is a federal regulation that stipulates that an airport must be listed in the National Plan of Integrated Airport Systems (NPIAS) in order to receive federal funds.

Several major differences exist between State Aid to Airports and block grant projects. First, the airport and the project must meet all eligibility requirements for federal funding. For example, State Aid to Airports can provide funds for terminals, but such buildings are ineligible for block grant funding. Airports receiving block grant funds must develop and implement a Minority Business Enterprise Program for construction (and for operations if the grant is large enough). And block grant recipients must agree to be bound by the FAA standard grant assurances that dictate a large range of requirements that must be adhered to in operations and maintenance.

Grant assurances

Almost all federal and state grant programs come with some measures of obligation by the airport to its funding source with respect to the operation of the airport. These obligations are known as grant assurances. Grant assurances provide the funding source that the funds will be used in accordance with the source's rules and regulations, design standards, and operational policies. In addition, most grant assurances include that the airport maintain overall standards of operation for a certain period of time (typically 20 years) following funding.

Airport financing

Since the mid-1990s the largest source of funding for capital improvements at airports has been through bond financing. In the years 1999–2001, for example, a total of $12 billion was funded for capital improvements at the nation's commercial service airports, $6.9 billion of which (59 percent) was issued through airport bonds.

The role of bond financing in overall investment varies greatly according to an airport's size and type of air traffic served. In terms of total dollar volume of bond sales, large and medium airports are by far the most prominent in the bond market. Of the total amount of municipal debt sold for airport purposes during the last two decades, 90 percent was for large and medium airports, in contrast to only 9 percent for small commercial airports. GA airports accounted for a little more than 1 percent of total airport bond sales.

General obligation bonds

General obligation bonds (GOB) are issued by states, municipalities, and other general-purpose governments for the purposes of financing large public works projects, including airport development. The payments (interest and principal) to bondholders are secured by the full faith, credit, and taxing power of the issuing government agency. An advantage of general obligation bonds is that, because of the community guarantee, they typically can be issued at a lower interest rate than can other types of bonds; however, most states limit the amount of general obligation debt that a municipality may issue to a specified fraction of the taxable value of all property within its jurisdiction. In addition, many states require voter approval before using general obligation debt.

Fiscal pressures on local governments for all manner of activities have been especially great in recent years. The need for school construction and other essential public works has required a considerable volume of general obligation bond financing. In numerous cases, local governments have reached statutory bond limits or desire to reserve whatever margin is left for more general functions of government. It is becoming increasingly difficult to obtain taxpayer approval for general obligation bond issues for airports.

Self-liquidating general obligation bonds are also secured by the full faith, credit, and taxing power of the issuing government body; however, there is adequate cash flow from the operation of the facility to cover the debt service and other costs of operation of the facility. In other words, they are self-liquidating (self-sustaining). The debt is not legally considered part of the community's debt limitation; however, because the credit of the local government bears the ultimate risk of default, the bond issue is still considered, for purposes of financial risk analysis, as part of the debt burden of the community; therefore, this method of financing generally means a higher rate of interest on all bonds sold by the community. The amount of interest rate generally depends in part upon the degree of "exposure risk" of the bond. Exposure risk occurs when there is insufficient net operating income to cover the level of debt service plus coverage requirements, and the community is therefore required to absorb the residual.

General airport revenue bonds

After World War II, larger airports began switching from general obligation bonds to revenue bonds as a method for financing new construction and improvements to existing fields. The first airport revenue bond in the United States was a $2.5 million issue sold in 1945 by Dade County, Florida, to buy what is now Miami International Airport from Pan American World Airways.

In the 1950s, the city of Chicago and the airlines that serve it worked out what has become the basic pattern for revenue bonds underwritten by airlines in the agreement that set up the financing for O'Hare International Airport. The airlines pledged that if airport income fell short of the total needed to pay off the principal and interest on the bonds, they would make up the difference by paying a higher landing fee rate. The historic O'Hare Agreement demonstrated that airports, backed up by the airlines that use them, could raise the money they need in the financial market without depending on general tax funds, and airport revenue bonding became the accepted way to raise money for construction and expansion.

The revenue bonds are usually issued for 25- or 30-year terms, in contrast to the customary 10- or 15-year terms for general obligation bonds. Interest rates run slightly higher on revenue bonds than on general obligation bonds.

A bond issue can be sold competitively, with the airport accepting bids and selling the issue to the bond house that offers to buy it for the lowest interest rate, or the interest rate can be negotiated between the seller and a single buyer. Often airport sponsors use the services of a bond counsel, who advises on the best way to market a particular bond issued. After a bond house buys a bond issue, it resells the bonds to commercial banks, insurance companies, pension funds, and other large investors.

In recent years, the vast majority of airport debt has been issued in the form of **general airport revenue bonds (GARB).** Used predominantly by large and medium-sized commercial airports, revenue bonds are secured solely by revenue generated by operations of the airport and are not backed by any additional governmental subsidy or tax levy.

Special facilities bonds

A special category of airport bonds is **special facilities bonds.** Although they are still issued by the airports' sponsors in order to obtain tax-exempt status, the special facility bonds are secured by the revenue from the indebted facility, such as a terminal, hangar, or maintenance facility, rather than the airport's general revenue. The annual amount of special facility bonds is more volatile than that for regular airport bonds because fewer special facility bonds are issued for larger amounts than regular airport bonds.

The perceived credit quality of an airport is the product of its performance in a number of analytical areas. Different analyses may place varying emphases on these issues but, generally, the following are considered: financial and operational comparables, nature of airline rates and charges, local economic base, airport current financial situation or debt level, strength of airport management, and airport layout.

Financial and operational factors

Standard financial ratios can be developed that represent median performance for airports of varying sizes, geographic locations, and passenger mix. Analysis of an airport's position with respect to these medians is a useful starting point for bond rating analysts. It develops a benchmark of airport financial and operational performance. The following is a representative list of ratios that might be analyzed in the development of a bond rating for a particular airport:

- Traffic ratios, such as total origin and destination (O&D) passengers to transfer passengers
- Annual increase in originating and transfer passenger traffic
- Annual increase in cargo traffic
- Aeronautical and nonaeronautical revenue per enplaned passenger

- Local per capita income, gross product, and total employment
- Debt per enplaned originating and transfer passenger
- Debt service coverage
- Percentage of traffic generated by the airport's two primary carriers

Airline rates and charges

Airline rates and charges generate a significant portion of total airport revenues. Because airport revenues are the sole backing for revenue bonds, the nature of airline rates and charges has a significant impact on an airport's credit rating. The fact that lease agreements can vary also makes analysis through comparisons of traditional financial ratios difficult, because these ratios do not indicate the relative flexibility of an airport's rate structure. Instead, analysts often consider whether the type of lease agreement seems appropriate, given local circumstances.

More important, however, is the fact that the rate-setting methodology affects the airport's control over its capital spending decisions. Under residual approaches in which the airlines assume the risk and guarantee revenues necessary to keep the airport operational, airlines can exercise control over capital spending through majority-in-interest (MII) lease provisions, which give signatory carriers the right of approval for airport capital spending. These provisions may allow existing carriers to resist capital projects designed to create facilities for new airlines. The debt of airports operating under such provisions is often considered less favorable by bond rating analysts.

Community economic base

The strength and diversity of the local economy in which the airport operates is a critical factor considered in airport bond rating. Economic strength results in greater demand for air transportation. Economic diversity protects the airport from economic fluctuations, resulting in more consistent enplanement levels. In addition, several nonaeronautical revenue sources such as parking and ground transportation (which contribute to the airport's financial viability) are closely linked to the economy in the local service area. These services represent a constant, dependable source of revenue (in that they are not subject to volatility in hubbing arrangements) as long as the local economy remains strong. Thus, airports located in economically booming areas may receive higher ratings than those in areas suffering from an economic downturn.

Current financial status and debt level

Credit analysts evaluate airports in the context of their capital plans and financial forecasts. An airport's overall level of indebtedness and need to generate future revenues affects its credit quality. However, the unique context in which each airport operates makes it difficult to develop simple comparative measures of the relative indebtedness of airports because of growth, changes in the air carrier industry, and varied service demand. Although there is usually a strong relationship between airport size and indebtedness, even this relationship can be skewed by the airport's stage in capital planning, debt issuance, and use of debt financing. Thus, although figures such as "debt per enplaned passenger" can be calculated, they are not always useful.

Airport management

Analysts review the managerial and administrative performance of airport operators and believe that well-run airports are generally better risks. Clearly, airport management's ability to negotiate favorable rates, charges, and tenant agreements is a positive indication of managerial control, as is general ability to manage financial and other resources during traffic declines. Both of these criteria may indicate the airport's likelihood of operating effectively in the future. Similarly, management's success in planning existing capital programs and implementing debt issuances demonstrates managerial quality.

Bond ratings

The major investor services (such as Moody's and Standard & Poor's) grade bonds according to investment quality. The top-ranked bonds are as follows:

1. *Best grade*. Bonds rated Aaa (by Moody's) or AAA (by Standard & Poor's) are graded best. Their exceptionally strong capacity to pay interest and repay principal offers the lowest degree of risk to investors in bonds.

2. *High grade*. Bonds rated Aa1 or Aa (by Moody's) or AA1 or AA (by Standard & Poor's) have very strong ability to pay interest and repay principal, but they are judged to be slightly less secure than best-grade bonds. Their margins of protection might not be quite so great, or the protective elements might be more subject to fluctuation.

3. *Upper-medium grade*. Bonds rated A1 or A (by Moody's) or A1, A, or A2 (by Standard & Poor's) are well protected, but the factors giving security to interest and principal are deemed more susceptible to adverse changes in economic conditions or other future impairments than for bonds in the best and high-grade categories.

4. *Medium grade*. Bonds rated Baa1 or Baa (by Moody's) or BBB1, BBB, or BBB2 (by Standard & Poor's) lack outstanding investment characteristics. Although their protection is deemed adequate at the time of rating, the presence of speculative elements might impair their capacity to pay interest and repay principal in the event of adverse economic conditions or other changes.

Although investors have considerable confidence in airport bonds, ratings vary between the top and medium grades. A medium grade means that rating firms see the investment as carrying a measure of speculative risk. General obligation bonds generally draw the best ratings. Under this form of security, ratings are determined by the economic vigor of the municipality or the entire state, and airports have little or no influence on the rating. Revenue bonds, on the other hand, draw ratings according to the fiscal vitality of the airport itself. Because more than 90 percent of all airport bonds (in terms of dollar volume) are secured with airport revenues, the criteria used by investor services to rate such bonds are central to the marketability of such bonds.

The final bond rating, which reflects the reliability of the bond, results from the airport's performance measured by these and other criteria. This rating determines the perceived risk potential investors associate with the bond issue and therefore affects the interest rate or terms attached to the debt issuance, which is important to the financial feasibility of the proposed project to be financed.

Interest costs

Interest costs represent the payments by airports to attract investors relative to what other municipal enterprises pay. The difference between interest costs paid by airports and by other public enterprises indicates that airports generally hold a strongly competitive position in the municipal bond market.

Like municipal bonds in general, airport bonds are sold and traded at prices that reflect both general economic conditions and the credit quality of the airport or (in the case of general obligation bonds) the creditworthiness of the issuing government. Rated revenue bonds are offered for sale in one of two ways. Under competitive bidding, the airport selects the lowest bid and thus obtains funds at the lowest cost of borrowing. Under a negotiated sale, the bond purchaser consents at the outset to purchase the bond issue at an agreed price. In either case, the entire bond issue is usually purchased by an underwriter (commonly, an investment brokerage company), or an underwriter team, which in turn markets the bond to institutional and individual investors.

In deciding the price of a particular bond issue, underwriters identify a "ballpark" interest rate on the basis of general market conditions and then refine this estimate according to the credit standing of the airport in question. General market conditions represent by far the most important determinant of interest costs on airport revenue bonds, and in this respect airports have little control over the cost of capital.

Within the range of interest costs dictated by market conditions, underwriters refine their bids on airport revenue bonds on the basis of the credit standing of the individual airport. Two factors have greatest importance here: the airport's basic fiscal condition (including its prospects for traffic growth and the strength of the local economic base) and the presence of special pressures on the airport to expand capacity, thereby necessitating extensive capital development. On average, larger airports pay lower interest costs than smaller airports, allowing for differences in types of security and average maturities of issues.

Defaults

The term *defaults* refers to the frequency with which a given type of enterprise has defaulted on a bond issue. This history of an enterprise, or of an entire industry, with regard to the number of defaults is an important index of investment value. By this measure, the record of airports is particularly strong. The airport industry has never suffered a single default, a fact noted by several credit analysts in citing the premium quality of airports as credit risks.

Private investment

In many instances, particularly internationally, airport capital projects have been funded by private investment. Many of these investments are focused on the construction of terminal and ground access facilities such as passenger and cargo terminal buildings, rental car facilities, and aircraft service facilities. Fewer private investments have been made on the construction of airfield facilities.

Many such investments are made through either public-private partnerships or complete privatization. Privatization can be structured in a number of ways. When assessing expanded private sector involvement in airports, two major financial issues must be examined:

- The profitability of the arrangement
- The ultimate costs of the arrangement and where the risks are borne

With regard to profitability, pro forma operating statement analysis can determine if these revenue streams can sustain the cost of the transaction and who will ultimately bear the burden of the transaction price and any new costs, including taxes. In assessing the costs of the arrangements, it may be found that government outlays and subsidies are not required, and therefore government costs are reduced and the government's exposure to financial risk is removed. The risk may not fall to zero if the public sector takes ultimate responsibility for the success of the venture.

Both internationally and domestically, privatization has become a popular way for government entities to finance new and existing infrastructure projects. In developing countries, governments are turning to the private sector as an alternative source of capital to build much-needed infrastructure or to improve the existing infrastructure. In more developed countries, the private sector is bringing efficiencies to traditionally government-run projects. Finally, government entities are turning to the private sector to provide innovation in service provision and operation. All of these factors have made privatization an attractive option for financing infrastructure projects, including the building and operating of airports.

Build, operate, and transfer (BOT) contracts

Most airports in the United States currently use private sector involvement to their advantage through some type of external contract for the construction and operations of facilities. Under a **build, operate, and transfer (BOT)** contract, private investment is used to construct and operate a facility for a period defined in the terms of the contract. At the end of the contract period, the ownership of the facility is transferred to the airport owner.

Lease, build, and operate (LBO) agreements

The lease arrangement allows a government entity to realize many of the benefits associated with complete privatization without losing control over the airport assets. In a long-term lease (usually lasting from 20 to 40 years), the government allows the private sector company or consortium to build and manage an airport facility, while leasing the property and facility from the airport. The private builder/operator has much of the authority over the facility, including operations, strategic decisions, and development.

In addition to a lease payment, the government is able to capture the efficiencies and innovation of the private sector. The private sector entity has the advantage of complete control over the airport; yet in many cases, the private sector firm also has the added benefits of access to tax-free financing and exemption from property taxes.

Full privatization

As illustrated in Fig. 9-3, the final basic form of airport **privatization** is the sale of the entire airport or partial interest in the airport. This form of privatization is prominent internationally but has not occurred domestically. Under the terms of complete sale, the government gives up all rights of ownership to the private entity; however, the government often maintains its regulatory authority.

Only one airport in the United States has operated under a fully privatized model. This airport, Newburgh's Stewart Airport in New York, was acquired by a public authority, the Port Authority of New York and New Jersey, operators of four other public-use airports in the New York metropolitan area, in 2007.

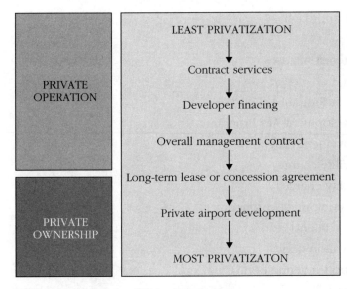

Figure 9-3. *Levels of Privatization.*
(Source: National Academies Transportation Research Board)

Attempts to fully privatize airports in the United States, however, continue, with the latest attempt in 2010 to privatize Chicago's Midway airport being revived after a failed attempt in 2008.

Concluding remarks

Financial planning of an airport is not a static activity. Continuous planning and management is required to adapt to the changing levels in demand, needs for maintaining and improving facilities, and especially the changing levels of revenues and other funding available to the airport.

Key terms

capital improvement expenses

O&M (operation and maintenance) costs

lump sum appropriation

appropriation by activity

line-item budget

zero-based budget

airport use agreement

residual cost approach

compensatory cost approach

signatory carrier

MII (majority-in-interest) clauses

grant programs

PGP (Planning Grant Program)

ADAP (Airport Development Aid Program)

AIP (Airport Improvement Program)

PFC (passenger facility charge)

CFC (customer facility charge)

F&E (facilities and equipment) program

LOI (federal letter of intent)

GOB (general obligation bond)

GARB (general airport revenue bond)

special facilities bond

bond rating

BOT (build, operate, and transfer) contracts

LBO (lease, build, and operate) agreements

privatization

Questions for review and discussion

1. What is airport accounting?

2. What are the different types of budget strategies found at airports?

3. What are the four categories of O&M expenses that exist at airports? What specific expenses lie within each category?

4. What are the differences between operating and nonoperating revenues?

5. What areas of airport operations are principal areas of potential litigation against airports?

6. What are the primary differences between the residual cost approach and compensatory approach?

7. What is a majority-in-interest clause? How do these clauses affect airport management?

8. How are airport facilities and services priced?

9. What grant programs exist on the federal level for airports? How are these programs funded? How may funds from these programs be used at airports?

10. What types of financing programs, or bond issues, are available to airports?

11. How do bond ratings affect the financial strategies of airports?

12. What forms of privatization exist at airports? How do each of these strategies differ?

Suggested readings

Airport Financing—Comparing Funding Sources with Planned Development. Washington, D.C.: U.S. General Accounting Office, March 1998.

Airport Financing—Funding Sources for Airport Development. Washington, D.C.: U.S. General Accounting Office, March 1998.

"Analysis of U.S. Airport Costs Incurred by Airlines," American Association of Airport Executives and Airports Council International—North America, September 1993.

Ashford, Norman, and Moore, Clifton. *Airport Finance.* The Loughborough Airport Consultancy, Leicestershire, United Kingdom, 1999.

Campbell, George E. *Airport Management and Operations.* Baton Rouge, LA.: Claitor's Publishing Division, 1972.

Cook, Barbara. "A New Solution to an Old Problem—What to Do When the AIP Well Dries Up?" *Airport Magazine.* November/December 1994.

Doganis, Rigas. *The Airport Business.* New York: Routledge, Chapman and Hall, 1992.

Eckrose, Roy A., and William H. Green. *How to Assure the Future of Your Airport.* Madison, Wis.: Eckrose/Green Associates, 1988.

Gesell, Laurence E. *The Administration of Public Airports,* 4th ed. Chandler, Ariz.: Coast-Aire Publications, 1999.

"Innovative Finance and alternative Sources of Revenue for Airports" *ACRP Synthesis 1*, Transportation Research Board, Washington, D.C., 2007

Hazel, Robert. "Airport Economics" *Handbook of Airline Economics*, 1st Edition, Chapter 17, McGraw-Hill, 1995.

Pino, Marc, and Fischbeck, Brian. "Airport Funding" *Handbook of Airline Economics*, 1st Edition, Chapter 16, McGraw-Hill, 1995.

Stanmeyer, Catherine, and Cote, Lorraine. "Airport Finance" *Handbook of Airline Economics*, 1st Edition, Chapter 15, McGraw-Hill, 1995.

10

The economic, political, and social role of airports

Outline

- Introduction
- The economic role of airports
  - Transportation role
  - Stimulating economic growth
- Political roles
  - Airport–airline relations
  - Airport–concessionaire relations
  - Airport–general aviation relations
- Environmental impacts of airports
  - Airport noise impacts
  - Measurement of noise
  - Air quality
  - Water quality
  - Hazardous waste emissions
  - Externalities
  - Economic and environmental sustainability practices
- Social responsibilities

Objectives

The objectives of this section are to educate the reader with information to:

- Understand the important economic role airports play within local communities.
- Describe how airport activity stimulates economic growth in a metropolitan region.

- Appreciate the complex relationships between airport management and the airlines that serve their airports.
- Understand the relationships airport management hold with concessionaires that serve the airport.
- Be familiar with the relationship between airport management and the general aviation community.
- Define the various measures used to determine the impact of noise around airports.
- Describe various noise abatement programs employed at airports.
- Be familiar with the impacts airport activity has on air and water quality.
- Describe methods used to make airports economically, environmentally, and socially sustainable.

Introduction

By nature of the fact that they are among the largest public facilities in the world, airports play significant roles in shaping the economic, political, and social landscape of the communities they serve. As such, airport management must assume the responsibility for leading the airport in positively contributing to the local economy, maintaining good working relations with the airport's users and surrounding community while minimizing the impacts that airports have on the surrounding natural environment. Maintaining this balance of roles is perhaps equally as challenging as maintaining the operations of the airport itself.

The economic role of airports

It is well understood that a viable and efficient transportation system is a fundamental and necessary component to the economy of any region. Transportation, by definition, provides the ability for people and goods to move between communities. This movement leads to trade and commerce between markets, which, in turn, leads to jobs, earnings, and overall economic benefit for a community's residents.

Transportation role

Even though there are a variety of transportation modes, such as automobiles, trucks, ships, and railroads, perhaps no other mode has as significant an impact on intercity trade and commerce than aviation. Travel in the aviation system allows for intercontinental travel of large volumes of passengers and cargo in relatively short periods of time. Access to markets around the world has resulted in the largest of communities reaping extraordinary economic benefit.

Airports are the gateways to the nation's aviation system, providing access to air transportation for the surrounding community. Commercial air carriers provide access to air transportation between many major metropolitan areas of the country. Thousands of smaller cities, towns, and villages have access to aviation by way of airports serving general aviation.

Stimulating economic growth

The airport has become vital to the growth of business and industry in a community by providing air access for companies that must meet the demands of supply, competition, and expanding marketing areas. Communities without airports or sufficient air service have limitations placed on their capacity for economic growth.

Airports and related aviation and nonaviation businesses located at the airport represent a major source of employment for many communities around the country. The wages and salaries paid by airport-related businesses can have a significant direct effect on the local economy by providing the means to purchase goods and services while generating tax revenues as well. Local payrolls are not the only measure of an airport's economic benefit to the community. In addition, employee expenditures generate successive waves of additional employment and purchases that are more difficult to measure but nevertheless substantial.

In addition to the local direct economic activity generated by the regular expenditures of resident employees, the airport also stimulates the economy indirectly through the use of local services for air cargo, food catering to the airlines, aircraft maintenance, and ground transportation on and around the airport. Regular purchases of fuel, supplies, equipment, and other services from local distributors inject additional income into the local community. Finally, earnings from direct and indirect economic generators further act to recycle money within the local community as dollars pass from one person to another. This *multiplier* effect operates in all cities as aviation-related dollars are channeled throughout the community.

Airports provide an additional asset to the general economy by generating billions of dollars per year in state and local taxes. These tax dollars increase the revenues available for projects and services to benefit the residents of each state and community. Whether the extra tax dollars improve the state highway system, beautify state parks, or help prevent a tax increase, airport-generated tax dollars work for everyone.

Cities with good airport facilities also profit from tourist and convention business. This can represent substantial revenues for hotels, restaurants, retail stores, sports and nightclubs, sightseeing, rental cars, and local transportation, among others. The amount of convention business varies with the size of the city, but even smaller communities show a sizable income from this area.

Beyond the benefits that an airport brings to the community as a transportation facility and as a local industry, the airport has become a significant factor in the determination of real estate values in adjacent areas. Land located near airports almost always increases in value as the local economy begins to benefit from the presence of the airport. Land developers consistently seek land near airports, and it follows inexorably that a new airport will inspire extensive construction around it.

Political roles

A major commercial airport is a huge public enterprise. Some are literally cities in their own right, with a great variety of facilities and services. Although the administration of these facilities is generally the responsibility of a public entity, such as a department of city government or aviation authority, airports also have a private character. Commercial airports must be operated in cooperation with the air carriers that provide air transportation service and all airports must work with tenants, such as concessionaires, FBOs, and other firms doing business on airport property. This combination of public management and private enterprise creates a unique political role for airport management.

Airport–airline relations

From the airlines' perspective, each airport is a point in a route system for the loading and transfer of passengers and freight. In order to operate efficiently, air carriers need certain facilities at each airport. These requirements, however, are not static; they change with traffic demand, economic conditions, and the competitive climate. Before airline deregulation in 1978, response to changes of this sort was slow and mediated by the regulatory process. Carriers had to apply to the Civil Aeronautics Board (CAB) for permission to add or to drop routes or to change fares. CAB deliberations involved published notices, comments from opposing parties, and sometimes hearings.

Deliberations could take months, even years, and all members of the airline-airport community were aware of a carrier's intention to make a change long before the CAB gave permission. Since the Airline Deregulation Act of 1978, air carriers can change their routes without permission and on very short notice. With these route changes, airline requirements at airports can change with equal rapidity.

In contrast to air carriers, which operate over a route system connecting many cities, airport operators must focus on accommodating the interests of a number of users at a single location. Changes in the way individual airlines operate might put pressures on the airport's resources, requiring major capital expenditures or making obsolete a facility already constructed. Because airports accommodate many users and tenants other than the airlines, airport operators must be concerned with the efficient use of landside facilities that are of little concern to the carriers, even though carriers' activities can severely affect (or be affected by) them.

Despite their different perspectives, air carriers and airport management have a common interest in making the airport a stable and successful economic enterprise. Traditionally, airports and carriers have formalized their relationship through airport use agreements. These agreements establish the conditions and methods for setting fees and charges associated with use of the airport by air carriers. Most agreements also include formulas for adjusting those fees from year to year. The terms of a use agreement can vary widely, from short-term monthly or yearly arrangements to long-term leases of 25 years or more. Within the context of these use agreements, carriers negotiate with the airport to get the specific airport resources they need for day-to-day operations. For example, under the basic use agreement, the carrier may conduct subsidiary negotiations for the lease of terminal space for offices, passenger lounges, ticket counters, and other necessities.

As with major airport planning decisions, negotiations related to the day-to-day needs of the carriers have traditionally been carried out between airport management and a negotiating committee made up of representatives of the scheduled airlines that are signatories to use agreements with the airport. In the past, negotiating committees have been an effective means of bringing the collective influence of the airlines to bear on airport management.

Since deregulation, the commercial air carrier environment has been characterized by competition rather than cooperation. Carriers might radically alter their routes, service levels, or prices on very short notice. They are reluctant to share information about their plans for fear of giving an advantage to a competitor. These factors make group negotiations more difficult. Some airport proprietors have complained that in this competitive atmosphere, carriers no longer give adequate advance warning of changes that might directly affect the operation of the airport.

As witnessed during the 1990s and 2000s, volatile economic climates have significant effects on airlines and hence on airports. Airlines very quickly attempt to adapt to changing economic environments, whether it be rapidly expanding services during strong economic times, or drastically reducing services during economic slowdowns. This volatility has significant impacts on airport management. As a result, airports must be flexible to modify their operations to accommodate the financial and operational health of their users, while by their very nature must plan in the long term when it comes to the construction and operation of facilities.

Airport–concessionaire relations

Services such as restaurants, bookstores, gift shops, parking facilities, car rental companies, and hotels are often operated under concession agreements or management contracts with the airport. These agreements vary greatly, but in the typical concession agreement, the airport extends to a firm the privilege of conducting business on airport property in exchange for payment of a minimum annual fee or a percentage of the revenues, whichever is greater. Some airports prefer to retain a larger share of revenues for themselves and employ an alternative arrangement called a management contract, under which a firm is hired to operate a particular service on behalf of the airport. The gross revenues are collected by the airport management, which pays the firm for operating expenses plus either a flat management fee or a percentage of revenues.

At a number of airports, the airport operator's share of parking and car rental fees (after concession or management fees are paid) represents the largest revenue source from the terminal area—and in some cases, larger than revenue from air carrier landing fees. At many locations, the parking and car rental firms operating at the airport are complemented by (or are in competition with) similar services operating at the airport.

Another important type of concessionaire is the Fixed Base Operator (FBO), who provides services for airport users lacking facilities of their own, primarily general aviation. Typically, the FBO sells fuel and operates facilities for aircraft service, repair, and maintenance. The FBO might also handle the leasing of hangars and rental of short-term aircraft parking facilities. Agreements between airports and FBOs vary. In some cases, the FBO constructs and develops its own facilities on airport property; in other cases the FBO manages facilities belonging to the airport. FBOs also provide service to some commuter and start-up carriers, especially those that have just entered a particular market and have not yet established (or have chosen not to set up) their own ground operations. The presence of an FBO capable of servicing small transport aircraft can sometimes be instrumental in a new charter carrier's decision to serve a particular airport.

In addition to concessionaires, some airport authorities serve as landlord to other tenants such as industrial parks, freight forwarders, and warehouses, all of which can provide significant revenue. These firms might lease space from the airport operator, or they might build their own facilities on the airport property.

Furthermore, nonaeronautical tenants that benefit from the proximity of airport activity play a major role in leasing airport property. Examples of such properties include hotels, restaurants, rental car agencies, and suppliers of goods that are associated with airport activity or the trade and commerce that flows through the airport. The relationship between airport management and these facilities is a true landlord-tenant relationship. Airport management leases the land, and often the associated facilities, based either on market-appraised land values, percentage of revenues earned from the property, or both. It is the responsibility of airport management to maintain fruitful relationships with all tenants, by ensuring reasonable lease fees, contract terms, and an overall mix of tenants that meet the needs of the airport and the public it serves.

Airport–general aviation relations

The relationship between airport operators and general aviation is seldom governed by the complex of use agreements and leases that characterize relationships with air carriers or concessionaires. General aviation (GA) is a diverse group. At any given airport, the GA aircraft will be owned and operated by a variety of individuals and organizations for a number of personal, business, or instructional purposes. Because of the variety of ownership and the diversity of aircraft type and use, long-term agreements between the airport and GA users are not customary. GA users often lease airport facilities, especially storage space such as hangars and tiedowns, but the relationship is usually that of landlord and tenant. There are instances when owners and operators of GA aircraft assume direct responsibility for capital development of an airport, but this is not common, even at airports where general aviation is a majority user.

Although GA activities make up about half the aircraft operations at FAA-towered airports, the average utilization of each aircraft is much lower than that of commercial aircraft. Only a small number, usually those operated by large corporations and flight schools, are used as intensively as commercial aircraft.

Thus, at the airport, the chief needs of general aviation are parking and storage space, along with facilities for fuel, maintenance, and repair. Whereas an air carrier might occupy a gate for an hour to load passengers and fuel, a general aviation user might need to park an aircraft for a day or more. At the user's home base, long-term storage facilities are needed, and the aircraft owner might own or lease a hangar or tiedown spot. In most parts of the country, the chief airport capacity problem for GA is a shortage of parking and storage space at popular airports. At some airports, waiting lists for GA parking spaces are several years long.

Some airport operators deal directly with their general aviation customers. The airport management might operate a GA terminal, collect landing fees, and lease tiedowns or hangars to users. At some airports, condominium hangars are available for sale to individual users. A corporation with an aircraft fleet commonly owns hangar space at its base airport. Often, however, at least some of this responsibility is delegated to the FBO, who thus stands as a proxy for the airport operator in negotiating with the individual aircraft owners for use of airport facilities and collecting fees.

Environmental impacts of airports

Although there is no doubt that the presence of an airport has great positive impacts on a surrounding community from an economic standpoint, the presence of an airport, much like any large industrial complex, unfortunately impacts the community and surrounding natural environment in what many consider a negative manner. These effects are a result of activity whose source is the airport itself and of vehicles, as well as both aircraft and ground vehicles, which travel to and from the airport. Regardless of which airport-related activities impact the surrounding environment, the burden of managing the impacts often lies with airport management. As such, it's vitally important for airport management to understand the types of environmental impacts that are associated with airport activity, the rules and regulations that govern environmental impact activity, and the political strategies that are available to airport management to satisfy the needs of the surrounding community while maintaining sufficient airport operations.

Airport noise impacts

Perhaps the most significant environmental impact associated with airports is that of the noise that emanates from aircraft movements to and from the airport. Citizens living around airports often complain that airport-related noise is annoying. Noise disturbs sleep, interferes with conversation, and generally detracts from the enjoyable use of property. There is increasing evidence that high exposure to noise has adverse psychological and physiological effects and that people repeatedly exposed to loud noises might exhibit high stress levels, nervous tension, and inability to concentrate.

Conflicts between airports and their neighbors have occurred since the early days of aviation, but airport noise became a more serious issue with the introduction of commercial jet aircraft in the 1960s. FAA estimates that the land area affected by aviation noise increased about sevenfold between 1960 and 1970. As a result of this increase in noise impacts, the FAA adopted federal regulations on noise levels emitted from jet engines that complied with newly created national environmental policies associated with the passing of the National Environmental Policy Act and with the creation of the Environmental Protection Agency 1969. The FAA adopted **Part 36—Certificated Airplane Noise Levels,** of the Federal Aviation Regulations, establishing noise certification standards for new design turbojet and transport category aircraft. In 1976, the Federal Aviation Regulations were amended, to provide U.S. operators until January 1, 1985, to quiet or retire the noisiest (stage 1) aircraft. In 1977, the Federal Aviation Regulations were again amended, defining three "stage" levels to categorize aircraft noise emissions and requiring aircraft certificated after March 3, 1977, to meet the more demanding stage 3 requirement. The federal program to encourage the use of quieter aircraft has been effective. The retirement of early model four-engine aircraft provided tremendous benefits, lowering the residential population exposed to incompatibly high noise levels from an estimated 7 million people in 1975 to 1.7 million people in 1995. This improvement is remarkable because it took place during a period of substantial growth in air transportation, with enplanements more than doubling.

The reduction of aircraft noise at the source, by using quieter aircraft, is supplemented by an ambitious program to encourage compatible land uses in areas around airports. **Part 150—Airport Noise Compatibility Planning,** of the Federal Aviation Regulations, adopted in January 1985, establishes the system for measuring aviation noise in the community and provides information about the land uses that are normally compatible with various levels of noise exposure. An FAA-approved Part 150 noise compatibility program clears the way for airports to obtain federal aid for noise abatement projects. A substantial amount of federal aid is available for preparing and implementing noise compatibility programs, with 10 percent of the annual Airport Improvement Program funds being reserved for this purpose.

Further significant improvements were assured by the **Aircraft Noise and Capacity Act of 1990,** which required the establishment of a national aviation noise policy, including a general prohibition against the operation of stage 2 aircraft of more than 75,000 pounds after December 31, 1999. In addition to the phase out of stage 2 aircraft, the act required the establishment of a national program for reviewing airport noise and access restrictions.

Measurement of noise

FAR Part 150 defines several methods that may be used to measure aircraft noise and its effect on a community. The level of sound can be measured objectively, but noise, unwanted sound—is a very subjective matter, both because the human ear is more sensitive to some frequencies than others and because the degree of annoyance associated with a noise can be influenced by psychological factors such as the hearer's attitude or the type of activity in which she or he is engaged. Techniques have been developed to measure single events measured in units such as dBA (A-weighted sound level in decibels) or EPNdB (effective perceived noise decibels). These measure the levels of noise in objective terms, giving extra weight to those sound frequencies that are most annoying to the human ear.

In some cases, annoyance is due not only to intensity of a single event, but also to the cumulative effects of exposure to noise throughout the day. Methods to measure this effect objectively include aggregating single-event measures to give a cumulative noise profile by means of such techniques as the **noise exposure forecast (NEF),** the **community noise equivalent level (CNEL),** and the **day/night average sound level (Ldn).** FAA uses EPNdB to measure single-event aircraft noise as part of its aircraft certification process. FAA has established dBA as the single-event unit and the Ldn system as the standard measure of cumulative noise exposure to be used by airports in the preparation of noise abatement studies. Ldn noise levels are calculated by considering the loudness of any one single aircraft operation, the altitude and flight path of the aircraft at the location of the noise measurement, the number of such events that occur throughout the day, and the number of such events that occur at night (typically considered between 10:00 P.M. and 7:00 A.M.). The Ldn system places additional emphasis on the noise burden of night operations on a community by adding 10 dB to the measured loudness reading of any operations occurring during night hours.

FAA has suggested, but not mandated, guidelines for determining land uses that are compatible with a given Ldn level. Ideally, residential uses should be located in areas below 65 Ldn. In the high noise impact areas (Ldn 80 to 85 or more), FAA suggests that parking, transportation facilities, mining and extraction, and similar activities are the most compatible. To identify locations surrounding an airport where different noise levels exist, the FAA suggests that a noise contour map be created. This map is created first by collecting noise data through field tests in select locations around the airport vicinity, then processing the data through the use of a noise contour modeling software program. The FAA's software, the **Integrated Noise Model (INM),** is one such program. Figure 10-1 illustrates a set of noise contours surrounding two airports in a metropolitan area.

Even though aircraft are the source of noise at airports, aircraft operators are not liable for damage caused by noise. The courts have determined that the sole legal liability for aircraft noise rests with the airport operator. Balancing their extensive exposure to liability claims, airport operators have some authority, albeit limited, to control the use of their airports in order to reduce noise. Basically, any restriction of operations at the airport must be nondiscriminatory. Further, no airport may impose a restriction that unduly burdens interstate commerce. The definition of "undue burden" is not precise, and restrictions at individual airports must be reviewed on a case-by-case basis. Restrictions must be meaningful and reasonable; a restriction adopted to reduce noise should actually have the effect of reducing noise. Finally, local restrictions must not interfere with safety or the federal prerogative to control aircraft in the navigable airspace.

Under FAR Part 150, airport operators can undertake noise compatibility studies to determine the extent and nature of the noise problem at a given airport.

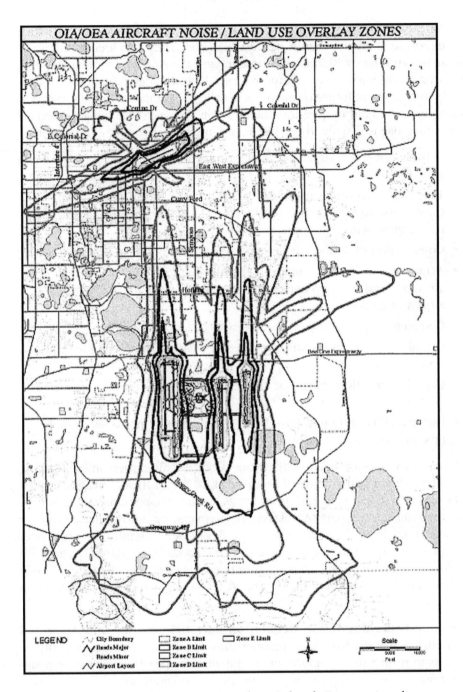

Figure 10-1. *Noise contours surrounding Orlando International and Orlando Executive airports.* (Figure courtesy Greater Orlando Airports Authority)

They can develop noise exposure maps indicating the contours within which noise exposure is greater than a permissible level. They can identify the non-compatible land uses within those contours and develop a plan for mitigating present problems and preventing future ones. Unfortunately, the airport operator's ability to prevent future problems is usually very limited. Unless the airport actually owns the land in question, the authority to make sure it is reserved for a compatible use is usually in the hands of a municipal zoning commission.

Many of these noise abatement programs allowed under current legislation are eligible for federal aid:

- Takeoff and landing procedures to abate noise and preferential runway use to avoid noise-sensitive areas (which must be developed in cooperation with and approved by FAA)
- Construction of sound barriers and soundproofing of buildings
- Acquisition of land and interests therein, such as avigation easements, air rights, and development rights to ensure uses compatible with airport operation
- Complete or partial voluntary curfews
- Denial of airport use to aircraft types or classes not meeting federal noise standards
- Capacity limitations based on the relative noisiness of different types of aircraft
- Differential landing fees based on FAA-certificated noise levels or on time of arrival and departure

The FAA provides assistance to airport operators and air carriers in establishing or modifying flight paths to avoid noise-sensitive areas. In some cases, aircraft can be directed to use only certain runways, to stay above minimum altitudes, or to approach and depart over lakes, bays, rivers, or industrial areas rather than residential areas. Procedures might be developed to scatter the noise over several communities through some "equitable" rotation program. These noise abatement procedures might have a negative effect on airport capacity. They might require circuitous routing of aircraft or use of a runway configuration that is less than optimum with respect to capacity.

In addition, **Federal Aviation Regulation Part 161—Notice and Approval of Airport Noise and Access Restrictions** provides guidelines for airport operators to restrict the operation of certain aircraft operations that have a significant adverse noise impact on the surrounding community.

Air quality

Although there is evidence that aircraft engine emission constitutes less than 1 percent of the total air pollutants in a typical metropolitan area, this facet of the environmental impact of airport operations cannot be overlooked in the development of the airport master plan. It is rather evident to the observer on the ground that exhaust smoke does exist and that contaminants are emitted into the environment.

Federal regulations concerning air quality date back to the **Clean Air Act of 1970,** established to protect the nation's air quality and protect the public's health. The act recognizes five major pollutants that require emissions regulation. They are sulfur dioxide (SO_2), suspended particulate matter, nitrogen oxide (NOx), carbon monoxide (CO), and volatile organic compounds such as hydrocarbons, and hazardous air pollutants such as asbestos, inorganic arsenic, beryllium, mercury, vinyl chloride, benzene, and radionuclides.

The majority of emissions that contribute to the reduction in air quality around airports are from aircraft engines as well as ground vehicles operating both on and to and from the airport property. In addition, industrial facilities and operations associated with airports, including generators, fossil fueled equipment, deicing materials, painting materials, paving operations, fuel dispensing operations, and construction activity all contribute to emissions of concern to air quality.

Reduction or mitigation of air quality impacts from airports is generally associated with raising the efficiency of airport operations. For example, more efficient aircraft taxi operations, which minimize the total time and distance aircraft are burning fuel on airport property, will reduce the volume of pollutants from engine operations. In addition, the use of mass transit systems, rather than private automobiles for travel to and from airport property, by passengers and employees, will contribute to reduced emissions generated from automobile use.

Water quality

An airport can be a major contributor to water pollution if suitable treatment facilities for airport wastes are not provided. Sources of water pollution are domestic sewage from airport facilities, industrial wastes such as fuel spills, and high temperature water degradation from various power plants at the airport. In addition, runoff from deicing operations contributes to the collection of pollutants in the surrounding water table.

In 1977, Congress passed the **Clean Water Act** as an amendment to the **1972 Federal Water Pollution Control Act.** This act authorized the issuance of regulations to prevent discharges of pollutants into navigable and nonnavigable waterways, rivers, streams, and creeks. In accordance with the Clean Water Act, airports are required to prevent the discharge of any contaminated runoff into any drainage system that empties into these water sources unless a specific permit is obtained by the Environmental Protection Agency or other authorized body.

Hazardous waste emissions

The Environmental Protection Agency defines waste as any solid, liquid, or contained gaseous material that is no longer used, and is either recycled, thrown away, or stored until enough is accumulated to treat or be disposed in another manner. Hazardous wastes are those that can cause injury or death to people or animals, or damage or pollute land, air, or water. Waste may also be considered hazardous if it exhibits any of four characteristics: ignitability, corrosiveness, reactivity, or toxicity.

Airports are sources of various emissions that may be considered hazardous waste, including fuel, deicing and other liquid runoff, used oil, corroded electric components, chemicals, paints, solvents, lavatory waste, and other solid, liquid, and gaseous materials.

Generators of hazardous waste are classified under the **Resource Conservation and Recovery Act of 1976.** The act classifies hazardous waste generators as follows:

- *Conditionally exempt small-quantity generators.* These generate less than 100 kg per month of hazardous waste.
- *Small-quantity generators.* Machines that generate more than 100 kg, but less than 1,000 kg, of hazardous waste per month.
- *Large-quantity generators.* Generators that generate more than 1,000 kg of hazardous waste per month.

Airports that accumulate hazardous waste must provide storage containment units that prevent the release of waste into the surrounding environment. Hazardous waste may be stored on airport property only temporarily, typically for no longer than 180 days, before it must be disposed of in a certified location off-site, or properly treated.

Externalities

In addition to environmental impacts generated directly from airport operations, airport management should also be concerned with the environmental impacts of activities that occur as a result of operations from other sources, as an indirect result of an airport's presence. These impacts are known as **externalities.** One example of an externality would be the environmental impacts resulting from the operation of a factory, which had been located in a region near the airport simply because of the airport's presence. Another example includes the increase of automobile traffic in the vicinity of the airport, created as a result of the operation of hotels, restaurants, gas stations, and other facilities that tend to appear near airports. Although these activities are hardly the responsibility of the airport manager and the airport manager does not have any authority over the operation of these activities, the airport operator typically is charged with these environmental issues in the form of externalities.

Careful and strategic negotiations with local facility operators, as well as with the local metropolitan planning organizations, may help manage external activity, which in turn may lead to the reduction in external environmental impacts.

Economic and environmental sustainability practices

In an effort to better manage the economic and environmental impacts of airports, airport management has begun to engage in active management of these impacts using what are known as **sustainability practices.**

Examples of economic sustainability practices include hiring a local workforce and contracting with local vendors, making contributions to the surrounding community, and investing in research and development that may lead to future improvements in the efficiency of the airport.

Environmental sustainability practices involve an active records management and analysis of environmental impacts. Findings from these analyses have led airports to develop sophisticated environmental management programs including waste management, recycling, stormwater collection and reuse, and alternative fuel usage for ground service vehicles. Terminal buildings themselves are being retrofitted with green building in mind; that is, buildings that are energy efficient minimize water usage and limit waste. Green buildings aim to become **LEED (Leadership in Energy and Environmental Design) certified.**

Social responsibilities

While airports are primarily considered transportation facilities, they do play a role in the social fabric of a community. Airports are representatives of their communities, and while being good representatives to visitors, airports must also be good neighbors of the community. As such, airports should be, in one way or another, active members of their communities.

Strategies that airports have used to contribute to their communities vary widely, but have included hosting community cultural projects and programs; providing educational programs to local schools, colleges, and universities; and sponsoring local events. Airport management must also take an active role in ensuring the well-being of all of its users, including providing a safe, healthy, and rich environment to work, travel, and visit.

Concluding remarks

Whether it be concerning economic, political, or environmental issues, airport management must be prepared to interact with the community that it serves, including tenants that provide air transportation, suppliers and service providers, nonaeronautical tenants, the public who use the airport, and those in the community who never even see the airport. The challenge for airport management is to understand all the rules, regulations, and policies governing each airport parties' concerns, and provide an environment that is economically and socially beneficial to all.

Airports that are successful in managing these roles are known to be significant positive contributors to their communities. It should be the goal of every airport management team to make such contributions, and hence receive community support for current airport operations and future airport planning.

Key terms

NEF (noise exposure forecast)

CNEL (community noise equivalent level)

Ldn (day/night noise level)

INM (Integrated Noise Model)

externalities

sustainability practices

LEED (Leadership in Energy and Environmental Design) certification

Key acts

1969—National Environmental Policy Act

1970—Clean Air Act

1972—Federal Water Pollution Control Act

1976—Resource Conservation and Recovery Act

1977—Clean Water Act

1990—Airport Noise and Capacity Act

Key federal aviation regulations

FAR Part 36—Certificated Airplane Noise Levels

FAR Part 150—Airport Noise Compatibility Planning

FAR Part 161—Notice and Approval of Airport Noise and Access Restrictions

Questions for review and discussion

1. How do airports contribute to the economic prosperity of the communities they serve?

2. What types of economic activity are said to be directly generated by airport activity?

3. What types of economic activity are said to be indirectly generated by airport activity?

4. What is a multiplier effect?

5. How have airport–airline relations changed since the years before airline deregulation?

6. How are contracts between airports and the concessions that serve the airport negotiated?

7. What issues arise between airport management and the general aviation community they serve?

8. When did airport noise become a major environmental issue?

9. What Federal Aviation Regulations are concerned with aircraft and airport noise issues?

10. What are the different methods of estimating noise impacts of airport activity?

11. What are the most common noise abatement strategies employed at airports?

12. How does FAR Part 161 aid airports in reducing noise impacts?

13. What are the most common pollutants affecting air quality emitting from airport activity?

14. How is hazardous waste classified?

15. What does the Clean Water Act do to reduce hazards to water quality from airport activity?

16. What are externalities?

17. What are airports doing to become more economically and environmentally sustainable?

Suggested readings

Air Quality Procedures for Civilian Airports & Air Force Bases. Washington, D.C.: Federal Aviation Administration, April 1997.

Integrated Noise Model, Federal Aviation Administration, available via the Internet at http://www.aee.faa.gov/Noise/inm/index.htm.

Regional Multipliers: A User Handbook for the Regional Input-Output Modeling System (RIMS II). Washington, D.C.: U.S. Department of Commerce, March 1997.

Airport Sustainability Practices, ACRP Synthesis 10, National Academies Transportation Research Board, Washington, D.C., 2008.

11

Airport planning

Outline

- Introduction
  - Defining the planning horizon
- Airport system planning
  - National-level system planning
  - Regional-level system planning
  - State-level system planning
- The airport master plan
  - Objectives of the airport master plan
  - Elements of the master plan
  - Inventory
  - Historical review of airports and facilities
    - Airspace structure and NAVAIDs
    - Airport-related land use
    - Aeronautical activity
    - Socioeconomic factors
- The airport layout plan
- Forecasting
  - Qualitative forecasting methods
  - Quantitative methods
  - Regression analysis
  - Forecasts of aviation demand
    - Civil airport users
    - Operational activity
- Facilities requirements
  - Aircraft operational requirements
  - Capacity analysis

- Design alternatives
  - Site selection
  - Runway orientation and wind analysis
    - Identifying the Airport Reference Code on the basis of critical aircraft
    - Analyzing historical wind data for the airfield
  - Airspace analysis
  - Surrounding obstructions
  - Availability for expansion
  - Availability of utilities
  - Meteorological conditions
  - Economy of construction
  - Convenience to population
    - Noise
  - Cost comparisons of alternate sites
  - Terminal area plans
    - Terminal area factors
    - Steps involved in determining space requirements
  - Airport access plans
- Financial plans
  - Economic evaluation
  - Break-even need
  - Potential airport revenue
  - Final economic evaluation
- Land use planning
  - Land uses on the airport
  - Land uses around the airport
- Environmental planning

Objectives

The objectives of this section are to educate the reader with information to:

- Define the various types of airport planning studies.
- Understand the concepts of national-, regional-, and state-level system planning.
- Describe the different elements of the airport master plan.

- Be familiar with an airport layout plan.
- Describe the various qualitative and quantitative forecasting methods used in airport planning.
- Understand runway orientation planning using wind rose analysis.
- Describe the factors that are considered in terminal area planning.
- Identify the considerations involved in financial planning of an airport.
- Describe the various processes involved with airport environmental planning.

Introduction

Along with the multitude of responsibilities and tasks associated with operating an airport on a day-to-day basis, airport management is also ultimately responsible for the significant responsibility of providing a vision for the future of the airport. On a larger scale, municipalities that are served by more than one airport, as well as individual states and even the United States as a whole, are handed the responsibility of strategically planning for a coordinated system of airports to best meet the future needs of the traveling public.

Airport planning may be defined as the employment of an organized strategy for the future management of airport operations, facilities designs, airfield configurations, financial allocations and revenues, environmental impacts, and organizational structures. There are various types of airport planning studies, including:

- **Facilities planning,** which focuses on future needs for airfield infrastructure such as runways, taxiways, aircraft parking facilities, associated lighting, communication and navigational systems, terminal buildings and facilities, parking lots, ground access infrastructure, and support facilities such as fuel farms, power plants, and nonaeronautical land uses such as office parks, hotels, restaurants, or rental car locations.
- **Financial planning,** which is concerned with predicting future revenues and expenses, budgeting resources, and planning for financial assistance through grant programs, bond issues, or private investment.
- **Economic planning,** which considers the future of economic activity, such as trade and commerce, and the activity of industries that exist on airport and off-airport property and are either a direct or indirect result of airport operations.
- **Environmental planning,** which concentrates on maintaining or improving existing environmental conditions in the face of changes in future airport activity. Environmental planning includes land

use planning, noise mitigation, wetland reclamation, and wildlife preservation.

- **Organizational planning,** which entails the management of future labor requirements and organizational structures for the airport administration, staff, and associated labor force.
- **Strategic planning,** which encompasses all other planning activities into a coordinated effort to maximize the future potential of the airport to the community.

Defining the planning horizon

The planning of airport operations, or any activities for that matter, is defined in part by the length of time into the future management considers in its planning. The length of time into the future that is considered is termed the **planning horizon.** Different planning efforts require different planning horizons. For example, the organizational planning of staffing levels per shift for airport operations may require a 3-month planning horizon, but certainly not a 20-year planning horizon. On the other hand, facilities planning of an airfield that may include runway construction requires at least a 5-year planning horizon, and certainly not a planning horizon of less than 1 year.

The various types of airport planning studies may be performed on a variety of different levels. Three such levels of planning include system planning, master planning, and project planning.

Airport system planning

Airport **system planning** is a planning effort that considers a collection of airports, either on a local, state, regional, or national level, expected to compliment each other as part of a coordinated air transportation system. Through airport system planning, the objectives of individual airports are set in accordance with the needs of the community by, for example, setting the mission of each airport to serve certain segments of the demand for aviation, such as targeting one airport in a region to handle international commercial air travelers and another airport to handle primarily smaller general aviation aircraft operations.

National-level system planning

Airport planning at the national level is the responsibility of the FAA, whose interests are to provide guidance for development of the vast network of publicly owned airports and to establish a frame of reference for investment of federal funds. These interests are set forth in the **National Plan of Integrated Airport Systems (NPIAS),** a document required under the Airport and Airway Improvement Act of 1982. The NPIAS is a 10-year plan that is revised every

2 years and is closely coordinated with the FAA's 10-year capital investment plan to improve the air traffic control system and airway facilities.

The NPIAS is not a plan in the fullest sense. It does not establish priorities, lay out a timetable, propose a level of funding, or commit the federal government to a specific course of action. Instead, it is merely an inventory of the type and cost of airport developments that might take place during the planning period at airports eligible for federal assistance. It is a tabular, state-by-state presentation of data for individual airports, listed in a common format, indicating location, role, type of service, and level of activity (enplanements and operations) currently and for 5 and 10 years in the future. Projected costs of airport needs in categories—land, paving, lighting, approach aids, terminal, and other—are shown, also at intervals of 5 and 10 years.

Estimates of need contained in NPIAS are developed by comparing FAA national and terminal area forecasts to the present capacity of each airport. Much of the initial determination of need and the regular updating is performed by FAA regional offices, which monitor changes and developments being carried out at the airports. NPIAS is not a simple compilation of local master plans or state airport system plans, although FAA does draw on these documents as sources in forming judgments about future needs and prospective airport improvements.

NPIAS is not a complete inventory of airport needs. The plan contains only "airport development in which there is a potential federal interest and on which federal funds may be spent under the current **Airport Improvement Program (AIP)** or former **Airport Development Aid Program (ADAP)** and the Planning Grant Program." There are two necessary conditions in the test of potential federal interest. First, the airport must meet certain minimum criteria as an eligible recipient for federal aid and, second, the planned improvement at that airport must be of a type that is eligible for federal aid. Eligible projects include such projects as land acquisition for expansion of an airfield, paving for runways and taxiways, installation of lighting or approach aids, and expansion of public terminal areas. Improvements ineligible for federal aid are not included in NPIAS: construction of hangars, parking areas, and revenue-producing terminal areas that airports are expected to build with private, local, or state funds.

NPIAS relates airport system improvements to three levels of need:

Level I. Maintain the airport system in its current condition

Level II. Bring the system up to current design standards

Level III. Expand the system

Maintaining the system includes such projects as repaving airfields and replacing lighting systems; bringing the system up to standards involves such projects as installing new light systems and widening runways; and expanding the system includes construction of new airports or lengthening runways to accommodate larger aircraft.

The classification system is somewhat misleading because it is not as hierarchical as it might appear, and the placement of a type of improvement at a particular program level does not necessarily reflect the priority that will be given a particular project. High-priority projects, those that FAA and a local sponsor agree must be carried out as soon as possible, might not necessarily correspond with level I needs in the NPIAS. An expansion project (level III) at an extremely congested and important airport might be more urgent than bringing a little-used airport up to standards (level II).

Regional-level system planning

In 2004, the FAA released Advisory Circular 150/5070-7—The Airport System Planning Process, marking the first formal guidance to regional and state agencies on the subject, and emphasizing the importance of system planning on regional levels. Figure 11-1 illustrates the elements of the typical airport regional-level system plan.

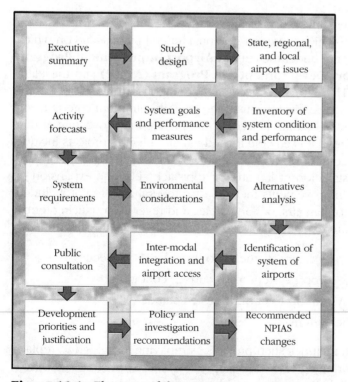

Figure 11-1. *Elements of the airport system plan.* (Source: FAA)

Regional system planning is concerned with air transportation for the region as a whole and must consider traffic at all the airports in the region, both large and small, as well as other modes of transportation used in the region to travel to and from, as well as within, the region. The practice of regional planning has been instituted to deal with questions of resource allocation and use that often arise when the airports in a region have been planned and developed individually and without coordination among affected jurisdictions. Regional planning seeks to overcome the rivalries and the jurisdictional overlaps of the various local agencies involved in airport development and operation. The goal is to produce an airport system that is optimum with respect to regionwide benefits and costs.

Thus, regional airport planning addresses one critical issue, the allocation of traffic among the airports in a region. This can be a sensitive subject. Questions of traffic distribution involve political as well as technical and economic issues, and they can greatly affect the future growth of the airports involved. One airport might be quite busy while another is underutilized. If traffic were to continue growing at the busy airport, new facilities would have to be constructed to accommodate that growth. On the other hand, if some of the new traffic were diverted to an underutilized airport, the need for new construction might be reduced and service to the region as a whole might be improved.

Although a planning agency might decide that such a diversion is in the interest of a metropolitan region and might prepare forecasts and plans showing how it could be accomplished, it might not necessarily have the power to implement these plans. Where airports are competitors, it is probably not reasonable to expect that the stronger will voluntarily divert traffic and revenues to the other. The planning agency would likely have to influence the planning and development process at individual airports so that they will make decisions reflecting the regional agency's assessment of regional needs.

Even where airports in a region are operated by the same authority, allocation of traffic between airports might still be difficult. For example, the Port Authority of New York and New Jersey can implement its planning decision to increase activity at Stewart Airport by creating financial incentives for air carriers, improved ground access, or other measures to increase use of that airport. Implementation of the policy, however, depends not just on control of airport development expenditures but also on the ability to influence the activities of private parties, the air carriers, and passengers.

Regional airport planning authorities may also, if they have planning responsibility for other transportation modes, plan for the airport as part of the regional transportation system. When multimodal planning responsibility resides in one organization, there is greater likelihood that the planning agency will consider airport needs in relation to other forms of transportation in the region. Also, the regional agency may try to improve coordination between the various modes so that, for example, airport developments do not impose an undue burden on surrounding highway facilities or so that advantage can be taken of opportunities for mass transit. For this to happen, however, two conditions are necessary: regionwide authority and multimodal jurisdiction.

State-level system planning

According to the National Association of State Aviation Officials (NASAO), all 50 state aviation agencies carry out some form of airport planning. A majority of these agencies are subdivisions of the state Department of Transportation; in the others, they are independent agencies. Several states have an aviation commission in addition to an aviation agency. The commissions are usually appointed by the governor and serve as policy-making bodies. State involvement in airport planning and development takes several forms: preparation of state airport system plans, funding of local master planning, and technical assistance for local planning.

Airport planning at the state level involves issues that are somewhat different from those of local or regional agencies. State governments are typically concerned with developing an airport system that will provide adequate service to all parts of the state, both rural and metropolitan. Development of airports is often seen as an essential tool for economic development or overcoming isolation of rural areas. Some state aviation agencies (in Ohio and Wisconsin, for example) have set a goal to develop at least one well-equipped airport in each county. Usually the allocation of traffic between airports serving the same community is not at issue. Rather, the issue is deciding how to allocate development funds among candidate communities and to maintain a balance between various parts of the state.

Before 1970, very few states conducted extensive or systematic airport planning. An important stimulus to state agencies to initiate comprehensive planning efforts was provided by the Airport and Airway Development Act of 1970, which set aside 1 percent of airport aid monies from the trust fund for this purpose each year. Most states applied for these funds promptly and typically spent from 1 to 4 years in developing **state aviation system plans (SASP)** under guidelines issued by the FAA, although a few took considerably longer. Most of the states sought assistance from outside consultants in some phase of the planning activity.

State plans typically encompass a planning period of 20 to 30 years. Planning periods are normally divided into short, medium, and long-term planning horizons (usually 5, 10, and 20 years, respectively). In each case, estimates of future needs have been developed by comparing existing facilities with projections of future traffic.

The major feature of the plans, and by far the bulk of each document, is a detailed listing of the actions planned by class of airport and type of improvement. The types of improvements most commonly cited are land acquisition (new sites or expansion of existing airports), pavement repair or improvement (runways, taxiways, aprons, roads, parking), installation of lighting and landing or navigation aids, and building construction (terminals, hangars, administrative facilities).

Although there are surface similarities, SASPs vary greatly in scope, detail, expertise, and planning philosophy. One state system plan might basically be a wish list, prepared primarily because planning funds were available and the state DOT required it. On the other hand, some state agencies regard the SASP as a valuable working document that is kept current and serves as a guide in programming and distribution of state funds.

In many states, programming of funds is somewhat separate from the system planning process. Although the SASP might have a long planning horizon of 20 years or more, the actual award of grants to complete particular projects is on a much shorter time scale. Some state agencies have developed methods for keeping current files on local airport projects planned for the near term, say, 3 years. When airports apply for state aid, or request state assistance in applying for federal aid, the SASP is used to assign priority for a grant award as funds become available. As a rule, only a fraction of the projects outlined in the SASP is undertaken.

Virtually all state plans estimate costs of recommended improvements and identify funding sources. Funding is the primary constraint in implementation of the SASPs. In all states, some sort of consultation, coordination, or review by persons outside the state aviation agency is part of the planning process. Often these are regional economic development or planning agencies created by state government. In many cases, airport planning is part of a general transportation planning process, but methods of interaction and feedback among the modal agencies vary considerably.

Some state agencies are involved in master planning activities for local airports, especially rural or small community airports that do not have the staff to carry out master planning on their own. State agencies might provide technical assistance or actually develop local master plans. Some states also participate in airport planning for major metropolitan areas, although most leave this responsibility with the local airport authority or a regional body. In recent years, state participation in planning at the larger airports has shown some increase, a trend that might be bolstered by current federal policy that earmarks a share of annual trust fund outlays for state aviation planning.

The airport master plan

At the local level, the centerpiece of airport planning is the airport **master plan,** a document that charts the proposed evolution of the airport to meet future needs. The magnitude and sophistication of the master planning effort depends on the size of the airport. At the largest commercial service airports, master planning is a formal and complex process that has evolved to coordinate large construction projects (or perhaps several such projects simultaneously) that can be carried out over a period of up to 20 years. At smaller airports, master planning might be the responsibility of a few staff members with other responsibilities, who depend on outside consultants for expertise and support. At very small airports, where capital improvements are minimal or are made infrequently, the master plan might be a very simple document, perhaps prepared locally but usually with the help of consultants.

An airport master plan presents the planner's conception of the ultimate development of a specific airport. It effectively presents the research and logic from which the plan was evolved and artfully displays the plan in a graphic and written report. Master plans are applied to the modernization and expansion of existing airports and to the construction of new airports, regardless of their size or functional role.

The typical airport master plan has a planning horizon of 20 years. The Federal Aviation Administration notes that for a master plan to be considered valid it must be updated every 20 years or when changes in the airport or surrounding environment occur, or when moderate and major construction may require federal funding.

Objectives of the airport master plan

The overall objective of the airport master plan is to provide guidelines for future development that will satisfy aviation demand and be compatible with the environment, community development, other modes of transportation, and other airports. Specific objectives within this broad framework are as follows:

- To provide an effective graphic presentation of the ultimate development of the airport and of anticipated land uses adjacent to the airport
- To establish a schedule of priorities and phasing for the various improvements proposed in the plan
- To present the pertinent backup information and data that were essential to the development of the master plan
- To describe the various concepts and alternatives that were considered in the establishment of the proposed plan

- To provide a concise and descriptive report so that the impact and logic of its recommendations can be clearly understood by the community the airport serves and by those authorities and public agencies that are charged with the approval, promotion, and funding of the improvements proposed in the airport master plan

Elements of the master plan

Although there is considerable variation in the content of the airport master plan and how it is used, its basic products are a description of the desired future configuration of the airport, a description of the steps needed to achieve it, and a financial plan to fund development. An airport master plan typically consists of the following elements: *inventory, activity forecasts, demand/capacity analysis, facilities requirements, design alternatives,* and *financial plans.* These elements provide a recipe for the airport in its effort to meet the demands of its users and the surrounding community over the airport's master plan. In addition, some master plans include environmental and economic assessments of plans associated with the future plans for the airport.

Inventory

The first step in the preparation of an airport master plan for an individual airport is the collection of all types of data pertaining to the area that the airport is to serve. This includes an inventory of existing airport facilities, area planning efforts that might affect the master plan, and historical information related to their development. This review will provide essential background information for the master plan report. It will also provide basic information for the development of forecasts and facility requirements.

Historical review of airports and facilities

The historical review traces the development of a community's airport facilities and the air traffic that they have served. A description of the airport and the date of construction or major expansion are included. Airport ownership is also mentioned.

The scope of the data collection is generally limited to the area that the master plan airport will serve and to national trends that will affect that area. The planner must carefully research and study data that are available from current sources such as state, regional, and national airport system plans and other local aeronautical studies. Existing airports and their configurations are shown on a base map. Included are all air carrier, general aviation, and military airports in the area.

Airspace structure and NAVAIDs

It is necessary to identify how the airspace is used in the vicinity of each airport and throughout the area, all air navigation aids and aviation communication facilities serving the area, and natural or manmade obstructions or structures that affect the use of the airspace.

The airway and jet-route structures have a significant effect on the utility of existing and future airport locations. The dimensions and configurations of the control zones and transition areas are noted. These segments of controlled airspace are designed to accommodate only specific instrument flight rules (IFR) requirements such as instrument approach, departure, holding, and transition flight maneuvers; thus, the inventory will show the current use of the area's IFR airspace and the balance of the airspace available for future use.

Additional maps or overlays showing the existing airspace structure are included in the inventory. Later in the planning process, proposed expansion of new airports can be related to the existing airspace structure and compatibility verified, or adjustments to the proposed development can be made.

Airport-related land use

An inventory of land uses in the vicinity of each existing airport is necessary so that later in the planning process a determination can be made on the feasibility of expansion and whether an expanded airport will be compatible with the surrounding area and vice versa. Current plans that show existing and planned land uses, highways, utilities, schools, hospitals, and so forth, are obtained from areawide agencies and transportation planning agencies that have jurisdiction over the area the master plan airport is to serve. Current land use is also displayed on a map to assist in later steps of the planning process. Also, if feasible, an estimate of the land values is made.

Normally when considering airport-related land use, a survey will be conducted of all ground travel entering or leaving the airport, including the air travelers, employees, suppliers, and visitors. Information might also be collected on parking and commodity movements. Sufficient data are obtained to establish the travel patterns of airport-oriented trips and to develop relationships that will be used to determine future travel patterns. Copies of zoning laws, building codes, and other regulations and ordinances that might be applicable to the development of an airport master plan are obtained. All of these have an effect on airport related land use.

Aeronautical activity

The principal determinant of future airport system requirements is the amount of aeronautical activity that will be generated in the metropolitan area. A record of current aviation statistics as well as a consideration of historical airport traffic data for such elements as passenger and air cargo traffic, aircraft movements, and aircraft mix is necessary to forecast aeronautical activity. The assessment of these aviation statistics, along with consideration of the socioeconomic attributes for the area, form the basis for forecasts of aeronautical activity for the metropolitan area. The forecasts of aeronautical activity, in turn, form the basis for facilities planning for future requirements.

Aeronautical statistical data include federal, state, and regional statistics as they relate to the master plan airport and the collection of as many local statistics as can be obtained. At the local level, surveys and questionnaires are used to supplement data on operations, frequency, and hours of use of aircraft and origins and destinations of travelers. The primary aviation statistics needed are taken up in this chapter's forecasting section.

Socioeconomic factors

The collection and analysis of socioeconomic data for a metropolitan area helps answer the basic questions regarding the type, volume, and concentration centers of future aviation activity in the region. Accordingly, the determinants (what causes a market to be the size it is) of a market for airports are established. What industries need air transportation? Do they have a need for better air transportation facilities? How many people will be available in the future who possess the income to make use of air service? Will the people and industries having the wherewithal to utilize the airport be there? Because people are associated with a multitude of income-earning and income-spending activities at any particular location from and to which they travel, transportation facilities are needed between those points where the future travel is expected to occur.

The primary forces that measure and help determine economic change and a general rationale for their use in determining air transportation demand follow.

Demography The size and structure of the area's population and its potential growth rate are basic factors in creating demand for air transportation services. The existing population along with its changing age and educational and occupational distributions can provide a primary index of the potential size of the aviation market and resultant airport employment over short-, medium-, and long-range forecast periods. Demographic factors influence the level of airport traffic and its growth, in terms of both incoming traffic from other states, regions, or cities, and traffic generated by the local or regional populations concerned.

Disposable personal income per capita This economic factor refers to the purchasing power available to residents in any one period of time, which is a good indicator of average living standards and financial ability to travel. High levels of average personal disposable income provide a strong basis for higher levels of consumer spending, particularly on air travel.

Economic activity and status of industries   This factor refers to situations within the area the airport serves that generate activity in business aviation and air freight traffic. A community's population, size, and economic character affect its air traffic–generating potential. Manufacturing and service industries tend to generate greater air transport activity than primary and resource industries, such as mining. Much will depend on established and potential patterns of internal and external trade. In addition, other aviation activities such as agricultural and instructional flying and aircraft sales are included in this factor.

Geographic factors   The geographic distribution and distances between populations and commerce within the area that the airport serves have a direct bearing on the type of transportation services required. The physical characteristics of the land and climatic differences are also important factors. In some cases, alternative modes of transportation might not be available or economically feasible. Furthermore, physical and climatic attractions assist in determining focal points for vacation traffic and tourism and help in establishing the demand for air services that they generate.

Competitive position   The demand for air service also depends on its present and future ability to compete with alternative modes of transportation. Also, technological advances in aircraft design and in other transportation modes, as well as industrial and marketing processes, can create transportation demands that have not previously existed.

Political factors   The granting of new traffic rights and routes for international air service will influence the volume of traffic at an airport. Demand for air transportation also depends on government actions such as the imposition of taxes and other fees. In addition, government might support other modes of transportation, which might result in changes in demand for air transportation services.

Community values   A very important factor in the airport master planning process is the determination of the attitude of the community toward airport development. Poor airport-community relations, unless they are changed, could influence the ability to implement an airport master plan. On the other hand, recognition by the community of the need for progress in the development of air transportation can have a positive influence in minimizing complaints; thus, it is necessary to place airport development in its proper perspective relative to community values.

The airport layout plan

Even though a narrative description of the airport environment is a necessary part of an airport master plan inventory, a graphical representation is also required. This graphical representation is known as the **airport layout plan,** or **ALP.**

The airport layout plan is a graphic presentation to scale of existing and proposed airport facilities and land uses, their locations, and the pertinent clearance and dimensional information required to show conformance with applicable standards. It shows the airport location, clear zones, approach areas, and other environmental features that might influence airport usage and expansion capabilities.

The airport layout plan also identifies facilities that are no longer needed and describes a plan for their removal or phaseout. Some areas might be leased, sold, or otherwise used for commercial and industrial purposes. The plan is always updated with any changes in property lines; airfield configuration involving runways, taxiways, and aircraft parking apron size and location; buildings; auto parking; cargo areas; navigational aids; obstructions; and entrance roads. The airport layout plan drawing includes the following items: the airport layout, location map, vicinity map, basic data table, and wind information.

The airport layout is the main portion of the drawing. It depicts the existing and ultimate airport development and land uses drawn to scale and includes as a minimum the following information:

- Prominent airport facilities such as runways, taxiways, aprons, blast pads, extended runway safety areas, buildings, NAVAIDs, parking areas, roads, lighting, runway marking, pipelines, fences, major drainage facilities, segmented circle, wind indicators, and beacons
- Prominent natural and man-made features such as trees, streams, ponds, rock outcrops, ditches, railroads, powerlines, and towers
- Outline of revenue-producing non-aviation-related property, surplus or otherwise, with current status and use specified
- Areas reserved for existing and future aviation development and services such as for general aviation fixed-base operations, heliports, cargo facilities, airport maintenance, and so forth
- Areas reserved for nonaviation development, such as industrial areas, motels, and so forth
- Existing topographic contours
- Fueling facilities and tiedown areas
- Facilities that are to be phased out
- Airport boundaries and areas owned or controlled by the sponsor, including avigation easements
- Airport reference point with latitude and longitude given on the basis of the U.S. Geological Survey grid system
- Elevation of runway ends, high and low points, and runway intersections

- True azimuth of runways (measured from true north)
- North point—true and magnetic
- Pertinent dimensional data—runway and taxiway widths and runway lengths, taxiway-runway-apron clearances, apron dimensions, building clearance lines, clear zones, and parallel runway separation

The **location map** shown in the lower-left-hand side of the airport layout plan drawing is drawn to scale and depicts the airport, cities, railroads, major highways, and roads within 25 to 50 miles of the airport.

The **vicinity map** shown in the upper-left-hand side of the airport layout plan drawing shows the relationship of the airport to the city or cities, nearby airports, roads, railroads, and built-up areas (Fig. 11-2).

The **basic data table** contains the following information on existing and ultimate conditions where applicable:

- Airport elevation (highest point of the landing areas)
- Runway identifications
- Percent effective runway gradient for each existing and proposed runway
- Instrument landing system (ILS) runway when designated, dominant runway otherwise, existing and proposed
- Normal or mean maximum daily temperature of the hottest month

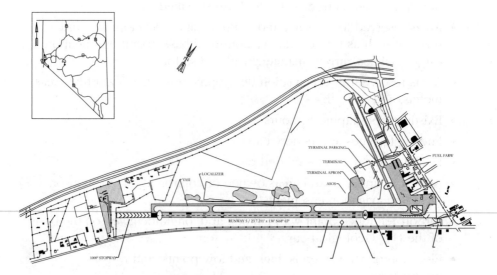

Figure 11-2. *Example of ALP layout of airfield and facilities.*

- Pavement strength of each runway in gross weight and type of main gear (single, dual, and dual tandem) as appropriate
- Plan for obstruction removal, relocation of facilities, and so forth

In addition, a wind rose (described in detail later in this chapter) is always included in the airport layout plan drawing with the runway orientation superimposed. Crosswind coverage and the source and period of data are also given. Wind information is given in terms of all-weather conditions, supplemented by IFR weather conditions where IFR operations are expected.

Airport layout plans also include to scale diagrams of all FAR Part 77 surfaces noise impacted areas, and detailed-to-scale drawings of major facilities at the airport, including terminal buildings, aircraft and automobile parking facilities, ground access roads, and public transit infrastructure, such as rail systems (Fig. 11-3).

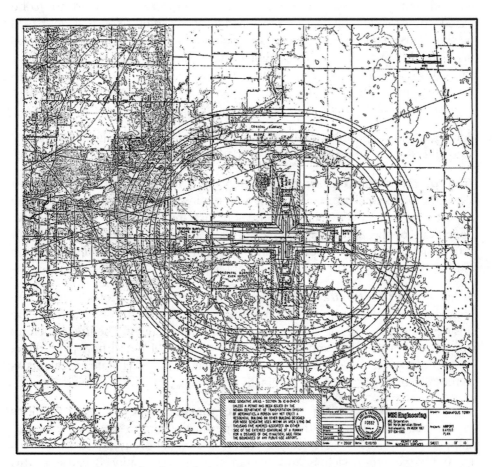

Figure 11-3. *Example of ALP layout illustrating FAR Part 77—surfaces.*

Forecasting

Airport master plans are developed on the basis of forecasts. From forecasts, the relationships between demand and the capacity of an airport's various facilities can be established and airport requirements can be determined. Short-, intermediate-, and long-range (approximately 5-, 10-, and 20-year) forecasts are made to enable the planner to establish a schedule of development for improvements proposed in the master plan.

Two types of forecasting methods are available to assist planners in the decision-making process: qualitative and quantitative.

Qualitative forecasting methods

Qualitative forecasting methods rely primarily on the judgment of forecasters based on their expertise and experience with the airport and surrounding environment. Judgmental predictions of future airport activity tend not to be based on historical data, but by the foresight that certain experts have, based on their knowledge of the current and potential future environment. Qualitative forecasts may almost be thought of as "educated guesses," opinions, or "hunches" of future activity, although they tend to be just as accurate as quantitative methods. Despite this, qualitative forecasts tend to require the support of some quantitative analysis to justify the forecasts to the public.

Four of the more popular qualitative methods include Jury of Executive Opinion, Sales Force Composite, consumer market survey, and the Delphi method.

Jury of Executive Opinion The **Jury of Executive Opinion method** seeks the predictions of management and administration of the airport and the airport's tenants. Given that these persons are the closest to the day-to-day operations of the airport, and typically have extensive experience in airport activity at this airport and perhaps others as well, the Jury of Executive Opinion tends to yield fairly accurate qualitative forecasts.

Sales Force Composite The **Sales Force Composite method** seeks the judgment of airport employees, and the employees of those firms that do business at the airport for their predictions of future activity. The theory behind this method is that the employees, or "sales force," of the airport have direct interaction with the users of the airport, and may provide accurate judgments as to future activity based on this interaction.

Consumer market survey A **consumer market survey** seeks the opinions of the consumer base of the airport, specifically airport passengers, cargo shippers, and users of aeronautically and nonaeronautically based businesses located in the airport vicinity and the surrounding community. Because it's this population that will actually partake in airport activity in the future, soliciting this population's judgment through a consumer market survey is a reasonable qualitative forecasting method.

Delphi method The **Delphi method** is a qualitative forecasting method originally developed by marketing researchers in private sector businesses. In the Delphi method, a group of experts in the field of interest is identified and each individual is sent a questionnaire. The experts are kept apart and are unknown to each other. The independent nature of the process ensures that the responses are truly independent and not influenced by others in the group. This forecasting method involves an iterative process in which all the responses and supporting arguments are shared with the other participants, who then respond by revising or giving further arguments in support of their answers. After the process has been repeated several times, a consensus develops.

Qualitative forecasting for the purposes of airport master planning may, and often do, use one or more of the above methods to derive initial forecast results.

Quantitative methods

Quantitative forecasting methods are those that use numerical data and mathematical models to derive numerical forecasts. In contrast to qualitative methods, quantitative methods are strictly objective. Because only numerical data are used, quantitative methods do not directly consider any judgment on the part of the forecaster. Quantitative methods are either used as stand-alone forecasting methods, or used to support forecasts made under qualitative methods.

Quantitative methods include *time-series* or *trend analysis models,* which forecast future values strictly on the basis of historical data collected over time, and *causal models,* which attempt to make accurate predictions of the future on the basis of how one area of historical data affects another.

Causal models use sophisticated statistical and other mathematical methods that are developed and tested by using historical data. The model is built on a statistical relationship between the forecasted (dependent) variable and one or more explanatory (independent) variables. A statistical correlation analysis is used as a basis for prediction or forecasting. Correlation is a pattern or relationship between two or more variables; the closer the relationship, the greater the degree of correlation.

A **causal model** is constructed by finding variables that explain, statistically, the changes in the variable to be forecasted. The availability of data on the variables, or more specifically their specific values, is largely determined by the time and resources the planner has available. For example, the number of aircraft operations forecast to occur at a general aviation airport may be statistically correlated to the strength of the economy, perhaps measured by the average income of residents in the area surrounding the airport.

The development of causal models that are hoped to make accurate predictions of future activity require significant amounts of research into all areas of the airport environment. Not until a comprehensive causal analysis is performed using a wide variety of potential explanatory characteristics will accurate forecast results be potentially achieved.

Another reasonably sophisticated statistical method of forecasting is time-series or trend analysis, the oldest and in many cases still the most widely used method of forecasting air transportation demand. **Time-series** models are based on a measure of time (months, quarters, years, etc.) as the independent or explanatory variable. This method is used quite frequently where both time and data are limited, such as in forecasting a single variable, for example, cargo tonnage, where historical data are obtained for that particular variable.

Forecasting by time-series or trend analysis actually consists of interpreting the historical sequence and applying the interpretation to the immediate future. It assumes that the rate of growth or change that has persisted in the past will continue. Historical data are plotted on a graph, and a trend line is drawn. Frequently a straight line, following the trend line, is drawn for the future; however, if certain known factors indicate that the rate will increase in the future, the line might be curved upward. As a general rule, there might be several future projections, depending upon the length of the historical period studied.

Airport authorities keep numerous records of data of particular concern to them (enplanements, aircraft movements, number of based aircraft, etc.), and when a forecast is needed, a trend line is established and then projected out to some future period. The accuracy of forecasting by historical sequence in time or trend analysis depends on good judgment in predicting those changing factors that might keep history from repeating.

The values for the forecasted variable are determined by four time-related factors: long-term trends, such as market growth caused by increases in population; cyclical variations, such as those caused by the business cycle; seasonal phenomena, such as weather or holidays; and irregular or unique phenomena, such as strikes, wars, special events, and natural disasters.

Regression analysis

The most widely used mathematical method for performing both time-series and causal quantitative forecasts is regression analysis. **Regression analysis** applies specific mathematical formulas to estimate forecast equations. These equations may then be used to forecast future activity by applying the equations to independent variables that may occur in the future. Regression equations come in many forms. The most common regression equation is one that represents a straight line. The method used to estimate the equation of a straight line that best represents either historical trends or causal relationships is known as ordinary least-squares (OLS) linear regression analysis.

Although based in sophisticated theories of statistics and calculus, OLS linear regression analysis tools are readily available on most personal computer spreadsheet software such as Microsoft Excel, Corel Quattro Pro, or IBM's Lotus 1-2-3. Other common statistical software tools available for personal computers include SPSS, SAS, and a variety programming languages that may be used to create custom regression models. All that is required of the forecaster is to collect appropriate data, enter the data into a software program, and apply the regression tool. Although applying data to today's regression tools is quite simple, proper interpretation and use of regression results require at least a fundamental knowledge of regression modeling from a theoretical perspective. It is suggested that anyone who will actively participate in performing or interpreting quantitative forecast results, such as those found from regression analysis, seek additional knowledge in statistical modeling (Fig. 11-4).

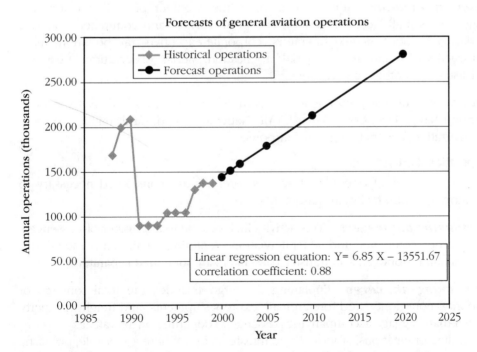

Figure 11-4. *Example of time-series forecast using OLS linear regression.*

Forecasts of aviation demand

Forecasts of aviation demand form the basis for facilities planning. There is a need to know the types of civil airport users, certificated air carriers, commuters, general aviation, and military services where applicable; the types and volume of operational activity, aircraft operations, passengers and cargo, based aircraft, and so forth; and the aircraft fleet mix, jet and large-capacity prop transport aircraft, smaller commercial, corporate, business, and pleasure aircraft, future vertical/short takeoff and landing (V/STOL) aircraft, and so forth.

Civil airport users

The FAA defines the various types of civil airport users as follows.

Air carriers   These airline aircraft operators holding Certificates of Public Convenience and Necessity issued by the former Civil Aeronautics Board and based on authorization from the Department of Transportation to perform passenger and cargo services. This general air carrier grouping includes the major, national, large, and medium regional air carriers.

Commuters   Commuters are noncertificated small regionals that perform scheduled service to smaller cities and serve as feeders to the major hub airports. They generally operate aircraft of less than 12,500 pounds maximum gross takeoff weight.

General aviation   General aviation is the segment of civil aviation that encompasses all facets of aviation except air carriers and commuters. General aviation includes air taxi operators, corporate-executive transportation, flight instruction, aircraft rental, aerial application, aerial observation, business, pleasure, and other special users.

Military   This category encompasses the operators of all military (Air Force, Army, Navy, U.S. Coast Guard, Air National Guard, and military reserve organizations) aircraft using civil airports.

Operational activity

Six major types of operational activity forecasts are considered necessary to determine future facility requirements.

Enplaning passengers   This activity includes the total number of passengers (air carrier, commuter, and general aviation) departing on aircraft at the airport. Originating, stopover, and transfer passengers are identified separately.

Enplaning air cargo   Enplaning air cargo includes the total tonnage of priority, nonpriority, and foreign mail, express shipments, and freight (property other than baggage accompanying passengers) departing on aircraft at an airport, including originations, stopover, and transfer cargo. Where applicable, domestic and international are identified separately.

Aircraft operations   Aircraft operations include the total number of landings (arrivals) at and takeoffs (departures) from an airport. Two types of operations—local and itinerant—are separately identified: *local operations,* performed by aircraft that operate in the local traffic pattern or within sight of the tower and are known to be departing for or arriving from flight in local practice and flight test areas located within a 20-mile radius of the airport and/or control tower and execute simulated instrument approaches or low passes at the airport; and *itinerant operations,* all aircraft arrivals and departures other than local operations described above. Where applicable, domestic and international itinerant operations should be identified separately.

Except for local training flights at some airports, air carrier aircraft movements are itinerant operations. The basic premise underlying the methodology for forecasting air carrier operations by an airport is that a relationship exists between the number of enplaned passengers and cargo shipments and the level of service provided. It is assumed that the number of aircraft seats for transiting and enplaning passengers and the number of flights by type of aircraft have been a function of the traffic demand and traffic characteristics of the community as well as the route structure and operating policies and practices of the individual carriers. It is also assumed that these same factors will continue to determine the level of operations in the future.

Based aircraft   Based aircraft is the total number of active general aviation aircraft that use or might be expected to use an airport as "home base." General aviation–based aircraft are separately identified as single-engine, multiengine, piston, or turbine, or vertical/short takeoff and landing (V/STOL) aircraft.

Busy-hour operations   Busy-hour operations is the total number of aircraft operations expected to occur at an airport at its busiest hour, computed by averaging the two adjacent busiest hours of a typically high-activity day. One definition of a typically high-activity day would be the average day of the busiest month of the year. The operations are identified by major user category, as applicable.

Aircraft fleet mix   Fleet mix is defined as the percentage of aircraft, by type or category, that operate or are based at the airport. Aircraft fleet mix is typically summarized as seating capacity groups for air carrier aircraft and operational characteristics groups for all four of the major airport user categories.

By performing a comprehensive forecast of the above measures of airport activity, using both quantitative and qualitative methods, the airport planner has the ability to incorporate into the master plan airport facilities that will accommodate forecast activity.

Facilities requirements

After an inventory of current facilities has been compiled and future aviation activity has been forecast, the next step in the airport master planning process is the assessment of facility requirements. The study of the demand/capacity relationship involves an estimation of the need to expand facilities and the cost of these improvements. This type of analysis is done in consultation with the airlines and the general aviation community. The analysis is applied to aircraft operations versus airfield improvements, passenger enplanements versus terminal building improvements, cargo tonnage versus cargo facility development, airport access traffic versus access roads and rapid transit facilities, and other improvements as might be appropriate. Airspace in the vicinity of the master plan airport is also analyzed.

Demand/capacity analysis is normally applied to short-, intermediate-, and long-range developments (approximately 5, 10, and 20 years). The analysis is only an approximation of facility requirements, their costs, and savings that will result from reduced delays to airport users as well as anticipated revenues that might be obtained from proposed improvements; thus, demand/capacity analysis will yield preliminary estimates of the number and configuration of runways, areas of apron, number of vehicle parking spaces, and capacities of airport access facilities. Preliminary estimates of economic feasibility may also be obtained. These approximations will provide a basis for developing the details of the airport master plan and for determining the feasibility of improvements considered in the plan.

Aircraft operational requirements

The forecasts of aviation activity will indicate the kinds of aircraft anticipated to use the master plan airport. The frequency of use, passenger/cargo load factors, and lengths of outbound nonstop flights will also be indicated. From this demand data, the planner can ascertain the required physical dimensions of the aircraft operational areas.

Although a capacity analysis provides requirements in terms of numbers of runways/taxiways and so forth, the analysis of aircraft operational requirements allows the determination of runway/taxiway/apron dimensions, strengths, and lateral clearances between airport areas. Of course, both of these analyses are interrelated and are accomplished simultaneously in order to determine system requirements.

Capacity analysis

An analysis of the existing air traffic capacity of the area the master plan airport is to serve will help determine how much additional capacity will be required at the master plan airport. Four distinct elements require investigation, namely, airfield and airspace capacity, terminal area capacity, and ground access capacity.

Airfield capacity is the practical maximum rate of aircraft movements on the runway/taxiway system. Levels of demand that exceed capacity will result in a given level of delay on the airfield (see Chap. 12 for a detailed analysis of airfield capacity). The proximity of airports to one another, the relationship of runway alignments, and the nature of operations [IFR or visual flight rules (VFR)] are the principal interairport considerations that will affect *airspace capacity* of the master plan airport. For example, it is not uncommon in a large metropolitan area to have major or secondary airports spaced so closely that they share one discrete parcel of airspace. In such cases there may be a reduction in the IFR capacity for the airports involved because of the intermixing of traffic within the common parcel of airspace. When this occurs, aircraft, regardless of destination, must be sequenced with the proper separation standards. This reduces the IFR capacity for a specific airport.

Terminal area capacity is the ability of the terminal area to accept the passengers, cargo, and aircraft that the airfield generates. Individual elements within terminal areas must be evaluated to determine overall terminal capacity. Terminal elements included in the analysis are airline gate positions, airline apron areas, cargo apron areas, general aviation apron areas, airline passenger terminals, general aviation terminals, cargo buildings, automobile parking, and aircraft maintenance facilities. The establishment of capacity requirements for the master plan airport will determine the capacity required for airport ground access. A preliminary examination of existing and planned highway and mass transit systems allows a judgment as to the availability of ground access capacity. In determining the volume of people, it is necessary for the planner to establish the percentage relationship between passengers, visitors, and airport employees. This can vary from one urban area to another and from one site to another.

Facility requirements are developed from information obtained in demand/capacity analysis and from FAA advisory circulars and regulations that provide criteria for design of airport components. Demand/capacity analysis yields the approximate number and configuration of runways, number of gates, square footage of terminal buildings, cargo facilities, number of public and employee parking spaces, types of airport access roads, and the overall land area required for the airport. From the mix of aircraft and the number of aircraft operations, general requirements for length, strength, and number of runways; spacing of taxiways; layout and spacing of gates; and apron area requirements can be determined.

Design alternatives

When planning for an airport's future, airport planners develop a series of design alternatives to accommodate forecast levels of demand. These design alternatives are then brought to airport management, the local government, the surrounding community, and often the Federal Aviation Administration to reach a consensus on the recommended design alternative.

The design alternatives for airports may include:

- The selection of an airport on a new yet undeveloped site
- The plans for design and operation of the airfield and local airspace
- The plans for design and operation of the terminal and ground access systems

Site selection

One of the design alternatives for the future of an airport may be to design a new airport on an open, or *greenfield,* site. If this is the case, the first and perhaps most important step in this process is that of proper site selection. The major factors that require careful analysis in the final evaluation of airport sites include runway orientation and wind analysis, airspace analysis, surrounding obstructions, availability for expansion, availability of utilities, meteorological conditions, noise impacts, and cost comparisons of alternative sites.

Runway orientation and wind analysis

Planning an airfield with respect to runway orientation is a nontrivial task. Runway orientation planning consists of three tasks:

1. Identifying the **Airport Reference Code (ARC)** on the basis of an airport's critical aircraft.
2. Analyze historical wind data for the airfield.
3. Apply the Airport Reference Code and historical wind data using a wind rose to find the appropriate orientation of the primary runway and any necessary crosswind runways.

Identifying the Airport Reference Code on the basis of critical aircraft

Every aircraft in use today is limited by the amount of crosswind that may exist in order to land or takeoff. This limit may be found in an aircraft's operating handbook. In general, however, aircraft with shorter wingspans, and slow approach speeds have lower crosswind tolerance limits.

The FAA categorizes aircraft for airports by their approach speeds and wingspans. The wingspan of any given aircraft puts the aircraft in to an "Airplane

Design Group." The approach speed of an aircraft denotes the airports "Aircraft Approach Category." Table 11-1 identifies specifications that determine an aircraft's Airplane Design Group and Aircraft Approach Category.

The combination of Airport Design Group and Aircraft Approach Category make up an Airport's Reference Code. For example, a Cessna 172 aircraft with a 36-foot wingspan and an approach speed of 65 knots would result in an Airport Reference Code of A-I.

For each Airport Reference Code, the FAA has determined the maximum allowable crosswinds for use in planning the orientation of runways at airports. The maximum allowable crosswinds for each reference code are:

- 10.5 knots for Airport Reference Codes A-I and B-I
- 13 knots for A-II and B-II
- 16 knots for A-III, B-III, C-I through C-III, and D-I through D-III
- 20 knots for A-IV through D-VI

The critical aircraft at an airport is the aircraft that operates at least 500 itinerant operations each year whose reference code represents the lowest maximum allowable crosswind. The planning and management of runways is, in fact, based on the maximum crosswind tolerances dictated by the Airport Reference Code associated with the critical aircraft.

Analyzing historical wind data for the airfield

At airports, wind is typically measured by its velocity (in knots), and direction (in degrees from north). Wind direction and velocity data have historically been recorded on an hourly basis at airports and other areas of interest by the National Oceanic and Atmospheric Administration.

Historical wind data are compiled, categorized, and illustrated by means of a graphical tool called a **wind rose.** A wind rose graphically represents wind speed, and direction by a series of concentric rings, which represent wind speed, and spokes, which represent direction. The center of the wind rose represents calm winds. Rings further out from the center of the wind rose represent winds of increasingly stronger velocity. The spokes represent direction from North. The percentage of time that the wind blows between certain directions and between certain speeds is placed in the cells created from the rings and spokes of the wind rose.

A wind rose is designed to provide the airport planner and manager a visual guide to the appropriate direction of the primary runway and any necessary crosswind runways. By overlaying a proposed runway direction over the wind rose, an airport planner can visually identify the direction of the prevailing winds and assess the approximate percentage of time that the

Table 11-1 Airport Reference Code

FAA Reference Code Element 1		FAA Reference Code Element 1	
Aircraft Approach Category	Aircraft Approach Speed (AS) in Knots	Airplane Design Group	Aircraft Wingspan (WS)
A	$AS < 91$	I	$WS < 49$ ft (15 m)
B	$91 \leq AS < 121$	II	49 ft (15 m) $\leq WS < 79$ ft (24 m)
C	15 m $\leq AS < 141$	III	79 ft (24 m) $\leq WS < 118$ ft (36 m)
D	$141 \leq AS < 166$	IV	118 ft (36 m) $\leq WS < 171$ ft (52 m)
E	$166 \leq AS$	V	171 ft (52 m) $\leq WS < 214$ ft (65 m)
		VI	214 ft (65 m) $\leq WS < 262$ ft (80 m)

Source: FAA.

runway orientation will provide less-than-maximum tolerable crosswinds. This is performed by adding up the percentages found within the "runway template." The "width" of the runway template is associated with the maximum crosswind component associated with the Airport Reference Code for the airport.

The necessity of a crosswind runway is determined on the basis of FAA regulations. The FAA requires that all runways at an airport should be oriented so that aircraft may use the airport at least 95 percent of the time with crosswind components not exceeding that of the critical aircraft.

Figure 11-5 illustrates an example of a wind coverage analysis using a wind rose. The wind rose estimates the wind coverage of a north-south runway designed for an aircraft with reference code C-II and a northeast-southwest runway designed for an aircraft with reference code A-I.

Airspace analysis

In major metropolitan areas, it is not uncommon for two or more airports to share common airspace. This factor might restrict the capability of any one airport to accept IFR traffic under adverse weather conditions. Airports too close to each other can degrade their respective capabilities and create a serious traffic control problem. It is important to analyze the requirements and future needs of existing airports before considering construction sites for a new airport.

Wind direction (blowing from)	Velocity between 10 and 16 kn	Velocity between 16 and 20 kn
North	8%	2%
North East	13%	11%
East	11%	8%
South East	6%	1%
South	5%	2%
South West	3%	4%
West	5%	3%
North West	4%	1%

Wind calm 15%

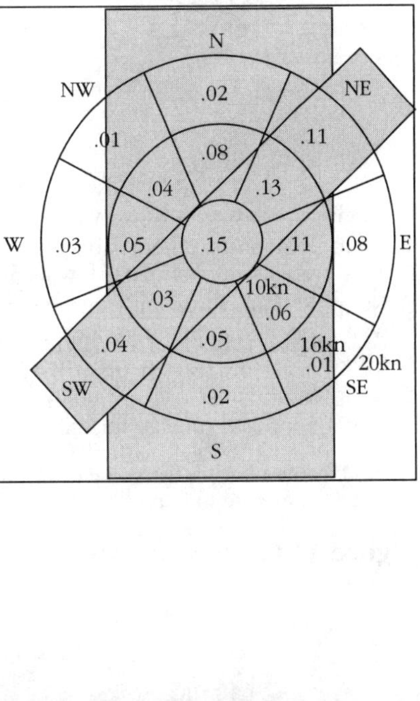

Figure 11-5. *Wind rose analysis.*

Surrounding obstructions

Obstructions in the vicinity of the airport sites, whether natural, existing, or proposed man-made structures, must meet the criteria set forth in Federal Aviation Regulations Part 77—Objects Affecting Navigable Airspace, as described in Chap. 4 of this text. The FAA requires that clear zones at the ends of runways be provided by the airport operator. **Runway protection zones (RPZs)** are areas immediately off the approach ends of runways. The dimensions of the RPZs are shown in Fig. 11-6.

The FAA requires that the airport owner have "an adequate property interest" in the RPZ area in order that the requirements of FAR Part 77 can be met and the area protected from future encroachments. Adequate property interest might be in the form of ownership or a long-term lease or other demonstration of legal ability to prevent future obstructions in the RPZ.

Availability for expansion

Available land for expansion of the airport is a major factor in site selection; however, it is not always necessary to purchase the entire tract at the start because adjacent land needed for future expansion could be protected by lease or option to buy. The Airport and Airway Development Act of 1970 first established funding for communities to acquire land for future airport development.

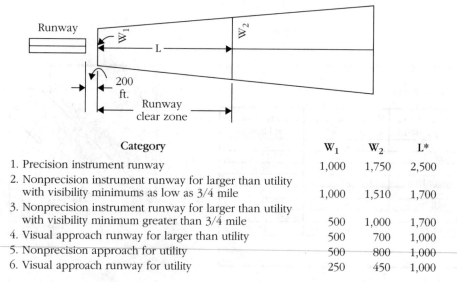

Category	W_1	W_2	L^*
1. Precision instrument runway	1,000	1,750	2,500
2. Nonprecision instrument runway for larger than utility with visibility minimums as low as 3/4 mile	1,000	1,510	1,700
3. Nonprecision instrument runway for larger than utility with visibility minimum greater than 3/4 mile	500	1,000	1,700
4. Visual approach runway for larger than utility	500	700	1,000
5. Nonprecision approach for utility	500	800	1,000
6. Visual approach runway for utility	250	450	1,000

*Length of clear zone is determined by distance required to reach a height of 50 ft. for the appropriate surface.

Figure 11-6. *Dimensions of runway protection zones.* (Source: FAA)

Availability of utilities

Consideration is always given to the distance that electric power, telephone, gas, water, and sewer lines must be extended to serve the proposed site. Cost of obtaining utilities can be a major influence on the site selection.

Meteorological conditions

Sites must be carefully investigated for prevalence of ground fog, bad wind currents, industrial smoke, and smog. A study of wind direction on a year-round basis is always made because prevailing winds will influence the entire design of the airport.

Economy of construction

Soil classification and drainage can have an effect on the cost of construction. Similarly, sites lying on submerged or marshy land are much more costly to develop than those on dry land. Rolling terrain requires much more grading than flat terrain. The site that is more economical to construct will be a deciding factor in the final selection.

Convenience to population

An airport must be convenient for the people who will use its facilities. Much in the same way that shopping centers derive their success from convenient access and parking, the airport too must be accessible in terms of time, distance, and cost of transportation. As a rule of thumb, the airport should be located no more than 30 minutes from the majority of potential users. Consideration in site selection is always given to the proximity of railroads, highways, and other types of transportation for movement and transfer of cargo and passengers.

Noise

Noise is the most predominant objection raised by opponents to new airports and airport expansion projects. Numerous efforts are being made by industry and government to seek new and better ways to reduce aircraft sound levels. Many of the older jet aircraft are now being retrofitted with noise kits that are designed to reduce noise. Engine manufacturers are exploring new engineering concepts and designs that will reduce this source of noise to an absolute minimum. Pilots of airliners are required to maintain certain power settings and to fly prescribed routes that reduce noise levels in the vicinity of takeoff and landing areas. Noise certification standards have been established by the FAA for new aircraft.

Cost comparisons of alternative sites

A quantitative and qualitative comparison of the aforementioned factors is made from the standpoint of cost. Quantitative analysis includes an evaluation of the costs of land acquisition and easements, site developments, major utilities, foundations, access facilities, ground travel for users, and effects on surrounding areas such as noise, air and water pollution, and safety. Qualitative evaluation considers accessibility to users, compatible land uses, expansion capabilities, and air traffic control compatibility.

Terminal area plans

The primary objective of the terminal area plans is to achieve an acceptable balance between passenger convenience, operating efficiency, facility investment, and aesthetics. The physical and psychological comfort characteristics of the terminal area should afford the passenger orderly and convenient progress from automobile or public transportation through the terminal to the aircraft and back again. A detailed description of airport terminal geometries, along with the facilities that are found in airport terminals, is provided in Chap. 6 of this text.

One of the most important factors affecting the air traveler is walking distance. It begins when the passenger leaves the ground transportation vehicle and continues on to the ticket counter and to the point at which he or she boards the aircraft. Consequently, terminals are planned to minimize the walking distance by developing convenient auto parking facilities, convenient movements of passengers through the terminal complex, and conveyances that will permit fast and efficient handling of baggage. The planner normally establishes objectives for average walking distances from terminal points to parked aircraft. Conveyances for passengers such as moving walks and baggage handling systems are also considered.

The functional arrangement of the terminal area complex with the airside facilities is designed so as to be flexible enough to meet the operating characteristics of the airline industry for handling passengers and for fast ground servicing of aircraft so that minimum gate occupancy time and maximum airline operating economy will be achieved.

The final objective of the terminal area plans is to develop a complex that provides all necessary services within an optimum expenditure of funds from the standpoints of capital investment and maintenance and operating costs. This takes into account flexibility and costs that will be required in future expansions of the terminal area.

Terminal area factors

In the selection of a terminal area concept, the following factors are taken into consideration by airport planners:

- Passengers
  - Adequate terminal area curb space for private and public transportation

- Minimum walking distance—terminal entrance to check-in facilities
- Minimum walking distance—check-in facilities to security screening and security screening to aircraft gates
- Ample space for passenger queuing at passenger processing facilities
- Passenger transportation—where long distances must be traversed
- Pedestrian walkways to aircraft—as backup to mechanical transportation systems for passengers
- Efficiency of passenger interline connection
- Baggage handling—enplaning and deplaning
- Convenient hotel-motel accommodations
- Efficient handling of visitors at the airport
- Passenger vehicles
  - Public automobile flow separation from service and commercial traffic
  - Public transportation to and from the airport
  - Public parking—long term (3 hours or more) and short term (less than 3 hours)
  - Airport employee parking
  - Airline employee parking
  - Public auto service area
  - Rental car parking and service areas
- Airport operations
  - Separation of apron vehicles from moving and parked aircraft
  - Passenger flow separation in the terminal building (departing and arriving)
  - Passenger flow separation from apron activities
  - Concession availability and exposure to public
  - Airfield security and prevention of unauthorized access to apron and airfield
  - Air cargo and freight forwarder facilities
  - Airport maintenance shops and facilities
  - Airfield and apron drainage
  - Airfield and apron utilities
  - Utility plants and heating and air-conditioning systems
  - Fire and rescue facilities and equipment

- Aircraft
  - Efficient aircraft flow on aprons and between terminal aprons and taxiways
  - Easy and efficient maneuvering of aircraft parking at gate positions
  - Aircraft fueling
  - Heliport areas
  - General aviation areas
  - Noise, fumes, and blast control
  - Apron space for staging and maneuvering of aircraft service equipment
- Safety
  - Enplaning and deplaning at aircraft
  - Elevators, escalators, stairs, and ramps as to location, speed, and methods of access and egress
  - People-mover systems as to location, speed, methods of access and egress
  - Road crossings as to protection of pedestrians
  - Provisions for disabled persons.

The terminal building is a complex major public-use facility serving the needs of passengers, air carriers, visitors, airport administration and operations, and concessionaires. Clearly, different objectives and space requirements are sought by each of these groups of users. Conflicts in objectives and space requirements often arise in planning passenger handling systems.

It has been recognized in airport planning that two sets of space criteria are needed. One is a set of criteria that can be used for general concept evaluation. This is a set of general considerations that the planner uses to evaluate and select among alternative concepts in a preliminary fashion prior to any detailed design and development. The other set of space criteria is the actual criteria for design and development. In this set, specific performance measures are needed in order to evaluate the likely operation of well-developed plans.

Although general concept evaluation criteria can be developed on the basis of experience and observation of existing terminal buildings, the more specific design and development criteria require the use of a number of analytic techniques for their generation. These include network models, critical path methods (CPM), queuing models, and simulation models.

The most important general concept evaluation criteria for space requirements are:

- Ability of the facility to handle expected demand
- Compatibility with the expected aircraft fleet mix
- Flexibility for growth and response to advances in technology
- Compatibility with ground access systems
- Compatibility with the airport master plan
- Minimal directional confusion caused by the physical layout of the building
- Cost considerations
- Sociopolitical and environmental considerations

The most important specific design and development criteria for space requirements are:

- Processing costs per passenger
- Walking distances for various types of passengers
- Passenger queuing and service rates at processing points
- Occupancy levels for lounges and corridors
- Aircraft maneuvering delays and costs
- Construction costs
- Operating and maintenance expenses
- Estimated revenues from concessionaires

Steps involved in determining space requirements

Once the sets of criteria have been established, the next determination is the actual space requirements for the various users. For planning purposes, the FAA has historically recommended the following steps to determine space requirements for facilities in the airport terminal:

1. *Estimate passenger demand levels.* This first step involves a forecast of the annual passenger volume. Next is a determination of the approximate hourly volume. Planners refer to this figure as the typical peak-hour passenger volume or design volume. The peak hour of an average day in the peak month is commonly used as the hourly design volume for terminal space. This figure is generally in the range of 3 to 5 percent of the annual volume.

   The type of passenger is broadly classified as domestic, international, or transfer. A further breakdown by type would include such items as (a) arriving or departing; (b) with or without checked baggage; (c) mode of access to or egress from the airport—automobile, bus, limousine, train, or helicopter; (d) scheduled, charter, or general aviation flight; and (e) any other characteristics that might be relevant to the particular airport.

2. *Estimate demand for particular facilities.* A matrix is developed matching passenger types and volumes with the various facilities in the terminal. These would include such areas as the ticket lobby, restrooms, baggage claim area, waiting rooms, eating facilities, and so forth. Areas for servicing international passengers would include public health, immigration, customs, agriculture, and visitor waiting areas. By summing the volume of passengers in rows corresponding to the facilities, it is possible to approximate the total load on each facility.

3. *Determine space requirements.* The actual space requirements are determined by multiplying the estimated number of passengers using each facility with an empirical factor to arrive at the approximate area or capacity of the facility required. The empirical factor or constant is based upon experience acquired by planners and contemplates a reasonable level of service and occupancy.

It should be noted that the above method of estimating space requirements in airport terminals is appropriate strictly for conceptual planning. To more accurately estimate the size and location of such facilities, a unique understanding of the flows of passengers and baggage for a specific airport terminal must be gained.

Terminal area planning takes into consideration expansion capabilities to accommodate increasing passenger volumes and aircraft gate positions. In addition, a proper balance between capital investment, aesthetics, operation, maintenance costs, and passengers and airport revenues is considered.

The FAA Advisory Circular 150/5360-13—*Planning and Design Guidelines for Airport Terminal Facilities* provides terminal area planning and design specifications based on recommended planning criteria for major terminal area components. Information for terminal requirements is obtained from the air carriers, general aviation interests, airport concessionaires, airport management, and special technical committees. The criteria are analyzed and agreed upon by all parties involved before they are incorporated in the master plan. It is essential that coordination with airport interests and users be effected before the final selection of a terminal area concept is made.

Airport access plans

The airport access plans are an integral part of the master planning process. These plans indicate proposed routing of airport access to the central business district and to points of connection with existing or planned ground transportation arteries. All modes of access are considered, including highways, rapid transit, and access by vertical and short takeoff and landing (V/STOL) aircraft. The estimated capacity requirement for the various modes considered is determined from forecasts of passengers, cargo, and aircraft operations. The airport access plans normally are general in nature because detailed plans of access outside the boundaries of the airport will be developed by highway departments, transit authorities, and comprehensive planning bodies.

Financial plans

The financial plan is an economic evaluation of the entire plan of development. It looks at the master plan activity forecasts from the point of view of revenues and expenditures, analyzing the airport's balance sheet over the planning period to ensure that the airport sponsor can afford to proceed. A corollary activity in this phase is the consideration of funding sources and financing methods for the proposed development. Questions to be addressed include which portions will be funded through federal grants-in-aid; the size and timing of bond issues; and the revenue from concessionaire rents, parking fees, landing fees, and so on. A more complete description of airport financial strategies may be found in Chap. 9 of this text.

Economic evaluation

Although the primary objective of the airport master plan is to develop a design concept for the entire airport, it is essential to test the economic feasibility of the plan from the standpoints of airport operation and individual facilities and services. Economic feasibility will depend on whether the users of the airport improvements programmed under the plan can produce the revenues (as might be supplemented by federal, state, or local subsidies) required to cover annual cost for administration, operation, and maintenance. This must be determined for each stage of development scheduled in the master plan. This consideration includes the cost of capital to be employed in financing the improvement, the annual operating costs of facilities, and prospective annual revenues.

This preliminary cost estimate for each of the proposed improvements provides the basic capital investment information needed for evaluating the feasibility of the various facilities. Estimated construction costs are adjusted to include allowance for architect and engineering fees for preparation of detailed plans and specifications, overhead for construction administration, allowance for contingencies, and allowance for interest expenses during construction. Estimated costs of land acquisitions as well as the costs of easements required to protect approach and departure areas are included. If the master plan provides for the expansion of an existing airport, the cost of the existing capital investment might be required to be added to the new capital costs.

The airport layout plan also indicates the stage development of the proposed facilities. The drawings are normally written with appropriate legends to indicate staging shown on the plan, either on single or separate sheets. Charts that show the schedule of development for various items of the master plan are developed for inclusion in the master plan report.

Break-even need

The annual amount that is required to cover cost of capital investment and costs of administration, operation, and maintenance can be called the break-even need. The revenues required to produce the break-even need are derived from user charges, lease rentals, and concession revenues produced by the airport as a whole. In order to make sure that the individual components of the airport are generating a proper share of the required annual revenues, the airport can be divided into cost areas to allow allocation of costs to such areas following generally accepted cost accounting principles. Carrying charges on invested capital include depreciable and nondepreciable items.

Nondepreciable investment items are those that have a permanent value even if the airport site is converted to other uses. Nondepreciable investment items include the cost of land acquisition, excavation and fill operations, and road relocations that enhance the value of the airport site. The annual cost of capital invested in nondepreciable assets depends in the first instance on the source of capital used. If revenue or general obligation bonds have been issued to acquire the asset, the total of the principal and interest payments and required reserves or coverage payments called for by the bonds is used. Assets acquired with airport operating surpluses of prior years, general tax revenues, or gifts do not ordinarily impose a cash operating requirement and the treatment of these investments will require a decision by the operator based upon legal considerations and financial operating objectives of the airport. Interest or depreciation charges are not required to be recovered on amounts secured by the airport under the Airport and Airway Improvement Act of 1982 or previous acts. Treatment of funds acquired under state grants-in-aid programs are governed by the terms of the act involved.

The annual cost of capital invested in plant and equipment (as distinguished from land) can be regarded as depreciable investment. The annual charge for depreciation depends on the useful life of the asset and the source of capital used in acquiring the asset. If payments of principal and interest on bonds issued to pay for the asset are required over a shorter period than the useful life of the asset, this schedule would govern and form the basis for depreciation charges unless other revenues are available to service the debt. Depreciation charges for capital assets acquired with operating surpluses of prior years, general tax revenues, or gifts do not ordinarily impose a cash operating requirement on the operator, and the treatment of this investment will require a policy decision by the operator. Interest or depreciation charges are not required to be recovered on amounts secured under the Airport and Airway Improvement Act of 1982 or previous acts. Funds obtained under state grants-in-aid are governed by the terms of the act involved.

Estimated expenses for administration, operation, and maintenance are developed for each airport cost area on the basis of unit costs for direct expenses. For nonrevenue areas, these expenses are forecasted separately and distributed to various airport operations. For utility expenses, the net amount expected to be owed from the utility purchase, after a sale of utility services, is forecast.

Potential airport revenue

The sum of the estimated annual carrying charges on invested capital and the estimated average annual expenses of administration, operation, and maintenance establishes the break-even need for each revenue-producing facility and for the airport as a whole. The next step in establishing economic feasibility is to determine if sufficient revenues (that might be supplemented by federal, state, and local subsidies) can be expected at the airport to cover the break-even needs; therefore, forecasts are prepared for revenue-producing areas. These areas include the landing area, aircraft aprons and parking areas, airline terminal buildings, public parking areas, cargo buildings, aviation fuel, hangars, commercial facilities, and other usable areas.

Landing area   This area includes runways and related taxiways and circulation taxiways. Flight fee revenue determination is distributed among scheduled airlines, other air carrier users, and general aviation. Flight fee amounts should provide sufficient revenues to cover the landing area break-even need.

Aircraft aprons and parking areas   Revenues to obtain the break-even need for airline terminal aprons and cargo aprons are assigned to the scheduled airlines. Those for general aviation ramps are assigned to private aircraft. Apron and parking area fees should provide sufficient revenues to cover the break-even needs for specific aircraft aprons and parking areas.

Airline terminal buildings   Revenues for concessionaires and ground transportation services are usually based on a percentage of gross income with a fixed-rate minimum for each type of service. Space for scheduled airlines and other users is paid for on a fixed-rental basis. In order to establish rental rates, forecasts of potential revenue from concessions and ground transportation must be established. Rental rates are based on the break-even need of the terminal building after giving credit for forecasted revenues from concessions and ground transportation.

Public parking areas   Public parking is usually operated on a concessionaire basis with revenues obtained from rentals based on a percentage of gross income with a fixed-rate minimum. The revenue amount required to meet break-even needs will depend on whether parking facilities are constructed by the airport owner or under provisions of the concessionaire contract. These revenues apply to public parking for both airline and general aviation terminals. Revenues in excess of the break-even need for public parking are allocated to the break-even need for the airport as a whole.

Cargo buildings   Rentals are usually charged on a rate per square foot and cover investments in employee parking and truck unloading docks, as well as in building space. Rates are established to meet break-even needs.

Aviation fuel   Fees charged to aviation fuel handling concessionaires are established to cover the costs of fuel storage areas and associated pumping, piping, and hydrant systems.

Hangars   Rentals are usually based on a rate per square foot and cover investments in associated aircraft apron space and hangar-related employee parking. Hangar office space is charged on a similar basis and covers office-related employee parking.

Commercial facilities   Airport office buildings, industrial facilities, and hotels are usually operated on a lessee-management basis with revenues obtained from rentals on a square foot basis. The facilities are often financed by private capital. Revenues in excess of the break-even need are allocated to the break-even need of the airport as a whole.

Other usable areas   Various uses of ground space for activities such as gasoline stations, service facilities for rental car operators, and bus and limousine operators usually obtain revenues on a flat-rate basis. Those facilities are often financed by private capital. Revenues in excess of the break-even need are allocated to the break-even need of the airport as a whole.

Final economic evaluation

After analysis of the break-even needs for individual components of the master plan has been made, economic feasibility is analyzed on an overall basis. The goal of overall analysis is to determine if revenues will equal or exceed the break-even need. This determination requires an evaluation of the scope and phasing of the plan itself in terms of the users' requirements and their ability to make the financial commitment necessary to support the costs of the program. If this review indicates that revenues will be sufficient, revisions in the scheduling or scope of proposed master plan developments might have to be made, or recovery revenue rates for airport cost areas might require adjustment. These factors are adjusted until the feasibility of the master plan is established; this is to say, airport revenues (as might be supplemented by federal, state, or local subsidies) will match capital investment throughout the master plan forecast period. When the economic feasibility of improvements proposed in the master plan has been established, capital budget and a program for financing those improvements is developed.

Land use planning

The airport land use plan shows on-airport land uses as developed by the airport sponsor under the master plan effort and off-airport land uses as developed by surrounding communities. The work of airport, city, regional, and state planners must be carefully coordinated. The configuration of airfield runways, taxiways, and approach zones established in an airport layout plan provides the basis for development of the land use plan for areas on and adjacent to the airport. The land use plan for the airport and its environment in turn is an integral part of an areawide comprehensive planning program. The location, size, and configuration of the airport need to be coordinated with patterns of residential and other major land uses in the area, as well as with other transportation facilities and public services. Within the comprehensive planning framework, airport planning, policies, and programs must be coordinated with the objectives, policies, and programs for the area that the master plan airport is to serve.

Land uses on the airport

The amount of acreage within the airport's boundaries will have a major impact on the types of land uses to be found on the airport. For airports with limited acreage, most land uses will be aviation oriented. Large airports with a great deal of land in excess of what is needed for aeronautical purposes might be utilized for other uses. For example, many airports lease land to industrial users, particularly those who employ business aircraft or whose personnel travel extensively by air carrier or charter. In many cases, taxiway access is provided directly to the company's facility. In some instances, railroad tracks serving the company's area, company parking lots, or low-level warehousing can be located directly under runway approaches (but free of clear zones). Companies that might produce electronic disturbances that would interfere with aircraft navigation or communications equipment or cause visibility problems because of smoke are not compatible airport tenants.

Some commercial activities are suitable for location within the airport's boundaries. Recreational uses such as golf courses and picnicking areas are quite suitable for airport land and might in effect serve as good buffer areas. Certain agricultural uses are appropriate for airport lands, but grain fields that attract birds are avoided.

Although lakes, reservoirs, rivers, and streams might be appropriate for inclusion within the airport's boundaries, especially from the standpoints of noise or flood control, care is normally taken to avoid those water bodies that have in the past attracted large numbers of waterfowl. Dumps and landfills that might attract birds are also avoided.

Land uses around the airport

The responsibility for developing land around the airport so as to maximize the compatibility between airport activity and surrounding activities, and minimize the impact of noise and other environmental problems, lies with the local governmental bodies. The more political entities that are involved, the more complicated the coordination process becomes.

In the past, the most common approach to controlling land uses around the airport was zoning. Airports and their surrounding areas become involved in two types of zoning. The first type of zoning is height and hazard zoning, which protects the airport and its approaches from obstructions to aviation while restricting certain elements of community growth. FAR Part 77—Objects Affecting Navigable Airspace is the basis for height and hazard zoning.

The second type of zoning is land use zoning. This type of zoning has several shortcomings. First, it is not retroactive and does not affect preexisting uses that might conflict with airport operations. Second, jurisdictions with zoning powers (usually cities, towns, or counties) might not take effective zoning action. This is partly because the airport might affect several jurisdictions and coordination of zoning is difficult. Or the airport might be located in a rural area where the county lacks zoning powers and the sponsoring city might not be able to zone outside its political boundaries. Another problem is that the interest of the community is not always consistent with the needs and interests of the aviation industry. The locality might want more tax base, population growth, and rising land values, all of which are not often consistent with the need to preserve the land around the airport for other than residential uses.

Another approach to land use planning around the airport is subdivision regulations. Provisions can be written into the regulations prohibiting residential construction in intense noise-exposure areas. These areas can be determined by acoustical studies prior to development. Insulation requirements can be made a part of the local building codes, without which the building permits cannot be issued.

Finally, another alternative in controlling land use around the airport is the relocation of residences and other incompatible uses. Often urban renewal funds are available for this purpose.

Environmental planning

For any proposed airport planning project, a review of how such expansion would affect the surrounding environment must be performed. This requirement was first established in the Airport and Airway Development Act of 1970 and the Environmental Policy Act of 1969. The National Environmental Policy Act of 1969 requires the preparation of detailed environmental statements for all major federal airport development actions significantly affecting the quality of the environment. The Airport and Airway Development Act of 1970 directed that no airport development project may be approved by the Secretary of Transportation unless he or she is satisfied that fair consideration has been given to the communities in or near which the project may be located. Environmental issues of concern to airport management, such as noise, air quality, and water quality impacts, are described further in Chap. 10 of this text.

For every proposed project, an **Environmental Impact Review (EIR)** is performed. The results of the review may find that there will be no significant impact to the surrounding environment as a result of the project. If this finding is realized then a **Finding of No Significant Impact (FONSI)** statement is issued. If the EIR reveals that there is the potential for significant environmental impact as a result of the project, then a more comprehensive **Environmental Impact Statement (EIS)** must be developed. The EIS states specifically the areas of the environment that will be impacted and the degree of impact on the environment, and, most important, requires a plan on the part of the airport to mitigate those impacts.

Studies of the impact of construction and operation of the airport or airport expansion upon accepted standards of air and water quality, ambient noise levels, ecological processes, and natural environmental values are conducted to determine how the airport requirements can best be accomplished. An airport is an obvious stimulus to society from the standpoints of economic growth and the services it offers to the public; however, this generation of productivity and employment might be negated by noise and air pollution and ecological compromises if compatibility between an airport and its environs is not achieved; thus, the airport master plan must directly contend with these problems identified in the studies of environmental qualities so that the engineering of airport facilities will minimize or overcome those operations that contribute to environmental pollution.

In line with the above guidelines and policy, an airport master plan (including site selection) must be evaluated factually in terms of any proposed development that is likely to:

- Noticeably affect the ambient noise level for a significant number of people
- Displace significant numbers of people
- Have a significant aesthetic or visual effect
- Divide or disrupt an established community or divide existing uses (e.g., cutting off residential areas from recreation or shopping areas)
- Have any effect on areas of unique interest or scenic beauty
- Destroy or detract from important recreational areas
- Substantially alter the pattern of behavior for a species
- Interfere with important wildlife breeding, nesting, or feeding grounds
- Significantly increase air or water pollution
- Adversely affect the water table of an area

The airport master plan is commonly thought of as a "living document" whose contents adapt to constant changes in community needs. A robust master plan is one that helps airport planners and management maintain and develop an airport that meets the needs of the community, surrounding environment, and the nation's aviation and transportation system overall.

Concluding remarks

Considering the high cost and long lead time for building or improving airports, planning is the key in determining what facilities will be needed and in creating programs for providing them in a timely manner, while making wise use of resources. Planning for airport development requires more than simply scheduling the capital improvements to be made. Airports are public entities whose managers interact with many other public and private organizations. Airport development plans affect other aspects of community life, such as the land dedicated to aviation use or the noise or automobile traffic that the airport generates. The need for aviation development must thus be weighed against other societal needs and plans. Planning cannot be done for one airport in isolation; each airport is part of a network that is itself part of the national transportation system.

Airport planning, as practiced today, is a formalized discipline that combines forecasting, engineering, and economics. Because it is performed largely by government agencies, it is also a political process, where value judgments and institutional relationships play as much a part as technical expertise. On the whole, airport planners have been reasonably successful in anticipating future needs and in devising effective solutions.

Airport planning at local, regional, state, and federal levels should be coordinated and integrated. This goal, however, is often difficult to achieve. To some extent, this arises naturally from different areas of concern and expertise. At the extremes, local planners are attempting to plan for the development of one airport, whereas the FAA is trying to codify the needs of several thousand airports that might request aid. Local planners are most concerned with details and local conditions that will never be of interest to a national planning body.

Airport Planners are assisted by a series of Advisory Circulars, published by the Federal Aviation Administration. Some such advisory circulars include:

AC Series 150/5050-3B—*Planning the State Aviation System.*

AC Series 150/5050-4—*Citizen Participation in Airport Planning.*

AC Series 150/5050-5—*The Continuous Airport System Planning Process.*

AC Series 150/5050-7—*Establishment of Airport AC Seriestion Groups.*

AC Series 150/5070-5—*Planning the Metropolitan Airport System.*

AC Series 150/5070-6A—*Airport Master Plans.*

AC Series 150/5360-13—*Planning and Design Guidelines for Airport Terminal Facilities.*

AC Series 150/5070-7—*The Airport System Planning Process.*

This lack of common goals and mutually consistent approach is also evident between federal and state planning. More than 30 years ago, the federal government recognized the need to strengthen state system planning and provided funds for this purpose under ADAP, and nearly all the state airport system plans have been prepared with federal funding; however, it does not seem that the FAA has always made full use of these products in preparing the NPIAS. The state plans contain many more airports than NPIAS, and the priorities assigned to airport projects by states do not always correspond to those of NPIAS. Although it is probably not desirable, or even possible, for NPIAS to incorporate all elements of the state plans, greater harmony between these two levels of planning might lead to more orderly development of the national airport system.

In addition, coordination between airport planning and other types of transportation and economic planning is vitally important. In many cases, such a lack of coordination in planning is evident in the case of land use, where airport plans are often in conflict with other local and regional developments. Even though the airport authority might prepare a thoroughly competent plan, lack of information about other public or private development proposed in the community (or failure of municipal authorities to impose and maintain zoning ordinances) allows conflicts to develop over use of the airport and surrounding land. This problem can be especially severe where there are several municipalities or local jurisdictions surrounding the airport property.

Just as there will always be a need for competent airport management, airport planners will always be needed to protect the viability of today's airport system for the aviation needs of tomorrow.

Key terms

facilities planning

financial planning

economic planning

environmental planning

organizational planning

strategic planning

planning horizon

system planning

NPIAS (National Plan of Integrated Airport Systems)

AIP (Airport Improvement Program)

ADAP (Airport Development Aid Program)

SASP (state aviation system plans)

master planning

ALP (airport layout plan)

qualitative forecasting

Jury of Executive Opinion method

Sales Force Composite method

consumer market survey

Delphi method

quantitative forecasting

causal forecast

time-series method

regression analysis

aircraft fleet mix

ARC (Airport Reference Code)

wind rose

runway protection zones

EIR (Environmental Impact Review)

FONSI (Finding of No Significant Impact)

EIS (Environmental Impact Statement)

Key FAA advisory circulars

Advisory Circular 150/5050-3B—*Planning the State Aviation System*

Advisory Circular 150/5050-4—*Citizen Participation in Airport Planning*

Advisory Circular 150/5050-5—*The Continuous Airport System Planning Process*

Advisory Circular 150/5050-7—*Establishment of Airport Action Groups*

Advisory Circular 150/5070-5—*Planning the Metropolitan Airport System*

Advisory Circular 150/5070-6A—*Airport Master Plans*

Advisory Circular 150/5360-13—*Planning and Design Guidelines for Airport Terminal Facilities*

Advisory Circular 150/5070-7—*The Airport System Planning Process*

Questions for review and discussion

1. What are some of the various types of airport planning studies. What is the focus of each type of study?

2. What is meant by an airport planning horizon? What is the typical planning horizon for an airport master plan?

3. Although the NPIAS is considered the national airport system plan, it often isn't a complete plan. Why is this?

4. What is the purpose of regional-level system planning? What are the most critical issues addressed by regional-level system planning of airports?

5. How does state-level system planning differ from regional-level system planning?

6. What are the primary objectives of the airport master plan?

7. What is described in the inventory section of the airport master plan?

8. What is an airport layout plan? What is included on an airport layout plan drawing?

9. What is the difference between qualitative and quantitative forecasting?

10. What are some of the more common qualitative forecasting methods?

11. What is the difference between causal models and time-series quantitative forecasting models?

12. How is regression analysis used in forecasting?

13. What elements of aviation demand are typically forecasted in airport planning studies?

14. What is the difference between local operations and itinerant operations?

15. What areas of the airport are commonly considered within an airport capacity analysis? What is the importance of airport capacity analysis?

16. How is an Airport Reference Code determined?

17. What is a wind rose?

18. How is the necessity of a crosswind runway determined?

19. What is meant by a runway clear zone?

20. What is the primary objective of airport terminal area planning?

21. What factors are commonly taken into consideration in planning the airport terminal area?

22. What steps are involved in estimating the space requirements in planning airport terminals?

23. What is an aircraft fleet mix?

24. What are some of the potential revenue-generating areas to be considered in airport planning?

25. What is the process involved with environmental planning for airport development?

Suggested readings

de Neufville, Richard. *Airport Systems Planning*. Cambridge, Mass.: MIT Press, 1976.

de Neufville, R., and A. Odoni. *Airport Systems: Planning, Design, and Management*. New York: McGraw-Hill, 2003.

Horonjeff, R., and F. McKelvey. *Planning and Design of Airports*. New York: McGraw-Hill, 1994.

Howard, George P. *Airport Economic Planning*. Cambridge, Mass.: MIT Press, 1974.

National Plan of Integrated Airport Systems (NPIAS), 2001–2005. Washington, D.C.: FAA, August 2002.

Schreiver, Bernard A., and William W. Siefert. *Air Transportation 1975 and Beyond: A Systems Approach*. Cambridge, Mass.: MIT Press, 1968.

Airport System Planning Practices, ACRP Synthesis 14, National Academies Transportation Research Board, Washington, D.C., 2009.

12

Airport capacity and delay

Outline

- Introduction
- Defining capacity
- Factors affecting capacity and delay
- Estimating capacity
- Illustrating capacity with a time-space diagram
- FAA approximation charts
- Simulation models
  - FAA's airport capacity benchmarks
- Defining delay
- Estimating delay
- Analytical estimates of delay: The queuing diagram
- Other measures of delay
- Approaches to reducing delay
  - Creating new airport infrastructure
  - Converting military airfields
- Administrative and demand management
  - Administrative management
  - Demand management

Objectives

The objectives of this section are to educate the reader with information to:

- Define the concepts of capacity, particularly as it relates to airport activity.
- Identify the factors of the airport environment that affect capacity and delay.

- Be familiar with the various runway configurations and their rules of operation that affect capacity.
- Describe the concept of LAHSO, as it relates to airport capacity.
- Estimate the capacity of an airfield on the basis of FAA approximation charts.
- Be familiar with the time-space diagram analytical tool used to estimate runway capacity.
- Describe various simulation models used to estimate airport capacity.
- Define the concepts of delay as it relates to airport activity.
- Be familiar with the queuing diagram as an analytical method of estimating delay.
- Discuss various strategies to reduce delay at airports.

Introduction

The efficient movement of aircraft and passengers between airports is highly dependent on two key characteristics of an airport's operations: the demand for service by aircraft operators and passengers and the capacity at the airport, both in airspace and local environment. A major concern of airport planning and management is the adequacy of an airport's airfield, specifically in relation to the layout of the airport's runways, to handle the anticipated demand of aircraft operations. If air traffic demand exceeds airport or airspace capacity, delays will occur, causing expense to air carriers, inconvenience to passengers, and increased workload for the FAA air traffic control system as well as airport employees and administrators.

At most airports in the United States, particularly those commercial service and general aviation airports serving smaller communities, the demand for service does not consistently exceed capacity. The Federal Aviation Administration estimates that airports with single runways have the capacity to serve approximately 200,000 operations annually, a demand level typically generated by metropolitan areas with approximately 350,000 inhabitants. Other airports, particularly those primary commercial service airports serving large metropolitan areas with populations exceeding 5,000,000 inhabitants, experience consistent periods of time where demand exceeds capacity, despite the fact that two or more such airports serve the area. Examples of these metropolitan areas include Chicago, Los Angeles, South Florida, the San Francisco Bay Area, and New York City.

Throughout the late 1990s and 2000s, the overall increase in demand for air transportation in the United States resulted in a growing number of airports that suffered from delays resulting from demand exceeding capacity. Within the entire system, over 550,000 air carrier operations suffered from at least 15 minutes of delay each during the year 2000, the year of second-greatest delays in the history of commercial aviation, next to 2007. The events of September 11, 2001, resulted in significant reductions in the overall demand for air travel for the year, which in turn resulted in fewer flight delays. However, the demand for air travel quickly rebounded, returning in 2005 to the highest number of commercial air carrier operations in history before again declining with the economic downturn beginning in 2008. After industry-wide consolidation, air travel demand was beginning to rebound again in 2010, and is expected to rebound in the long term, just as the industry has throughout its history (Fig. 12-1). With such growth, associated flight delays will increase as well, unless the capacity of the nation's airports and airspace system increases to accommodate the demand.

There are a number of potential specific reasons for any given aircraft to experience delay. The most common factors that might cause an aircraft to experience delay are weather, aircraft mechanical issues, or simply operating at a time when overall demand for operations exceeds capacity. The FAA estimates that the majority of flight delays occur because of adverse weather. Other delays are attributed to equipment, runway closures, and excessive volume or demand (Fig. 12-2).

The task of managing airports in particular, and the aviation system in general, with the goal of minimizing delays, either by providing sufficient capacity to handle demand, or managing the demand itself, is a challenging one. To

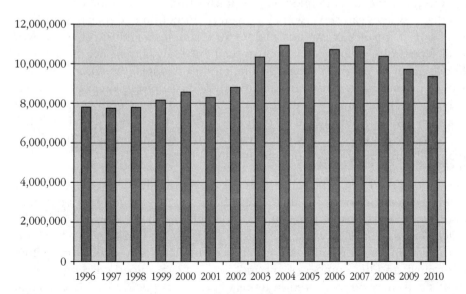

Figure 12-1. *Annual Systemwide Departures, U.S. Air Transportation.* (Source: RITA)

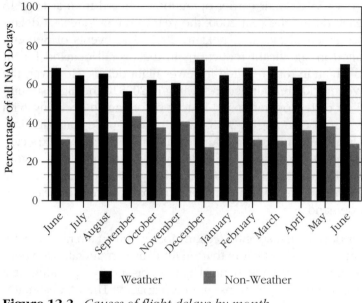

Figure 12-2. *Causes of flight delays by month, June 2009–June 2010.* (Source: Bureau of Transportation Statistics)

effectively achieve this task, a fundamental understanding of capacity, demand, and associated delays is needed.

Defining capacity

Capacity, in general, is defined as the practical maximum number of operations that a system can serve within a given period of time. Capacity is, in fact, a rate, similar to velocity. An automobile for example, might travel at a rate of 50 miles per hour, meaning that over an hour, traveling at this rate, the automobile will travel 50 miles. Traveling this rate for 30 minutes, the automobile will travel 25 miles, and so forth. Airport capacity is measured in aircraft operations per hour. A single runway at an airport, might have an operating capacity of 60 operations per hour, meaning, over the course of an hour, the airport will be able to serve approximately 60 aircraft takeoffs and landings; in 30 minutes, the airport can serve 30 such operations, and so forth.

It should be noted that although airport capacity typically refers to the capacity to handle aircraft operations, there are other areas of operation at an airport where other measures of capacity are equally important. For example, the efficient movement of passengers through points within an airport terminal is determined, in part, by the passenger processing capacity of locations within the terminal, and the number of automobiles that can unload passengers at an airport's curb may be measured in terms of *vehicle capacity*. The theories that govern capacity and delay are similar regardless of location. With this in mind, however, the focus of airport capacity will heretofore feature airport capacity in its traditional definition, that of aircraft operating capacity.

There are actually two commonly used definitions to describe airport capacity: throughput capacity and practical capacity. **Throughput capacity** is defined as the ultimate rate at which aircraft operations may be handled without regard to any small delays that might occur as a result of imperfections in operations or small random events that might occur. Throughput capacity, for example, does not take into account the small probability that an aircraft will take longer than necessary to take off, or a runway must close for a very short period of time because of the presence of small debris. Throughput capacity is truly the theoretical definition of capacity and is the basis for airport capacity planning.

Practical capacity is understood as the number of operations that may be accommodated over time with no more than a nominal amount of delay, usually expressed in terms of maximum acceptable average delay. Such minimal delays may be a result of two aircraft scheduled to operate at the same time, despite the fact that only one runway is available for use, or because an aircraft must wait a short time to allow ground vehicles to cross. The FAA defines two measures of practical capacity to evaluate the efficiency of airport operations. **Practical hourly capacity (PHOCAP)** and **Practical annual capacity (PANCAP)** are defined by the FAA as the number of operations that may be handled at an airport that results in not more than 4 minutes average delay during the busiest, known as the peak, 2-hour operating period, hourly and annually, respectively.

Factors affecting capacity and delay

The capacity of an airfield is not constant. Capacity varies considerably based on a number of considerations, including the utilization of runways, the type of aircraft operating, known as the fleet mix, the percentage of takeoff and landing operations being performed, ambient climatic conditions, and FAA regulations which prescribe the use of runways based on these considerations. When a specific number is given for airfield capacity at an airport, it is usually an average number based either on some assumed range of conditions or on one specific set of conditions.

An understanding of the variability of capacity, rather than its average value, however, is crucial to the effective management of an airfield. Much of the strategy for successful management of an airfield involves devising ways to compensate for a number of factors that, individually or in combination, act to reduce capacity or induce delays.

The physical characteristics and layout of runways, taxiways, and aprons, for example, are basic determinants of the ability to accommodate various types of aircraft and the rate at which they can be handled. Also important is the type of equipment, particularly the presence of instrument landing systems, installed on the airfield as a whole or on a particular segment.

One of the characteristics that affects an airport's capacity is the configuration of its runway system. Although every airport is different, the configurations of airport runways may be placed in the following categories: single runway, parallel runways, open-V runways, and intersecting runways. Although every runway configuration has a uniquely different capacity that is determined by a variety of factors, the FAA has established some basic estimates of capacity by runway configuration.

The single runway, for example, the simplest of runway configurations, can accommodate up to 99 operations per hour for smaller aircraft and approximately 60 operations per hour for larger commercial service aircraft during fair weather conditions, known as visual meteorological conditions (VMC), or operating under visual flight rules (VFR). Under poor weather conditions, known as instrument meteorological conditions (IMC), when aircraft fly according to instrument flight rules (IFR), the capacity of a single runway configuration is reduced to between 42 and 53 operations per hour, depending primarily on the size of the aircraft using the runway and any *navigational aids* that may be available.

In general, airport capacity is usually greatest in VMC, whereas IMC, in the form of fog, low cloud ceilings, or heavy precipitation, tends to result in reduced capacity. In addition, strong winds or significant accumulations of snow or ice on a runway can significantly reduce capacity or close an airport to aircraft operations altogether. Even a common occurrence such as a wind shift can reduce operating capacity while air traffic is appropriately rerouted.

The parallel runway configuration, characterized by two or more runways aligned parallel to each other, increases runway capacity at an airport over single runway configurations depending primarily on the distances between the parallel runways, specifically their lateral separation, defined as distance between the centerlines of each runway. Parallel runway configurations are typically found at airports that require increased levels of runway capacity but do not require any crosswind runways. For two parallel runways separated by at least 4,300 feet, total runway capacity is double that of the capacity of a single runway. However, if the lateral separation is less than 4,300 feet, then under IFR operations, operations on each runway must be highly coordinated, effectively reducing capacity. If the parallel runways are separated by less than 2,500 feet, the airfield must operate as a single runway configuration under IFR (Fig. 12-3).

The open-V runway configuration describes two runways that are not aligned in parallel with each other, yet do not intersect each other at any point on the airfield. The runway oriented into the prevailing winds is known as the primary runway. The other runway is identified as the crosswind runway. During low wind conditions, both runways may be used simultaneously. When aircraft

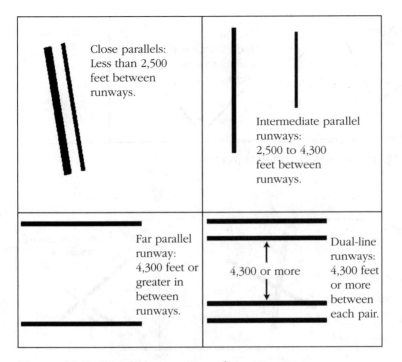

Figure 12-3. *Parallel runway configurations.* (Source: NASA)

operate outwardly from the V, the runway configuration is said to be used for diverging operations. Typically, takeoffs are allowed simultaneously during divergent operations. When the runway configuration is used in a converging manner, landings tend to be handled simultaneously. Runway capacity is typically greater when operations are performed under divergent operations. Under divergent operations total runway capacity can reach nearly 200 operations per hour for smaller aircraft and 100 operations per hour for commercial service aircraft. Under convergent operations, capacity rarely exceeds 100 operations per hour for smaller aircraft and frequently less than 85 operations per hour for commercial service aircraft. When winds are sufficiently strong or when IFR operations are in effect, only one runway in the open-V configuration is typically used, reducing capacity to that of a single runway configuration (Fig. 12-4).

The intersecting runway configuration describes two runways that are not aligned in parallel with each other and intersect each other at some point on the airfield. As with open-V runways, the runway oriented into the prevailing winds is known as the primary runway. The intersecting runway is identified as the crosswind runway. During low wind conditions and operating under VFR, both runways may be used simultaneously, yet in a highly coordinated manner, so as to avoid any *incursions* between two aircraft (Fig. 12-5).

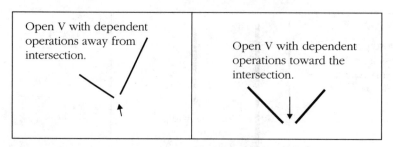

Figure 12-4. *Open-V runway configurations.* (Source: NASA)

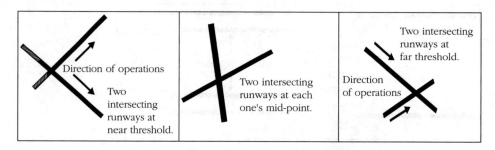

Figure 12-5. *Intersecting runway configurations.* (Source: NASA)

Under certain specific conditions, aircraft may land simultaneously and independently on intersecting runways. These operations, known as **LAHSO (land and hold short operations),** may be conducted with approval from the FAA and only when there is sufficient runway length on each runway before the intersection of the two runways for each aircraft to land and stop before reaching the intersection (Fig. 12-6).

LAHSO is one example of a spectrum of rules and procedures employed by air traffic control, intended primarily to ensure safety of flight that fundamentally determines airport capacity. Other such rules governing, for example, the speeds of aircraft, separations between aircraft, runway occupancy, and prescribed routes to be followed by aircraft on departure from, approach to, and en route between airports all have an effect on the capacity of an individual airport.

Another significant factor in determining airport capacity is the consideration of the volume of demand and characteristics of the aircraft that wish to use the airport during any given period of time. For any given level of demand, the varying types of aircraft with respect to speed, size, flight characteristics, and even pilot proficiency will in part determine the rate at which they can

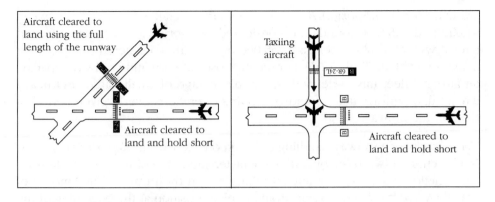

Figure 12-6. *LAHSO (land and hold short operations) on intersecting runways.* (FAA AIM)

perform operations. In addition, the distribution of arrivals and departures to the extent to which they are bunched rather than uniformly spaced, as well as the sequence of such operations, also play a part in determining an airport's operating capacity. In part, the tendency of traffic to *peak* in volume at certain times is a function of the flight schedules of commercial air carriers using an airport. For example, at airports that serve as **hubs** for major air carriers, high volumes of aircraft all arrive in *banks* and all depart a short time later, after passengers have transferred from one flight to another to complete their travel. Arrival banks of aircraft result in one level of airport capacity, whereas departure banks result in another level of capacity, merely by the different operating characteristics of aircraft arrivals and departures. Finally, the volume of demand for surrounding airspace is a critical component of determining airport capacity, particularly in large metropolitan regions with multiple airports in close proximity, such as the New York Metropolitan area.

Estimating capacity

The art of estimating the capacity of an airfield is a challenging and important one. In 2007, it was estimated that delays in air carrier operations cost more than $41 billion in terms of wasted fuel and labor, and lost opportunities for air carriers and their passengers. Significant investments are made on the part of airport management, the FAA, and air carriers to estimate as accurately as possible the capacity of an airfield under varying conditions and operating characteristics. However, if the basic fundamentals of aircraft operations are understood, initial estimates of runway capacity may be established with little effort.

The FAA categorizes the wide variety of aircraft types by their maximum certified takeoff weights (MTOW). Aircraft with MTOW less than 41,000 pounds are considered *category A/B* or *small* aircraft, aircraft with MTOW between 41,000 and 255,000 pounds are considered *category C* or *large* aircraft, and aircraft with MTOW greater than 255,000 pounds are considered *category D* or *heavy* aircraft (see Table 12-1). For the purposes of estimating runway capacity, an airport's fleet mix is defined by the percentage of small, large, and heavy aircraft that perform takeoff and/or landing operations over a given period of time on the runway.

The capacity of a runway handling only takeoffs, known as *departure capacity,* is a function of two basic operating characteristics. One of these characteristics is a function of the type of aircraft taking off on the runway. The amount of time the aircraft requires to start from an initial position at the beginning of the runway to the time it in fact leaves the runway environment allowing another aircraft to depart is called an aircraft's **runway occupancy time (ROT).** The shorter an aircraft's ROT, the greater the number of aircraft that can use the runway over time, and hence the greater the capacity of the runway. ROT of a given aircraft is a function of the performance specifications of the aircraft. In general, smaller and lighter aircraft (fleet mix categories A and B) tend to require smaller ROT for takeoff than larger or heavier aircraft (fleet mix categories C and D). As a result, the operating capacity of a given runway is higher when accommodating smaller aircraft departures than it is when accommodating departures of larger aircraft. ROTs for departing aircraft range from approximately 30 seconds for small aircraft to approximately 60 seconds for larger and heavier aircraft.

Witnessing a runway operating at departure capacity is clear. When a runway is constantly occupied by departing aircraft, that is, the runway is never empty or *idle,* the runway is operating at capacity.

The capacity of a runway handling only landings, known as *arrival capacity,* is similarly a function of the ROT of arriving aircraft. In addition, the velocity at which the aircraft travels while on approach to the runway (known as an aircraft's *approach speed*), as well as FAA regulations requiring that aircraft remain at least a given distance behind one another while on approach to landing

Table 12-1 Aircraft Fleet Mix Categories

Aircraft Fleet Mix Category	Maximum Takeoff Weight
A, B (small)	<12,500 lb
C (large)	12,500–300,000 lb
D (heavy)	>300,000 lb

Source: FAA AC 150-5060/-5.

(known as *longitudinal separation*), are determining factors in arrival capacity. In general, smaller and lighter aircraft tend to travel at lower approach speeds than larger and heavier aircraft. However, larger aircraft create the need for greater longitudinal separations. As a result of these characteristics, estimating arrival capacity becomes a nontrivial analysis of the various types of aircraft, known as the **fleet mix,** that wish to land over a given period of time. An airport's fleet mix is defined as the percentage of aircraft operations by aircraft type that occurs at the airport over a given period of time.

When two aircraft are on approach to a runway, the longitudinal separation required between the two aircraft is determined by the weight categories of the aircraft in front, known as the *lead aircraft,* and the aircraft following, known as the *lag aircraft.* Table 12-2 illustrates the FAA's required longitudinal separations, in nautical miles. As long as both aircraft are airborne on approach, these longitudinal separations must be maintained. The only exception to this rule is when operating under visual flight rules (VFR), small aircraft are required to maintain sufficient separation so that the lag aircraft does not touch down on the runway before the lead aircraft has landed and cleared the runway. In most airport environments, especially those airports with an operating air traffic control tower and those airports serving commercial air carriers, the longitudinal separation standards given in Table 12-2 are maintained. The primary reason for these standards is to prevent lag aircraft from experiencing severe *wake turbulence* as a result of very rough airflow emanating from the lead aircraft's wings.

As implied in Table 12-2, the arrival capacity of a runway can be significantly affected by the airport's fleet mix. In general, the more distributed the mix, that is, the more variability in aircraft sizes, the lower the arrival capacity. As a result, airports with multiple runways often attempt to separate the arrivals of aircraft by size onto separate runways.

Contrary to that of departure capacity, witnessing the arrival capacity of a runway is not as visually intuitive. Even though a runway is operating at its arrival capacity, there are often periods of time when the runway itself is empty of aircraft. This is due to the fact that the required longitudinal separation of

Table 12-2 Required Longitudinal Separations for Arriving Aircraft to a Single Runway When Performing under IFR (Distances in Nautical Miles)

Lead/Lag	Small	Large	Heavy
Small	3	3	3
Large	4	3	3
Heavy	6	5	4

aircraft, which results in the lead aircraft often landing and exiting the runway prior to the lag aircraft touching down, prevents arriving aircraft from reaching the runway any sooner.

The capacity of a runway handling both landings and takeoffs is called the runway's *mixed-use operating capacity*. In general, a runway's mixed-use operating capacity is determined first by estimating arrival capacity, and then, taking advantage of the times that the runway is idle because of longitudinal separation requirements, allowing departures to occur.

Illustrating capacity with a time-space diagram

Accurately estimating the capacity of a runway is a challenge, particularly when considering all the variations in aircraft and pilot performance, external conditions, and regulatory policies. However, to find basic estimates of runway capacity, a fundamental graphical analysis using what is known as a **time-space diagram** may be used.

A time-space diagram is a two-dimensional graph that may be used to represent the location of any particular object, such as an arriving or departing aircraft, at a given point in time. With a time-space diagram, visual representations of aircraft movements, based on performance characteristics and FAA regulations may be made.

For example, to illustrate the movement of departing aircraft on a runway, the time-space diagram illustrated in Fig. 12-7 may be used. Figure 12-7 represents

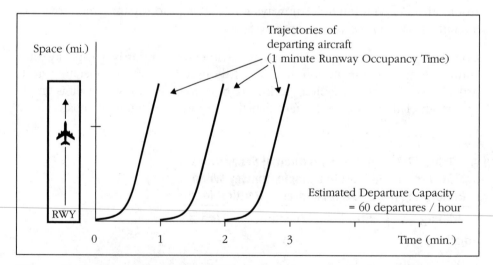

Figure 12-7. *Time-space diagram illustrating departure capacity of a runway serving departing aircraft with a runway occupancy time of 1 minute.*

the departure of aircraft along a runway. Each aircraft has a 60-second ROT. The diagram illustrates the fact that only one aircraft may be present on the runway at any given period of time. The trajectory of each aircraft is represented by a curve, which represents the aircraft's increase in velocity until it reaches takeoff speed. From the diagram, it can easily be seen that the departure capacity of the runway is one departure per minute, or 60 departures per hour.

Figure 12-8 illustrates the arrival capacity of a runway being used by small arriving aircraft under IFR conditions. Each aircraft has a velocity, known as the *approach speed,* of 60 nautical miles per hour and requires a 30-second runway occupancy time upon landing. In addition there is a required longitudinal separation of 3 nautical miles between lead and lag aircraft. From this diagram, it is realized that aircraft arrive to the runway at the rate of one landing every 3 minutes, resulting in an arrival capacity of 20 landings per hour.

Figure 12-9 illustrates the mixed-use capacity of the runway. Departures are allowed to occur during idle runway times between landings. Because departures require 1-minute runway occupancy times, it is possible to allow two departures between landings, resulting in a mixed-use departure capacity of 40 departures per hour. When combined with the 20 arrivals per hour arrival capacity, this runway is estimated to have a mixed-use operating capacity of 60 operations per hour.

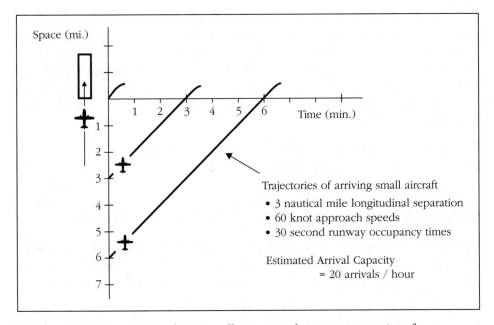

Figure 12-8. *Time-space diagram illustrating departure capacity of a runway serving arriving small aircraft.*

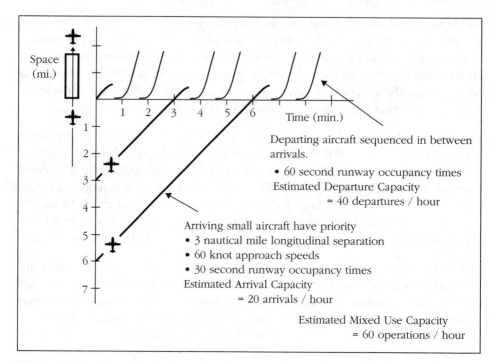

Departing aircraft sequenced in between arrivals.

- 60 second runway occupancy times
 Estimated Departure Capacity
   = 40 departures / hour

Arriving small aircraft have priority
- 3 nautical mile longitudinal separation
- 60 knot approach speeds
- 30 second runway occupancy times
 Estimated Arrival Capacity
   = 20 arrivals / hour

Estimated Mixed Use Capacity
   = 60 operations / hour

Figure 12-9. *Time-space diagram illustrating mixed-use operating capacity of a runway serving arriving and departing small aircraft.*

Applying the above simple analyses to aircraft with different performance characteristics reveals the interesting behavior of runway capacity. For example, the arrival capacity of large aircraft, which typically have approach speeds of approximately 90 nautical miles per hour and require longitudinal separations of 3 nautical miles between lead and lag aircraft, may be illustrated on a time-space diagram to yield a capacity of 30 arrivals per hour and a departure capacity of 30 departures per hour (Fig. 12-10).

In addition, a time-space diagram may be used to illustrate the degrading effects on capacity as a result of allowing aircraft of mixed sizes to use the same runway. As Fig. 12-11 illustrates, large longitudinal separation requirements for larger aircraft in front of smaller aircraft, along with the slower speeds of smaller arriving aircraft, result in significantly reduced arrival capacity.

One technique often used by air traffic controllers to increase capacity is to carefully sequence arriving aircraft to minimize the amount of time smaller aircraft are following larger aircraft. One example of sequenced arrivals may be illustrated in Fig. 12-12.

Even though the time-space diagram is an excellent tool for estimating runway capacity on the basis of fundamental principles, it quickly becomes cumbersome

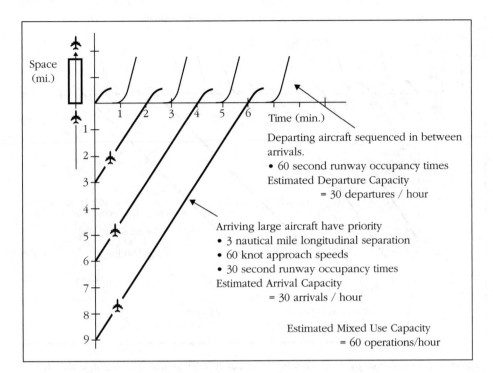

Figure 12-10. *Time-space diagram illustrating mixed-use operating capacity of a runway serving arriving and departing large aircraft.*

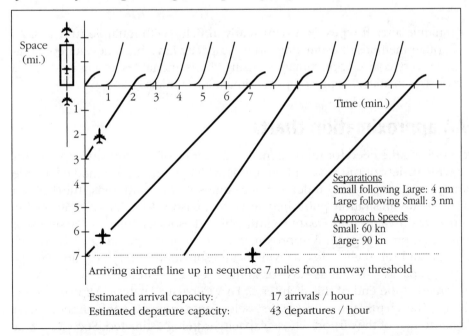

Figure 12-11. *Time-space diagram illustrating mixed-use operating capacity of a runway serving arriving and departing large and small aircraft.*

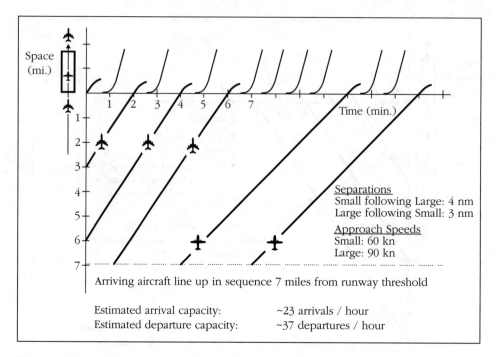

Figure 12-12. *Time-space diagram illustrating mixed-use operating capacity of a runway serving arriving and departing large and small aircraft sequenced to maximize arrival capacity.*

for multiple aircraft types, and particularly airfields with multiple runways. As such, other methods of estimating airfield capacity have been developed, ranging from very coarse approximations using charts and tables, to highly detailed complex estimations using computer simulation models.

FAA approximation charts

Recognizing the need for fundamental approximations of runway capacity for the wide variety of runway configurations which in turn accommodate a wide variety of aircraft types under varying atmospheric conditions, the Federal Aviation Administration published Airport Advisory Circular AC 150/5060-5 —*Airport Capacity and Delay*. Within this advisory circular are a series of charts that provide general approximations of hourly operating capacity under VFR and IFR flight rules, as well as an estimation of a typical annual operating capacity, known as ASV, or annual service volume. These charts are presented at the end of this chapter as FAA Capacity Analysis Approximation Charts. The charts are used by first selecting a runway configuration most similar to that of the airport whose capacity is being estimated. Second, a *mix index* is calculated, estimated by adding the percentage of large aircraft to the percentage of heavy aircraft multiplied by 3. The mathematical formula used to calculate mix index is MI 5 C + 3D. By applying the calculated mix index to the selected runway configuration, estimated VFR and IFR hourly capacities and ASV may be read from the chart.

For example, consider an airport with a single runway and a fleet mix of 80 percent small aircraft, 18 percent large aircraft, and 2 percent heavy aircraft. Such an airport is typical of a small general utility general aviation airport. The airfield configuration depicted in Fig. A1-1A, which best represents this airport is configuration 1. The mix index of the airport is 18 + 3(2) = 24. From reading the chart, it can be seen that the estimated hourly capacity of the airport is approximately 98 operations per hour under VFR, and 59 operations per hour under IFR (not coincidentally very close to the capacity estimated using a time-space diagram), and an annual service volume of approximately 230,000 operations per year.

A large commercial airport with two sets of closely separated parallel runways, each set separated by at least 3,500 feet with a fleet mix of 5 percent small aircraft, 80 percent large aircraft, and 15 percent heavy aircraft, might be represented in Fig. A1-B by configuration and a mix index of 80 + 3(15) = 125, resulting in an hourly VFR capacity of 189 operations per hour, an hourly IFR capacity of 120 operations per hour, and an annual service volume of 675,000 operations per year.

Simulation models

Although the charts described in AC 150/5060-5 are appropriate for coarse approximations of airport capacity, they do not, in fact, provide the actual operating capacity that may be occurring at an airport at any given period of time. One of the few methods of estimating capacity, particularly as a function of a constantly changing airport environment, is computer simulation (Fig. 12-13).

The Airport and Airspace Simulation Model (**SIMMOD**™), validated by the FAA, is an industry standard analysis tool used by airport planners and operators, airlines, airspace designers, and air traffic control authorities for conducting high-fidelity simulations of current and proposed airport and airspace operations. SIMMOD™ is designed to "play out" airport and airspace operations on the computer and calculate the real-world consequences of potential operating conditions. It has the capability and flexibility to address a wide range of "what-if" questions related to airport and airspace capacity, delay, and efficiency, including questions associated with:

- Existing or proposed airport facilities (e.g., gates, taxiways, runways, pads)

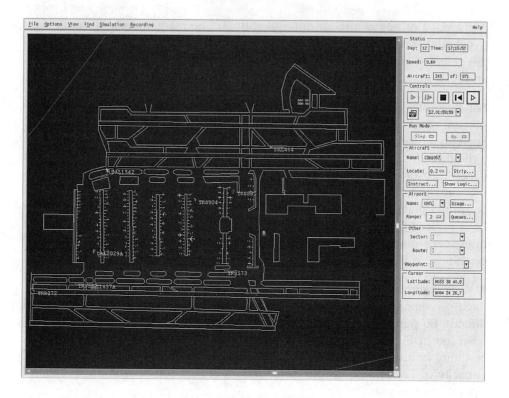

Figure 12-13. *Example of airfield simulation using computer software.*

- Airport operating alternatives (e.g., taxi patterns, runway use, departure queuing)
- Existing or proposed airspace structures (e.g., routes, procedures, sectors)
- Air traffic management/control technologies, procedures, and policies
- Aircraft separation standards parameters (e.g., weather, aircraft type, flight state)
- Airline operations (e.g., flight schedule, banking, gate use, and service times)
- Current and future traffic demand (e.g., volume, aircraft mix, new aircraft types)

Based upon a user-input scenario, SIMMOD™ tracks the movement of individual aircraft through an airport/airspace system, detects potential violations of separations and operating procedures, and simulates air traffic control actions required to resolve potential conflicts. The model properly captures the interactions within and between airspace and airport operations, including interactions among multiple neighboring airports.

As SIMMOD™ simulates airport and airspace operations, it computes and records detailed information on the activities and events associated with the operation of each aircraft at the airport and within the airspace. These results are provided as outputs that are available to the user for evaluating alternatives, including aircraft travel time, delay and operating costs, and system capacity, throughput and traffic loading.

In recent years, **TAAM**™ (Total Airport and Airspace Modeler) simulation modeling software has become another standard accepted computer application for estimating the capacity of an airport and associated airspace. The makers of TAAM also offer specialized products to estimate the capacity of airport terminals and baggage processing centers, as well as software for use of the administrative staff planning airports.

FAA's airport capacity benchmarks

Beginning in 2001, the FAA has developed models to estimate the airfield capacity at 35 of the nation's busiest and often most congested airports. The models estimate the capacity of these airfields under optimal (VFR) weather conditions, instrument (IFR) conditions, and marginal weather conditions (conditions IFR are not applied but visual approaches to runways are impractical) using the most common runway use configuration for a given airport.

Table 12-3 lists the estimated airfield capacity, defined in terms of total operations (takeoffs plus landings per hour), for the estimated airports.

It should be noted that some airports lose a significant amount of operating capacity under IFR, as compared to optimal conditions. This is most often due to the runway configurations at these airports, as under some configurations the use of certain runways such as closely separated parallel runways or intersecting runways are significantly reduced.

The FAA applies these estimates to the analysis of potential levels of airfield delays, in the face of varying levels of demand, and uses their findings to determine priorities and strategies for expanding capacity at these airports.

Defining delay

Delay is defined as the duration between the desired time that an operation occurs and the actual time the operation occurs. When aircraft depart and arrive "on time," according to their respective schedules, for example, the

Table 12-3 Estimated Airfield Capacity, Defined in Terms of Total Operations (Takeoffs Plus Landings) per Hour, for the Estimated Airports.

	Airport	Optimum	Marginal	IFR
ATL	Atlanta Hartsfield-Jackson International	180-188	172-174	158-162
BOS	Boston Logan International	123-131	112-117	90-93
BWI	Baltimore-Washington International	106-120	80-93	60-71
CLE	Cleveland Hopkins	80-80	72-77	64-64
CLT	Charlotte/Douglas International	130-131	125-131	102-110
CVG	Cincinnati/Northern Kentucky International	120-125	120-124	102-120
DCA	Ronald Reagan Washington National	72-87	60-84	48-70
DEN	Denver International	210-219	186-202	159-162
DFW	Dallas/Fort Worth International	270-279	231-252	186-193
DTW	Detroit Metro Wayne County	184-189	168-173	136-145
EWR	Newark Liberty International	84-92	80-81	52-56
FLL	Fort Lauderdale-Hollywood International	60-62	60-61	61-66
HNL	Honolulu International	110-120	60-85	58-60
IAD	Washington Dulles International	135-135	114-120	105-113
IAH	Houston George Bush International	120-143	120-141	108-112
JFK	New York John F. Kennedy International	75-87	75-87	64-67
LAS	Las Vegas McCarran International	102-113	77-82	70-70
LAX	Los Angeles International	137-148	126-132	117-124
LGA	New York LaGuardia	78-85	74-84	69-74

Code	Airport			
MCO	Orlando International	144-164	132-144	104-117
MDW	Chicago Midway	64-65	64-65	61-64
MEM	Memphis International	148-181	140-167	120-132
MIA	Miami International	116-121	104-118	92-96
MSP	Minneapolis-St Paul International	114-120	112-115	112-114
ORD	Chicago O'Hare International	190-200	190-200	136-144
PDX	Portland International	116-120	79-80	77-80
PHL	Philadelphia International	104-116	96-102	96-96
PHX	Phoenix Sky Harbor International	128-150	108-118	108-118
PIT	Greater Pittsburgh International	152-160	143-150	119-150
SAN	San Diego International - Lindbergh Field	56-58	56-58	48-50
SEA	Seattle-Tacoma International	80-84	74-76	57-60
SFO	San Francisco International	105-110	81-93	68-72
SLC	Salt Lake City International	130-131	110-120	110-113
STL	Lambert-St. Louis International	104-113	91-96	64-70
TPA	Tampa International	102-105	90-95	74-75

aircraft is said to have experienced no delay. If, however, an aircraft actually departs an hour after its scheduled departure time, that aircraft is said to have suffered 1 hour of delay. This delay may have been the result of any number of factors. A mechanical repair may have been required, luggage may have been slow in being loaded, weather may have required the aircraft wait until conditions improve, or perhaps the aircraft was one in a large number of aircraft that were scheduled to depart during a high-demand period time of day when the capacity of the airfield was insufficient to accommodate such high demand.

Figure 12-14 illustrates the relationship between demand, capacity, and delay. As Fig. 12-14 illustrates, some amount of delay is often experienced by aircraft, even when levels of demand are significantly less than capacity. These delays are usually nominal, created as the result of sparse instances of two aircraft wishing to operate within very close intervals of time, or minor operational anomalies. As demand nears capacity, delays tend to increase exponentially as the potential for such anomalies and scheduling conflicts increase.

The FAA defines the maximum acceptable level of delay as the level of demand, in relation to throughput capacity, that will result in aircraft delays of no more than 4 minutes per operation. Congestive delay occurs when demand is sufficiently close to throughput capacity to result in an average of nine or more minutes of delay per aircraft operation. As demand asymptotically reaches throughput capacity, delays can reach several hours per operation. During

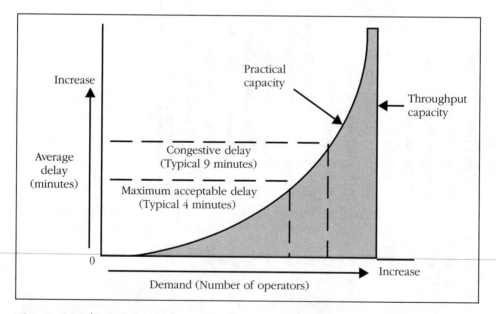

Figure 12-14. *Delay as a function of capacity and demand.*
(Source: FAA Office of Technological Assessment)

extreme periods, when both demand is at its highest and capacity is significantly reduced because of weather or any other adverse condition, scheduled aircraft operations may be delayed for several days, if not canceled.

How much delay is acceptable is, in fact, based on judgment involving three concepts. First is the concept that some delays are unavoidable because of factors beyond human control, such as changing meteorological conditions. Second, some delays, though avoidable, might be too expensive to eliminate. For example, the cost of building a new runway only to reduce delays by perhaps a few seconds per operation may be excessive. Third, even with the most vigorous effort, because aircraft operations are demanded on a somewhat random time frame (e.g., even though an air carrier may be scheduled to land at 12:00 noon, it might actually wish to depart at some random period between 11:58 and 12:03, depending on winds or other factors that determine an aircraft's travel time to its destination), there always exists the probability that some aircraft will encounter delay greater than some "acceptable" amount. Thus, acceptable delay is essentially a policy decision about the tolerability of delay being longer than some specified amount, taking into account the technical feasibility and economic practicality of available remedies.

Estimating delay

As with estimating capacity, various methods exist to estimate delay from fundamental analytical models, FAA tables and graphs, and computer simulation models. Similar to capacity estimation, analytical models allow an airport planner to estimate delays using fundamental estimations for aircraft demand and airport capacity. FAA tables provide cursory estimations of delay for more complex operating conditions, whereas computer simulation models provide detailed estimations of delay under a full variety of operating conditions, from the very simple to the highly complex.

A common analytical tool used to estimate delay over a period of time for a given airport capacity is the cumulative arrival diagram, also known as a queuing diagram. The diagram is based on the highly developed science of queuing theory, formed originally to estimate queues and delays for automobile traffic. Queuing theory may be applied to any environment where queues, and hence delays, occur, from toll booths, to grocery stores, to airports.

One example of a situation where a cumulative arrival diagram is particularly useful is a period of time where the demand changes while airport capacity remains essentially the same. This situation occurs often within airports. Periods of high demand, known as *peak* periods, tend to occur during morning and evening commute hours, at airports acting as hubs for major air carriers, and in periods of arrival or departure. Although the time and duration of peak periods of each individual airport is unique, peak periods, in general, are common to virtually all airports. Periods of time that experience less demand are known as *off-peak* periods.

Similar to capacity, demand is a rate, measured in operations per hour. Whereas capacity is the maximum number of operations that can be handled within an hour, demand is the number of operations that wish to occur over an hour. By definition, then, if a demand is less than capacity, the airport is said to be operating under capacity, and suffers minimal delays; as demand reaches capacity minor delays increase. When demand reaches, or exceeds capacity, the airport is said to be *saturated,* operating at capacity, but suffering large delays.

Analytical estimates of delay: The queuing diagram

Consider an airport with a single runway, having a capacity of 60 operations per hour. This airport may accommodate a demand for operations during most of the day, the off-peak period, of 30 operations per hour. During a 2-hour peak period, for example between 6:00 and 8:00 A.M., demand is 75 operations per hour. On the basis of this information, a **queuing diagram** may be constructed to illustrate aircraft demand, airport capacity, and overall delay that

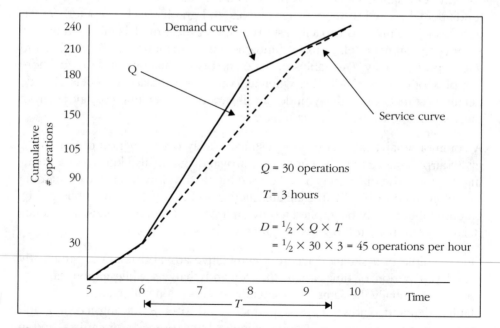

Figure 12-15. *Example of queuing diagram.*

occurs as a result of the relationship between the two over time. Figure 12-15 represents such a queuing diagram.

The solid line in Fig. 12-15, known as the *demand curve,* represents the cumulative number of operations that have been scheduled to occur by any given time of day. In this example, the first operation of the day is scheduled for 5:00 A.M. The steady rise in the demand curve between 5:00 A.M. and 6:00 A.M. represents a schedule of operations that occur steadily at the off-peak rate of 30 operations per hour. The period between 6:00 A.M. and 8:00 A.M. is the *peak period,* where an additional 75 aircraft per hour, for a total of 150 aircraft, are scheduled to operate. At the end of the peak period, a cumulative total of 180 operations were scheduled to occur since the beginning of the operating day. After the end of the peak period, the demand curve illustrates the return to off-peak demand levels, resulting in 210 cumulative operations scheduled by 9:00 A.M., 240 cumulative operations scheduled by 10:00 A.M., and so forth.

The dashed line in Fig. 12-15, known as the *service curve,* represents the cumulative number of operations that will be served by any given time of the day. During the first hour of operation, between 5:00 A.M. and 6:00 A.M., the service curve is the same as the demand curve, illustrating the fact that during this relatively low demand period, when demand is less than capacity, aircraft, are served at the time they are scheduled. This portion of the cumulative arrival curve illustrates the fact that no significant delays occurred during this period.

During the peak period, however, the demand for service is greater than the service capacity. As such, the service curve illustrated during the peak period represents the maximum number of operations that could be served during this period, that is, the capacity of the system. Because the capacity of the system is only 60 operations per hour, by 8:00 A.M., the end of the peak period, only 150 cumulative operations could be served, less than the 180 that were scheduled.

The vertical distance between the demand and service curves represents the number of aircraft that are demanding service, but have not yet been served. That is, the number of aircraft *in queue* for service. The increasing vertical distance between the two curves during the peak period represents the ever-growing queue during this time. The end of the peak period bears witness to the longest queue, Q, during this analysis. In this case, Q equals 30 aircraft.

All aircraft that arrive during the peak period wait in a queue of some length, and thus experience an amount of delay. The amount of delay experienced by any given aircraft is illustrated by the horizontal distance between the demand curve and the service curve. For example, the ninetieth aircraft observed in this system was scheduled to operate at 6:45 A.M., but was not served until 7:00 A.M., resulting in 15 minutes of delay time for this aircraft.

As illustrated on the queuing diagram, queues and delays do not end at the end of the peak period. At 8:00 A.M., the system continues to operate at capacity. Because the scheduled demand is less than the capacity after the peak period ends, the queue is able to decrease until it ultimately disappears. In this case, the queue is dissipated at 9:00 A.M., a full hour after the end of the peak period. Thus, between 6:00 A.M. and 9:00 A.M., which includes the period of peak demand, and part of the off-peak period, which occurs after the peak period ends, aircraft scheduled to operate experience some amount of delay. This time period, T, is known as the period of time the system is in a delayed state.

On a queuing diagram, the amount of delay experienced by all aircraft in the system is defined by the area that exists between the demand and service curves. In this example, this area may be represented by calculating the area of the triangle defined by length T and height Q. Thus the total delay experienced by this system is defined by the area of the triangle, calculated as $1/2\ QT$. In this case, $1/2(30)(3) = 45$ operating hours of delay. Averaged over the 180 aircraft that experienced some delay, each aircraft experienced an average of 0.25 hours, or 15 minutes, of delay.

The queuing diagram illustrated in Fig. 12-15 is a very simple application of queuing theory as used to estimate airport capacity. For a more comprehensive description of this very rich methodology, the reader is referred to Gordon Newell's text, *Applications of Queuing Theory*.

Other measures of delay

Traditionally, the FAA has gathered aircraft delay data from two different sources. The first is through the **Air Traffic Operations Network System (OPSNET),** in which FAA personnel record aircraft that are delayed 15 or more minutes by specific cause (weather, terminal volume, center volume, closed runways or taxiways, and NAS equipment interruptions). Aircraft that are delayed by less than 15 minutes are not recorded in OPSNET.

The delay data reported through OPSNET is not without its problems. It reports only delays of 15 minutes or more; it aggregates flight delays, thus making it impossible to determine if a particular flight was delayed; and it reports only flight delays that are due to an air traffic problem (i.e., weather, terminal, volume, center volume, closed runways or taxiways, and NAS equipment interruptions). OPSNET is based on controller reports, and the quality and completeness of reporting vary considerably with controller workload. In addition, it measures delay against the standard of flight times published in the *Official Airline Guide* (OAG). This, in all probability, results in an overestimation of delay because there is wide variation in the "no delay" time from airport to airport and, at a given airport, among various runway configurations. Many operations, when measured against a single nominal standard, are counted as delays but are within the normal expectancy for a given airport under given circumstances. There might also be a distortion in the opposite direction. Most airline schedules, especially for flights into and out of busy airports, have a built-in allowance for delay. In part, this is simply realistic planning, but there is also a tendency to inflate published flight times so as to maintain a public image of online operation. Finally, OPSNET incorporates whatever delay might be experienced en route. Delays en route might not be attributable to conditions at the airport, and including them in the totals for airports probably leads to overestimation.

The second source of delay data is through the **Consolidated Operations and Delay Analysis System (CODAS).** CODAS is a newer FAA database and reporting system containing delay information by phase of flight for U.S. domestic flights. CODAS was developed by merging the former **airline service quality performance (ASQP)** database with the FAA's **Enhanced Traffic Management System (ETMS).** In addition, CODAS contains flight schedule information from the OAG and weather data from the National Oceanic and Atmospheric Administration (NOAA). CODAS contains actual times for gate out, wheels off, wheels on, and gate in. From this information, gate delays, taxi-out delays, airborne delays, and taxi-in delays as small as 1 minute are computed. CODAS measures a delay where it occurs, not where it is caused. The principal purpose of CODAS is to support analytical studies and not the day-to-day management of the ATC (air traffic control) system.

Approaches to reducing delay

Many commercial service airports, particularly those in large metropolitan areas, have experienced significant operating delays on their airfields, and also in their terminals, and on ground access systems around the airports, as well as between the airport and the associated area. The strategies that may be employed to reduce delays fall into two categories: increasing system capacity and managing system demand. Increasing capacity includes the addition of new infrastructure, such as additional runways, terminal facilities, and ground access roads. Increasing capacity also includes provision of technologies and policies to make existing infrastructures operate more efficiently. For example, it reduces the amount of processing time required at any given facility to allow more operations over a given period of time. Managing demand focuses more on changing the behavior of system users that in turn will lead to better use of existing system capacity.

Creating new airport infrastructure

Historically, the development of new airports and the construction of new runways and runway extensions at existing airports have offered the greatest potential for increasing aviation system capacity. The new Denver International Airport (DIA), completed in 1995, increased capacity and reduced delays not only in the Denver area, but also throughout the aviation system. However, at a cost of over $5 billion for an airport like Denver International, it will remain a challenge to finance and build others.

These options for achieving major capacity increases have become more difficult because of surrounding community development, environmental concerns, shortage of available adjacent property and funding required, lack of public support, rival commercial and residential interests, and other competing requirements.

Between 1997 and 2010, more than 25 new major commercial airport runways in the United States were opened. The additional runway capacity at the largest hubs in the United States, including Atlanta's Hartsfield-Jackson International Airport, provided significant increases to the nation's airport capacity, and significantly reduced delays in the system. The additional capacity also allowed air carriers to add flights during peak periods, which in some cases has begun to increase delays at these facilities.

Also, the modification of runway configurations, in particular converting interesting runways to parallel configurations, or lengthening shorter runways to accommodate larger aircraft, have been recent capacity-enhancing strategies at airports such as Chicago's O'Hare field and Fort Lauderdale–Hollywood International Airport.

In addition, the increase in properly located taxiways, particularly those that provide egress from runways, has an important effect on reducing runway occupancy time. The placement of exit taxiways, where landing aircraft turn off the runways, and the angle at which these taxiways intersect the runways can be crucial. Poorly placed exit taxiways prolong runway occupancy by forcing arriving aircraft to taxi at low speed for an excessive distance before clearing the runway. Taxiways that leave the runway at right angles force the aircraft to reduce to very slow speeds in order to safely turn off the runway. The addition of taxiways in strategic locations along runways can contribute to the minimization of runway occupancy times, leading to increased runway capacity.

Converting military airfields

In a somewhat similar manner to the transfer of military air bases to municipal civil aviation airports after World War II, the downsizing of United States military facilities in the 1990s contributed to an increase in commercial aviation system capacity by allowing the conversion of closed military airfields to civilian use. Most of the military airfields that have been considered for conversion are already designed to accommodate heavy aircraft, with runways up to 13,000 feet in length. Many of these airfields are located in the vicinity of congested metropolitan airports where the search for major new airports has been under way. A prime example of military airfield conversion is the Austin-Bergstrom International Airport, in Austin, Texas, located on the site of the Bergstrom Air Force Base.

In addition to military airfield conversions to civil airports, there are a number of military airfields now in operation accommodating joint civil and military use. For the most part, these joint-use airfields provide primary service to the communities and have a modest impact on system capacity.

To assist in the transition of military airfields to civilian and joint-use airports, the Military Airport Program (MAP), established as a funding set-aside under the Airport Improvement Program (AIP), provides grant funding of airport master planning and capital development. The MAP allows the secretary of transportation to designate current or former military airfields for participation in the program. To participate, eligible airport sponsors apply to the FAA. In determining whether or not to designate a facility, the FAA considers (1) proximity to major metropolitan air carrier airports with current or projected high levels of delay, (2) capacity of existing airspace and traffic flow patterns in the metropolitan area, (3) the availability of local sponsors for civil development, (4) existing levels of operation, (5) existing facilities, and (6) any other appropriate factors.

Administrative and demand management

Two basic approaches to managing demand have the same objective: to ease congestion by diverting some traffic to times and places where it can be handled more promptly or efficiently. This might be done through administrative management; the airport authority or another governmental body might allocate airport access by setting quotas on passenger enplanements or on the number and type of aircraft operations that will be accommodated during a specific period. The alternative approach is economic—to structure the pricing system so that market forces allocate scarce airport facilities among competing users; thus, demand management does not add capacity, it promotes more effective or economically efficient use of existing facilities.

Any scheme of demand management denies some users free or complete access to the airport of their choice. This denial is often decried as a violation of the traditional federal policy of freedom of the airways and the traditional "first-come, first-served" approach to allocating the use of airport facilities. Economists reject this argument on the grounds that it is a distortion of the concept of freedom to accord unrestricted access to any and all users without regard to the societal costs of providing airport facilities. Attempts to manage demand are also criticized for adversely affecting the growth of the aviation industry and the level of service to the traveling public. Nevertheless, as growth in traffic has outstripped the ability to expand and build airports, some forms of demand management have already come into use, and many industry observers have taken the position that some form of airport use restrictions will become increasingly important in dealing with delay and in utilizing existing airport capacity efficiently.

Administrative management

Several administrative management approaches are being adopted to manage demand at individual airports or for a metropolitan region. Among these are required diversion of some traffic to reliever airports, more balanced use of metropolitan air carrier airports, restriction of airport access by aircraft type or use, establishment of quotas (either on the number of operations or on passenger enplanements), and "rehubbing" or redistributing transfer traffic from busy airports to underused airports.

The best regionwide solution to the problem of delay at a major airport might be to divert some traffic away from the busy airport to either a general aviation reliever airport or a lightly used commercial airport. To some extent, this can occur as a result of natural market forces. When delays become intolerable at the busy airport, users begin to divert of their own accord. Even though those who choose to move to a less crowded facility do so for their own benefit, they also reduce somewhat delays incurred by users that continue to operate at the crowded airport. Public policy might encourage this diversion through administrative action or economic incentives before traffic growth makes conditions intolerable or necessitates capital investment to accommodate peaks of demand at the busy airport.

Diversion of general aviation from busy air carrier airports is often an attractive solution. GA traffic, because it consists mostly of small, slow-moving aircraft, does not mix well with faster, heavier air carrier traffic. GA operators—especially those flying for recreational or training purposes—want to avoid the delays and inconveniences (and sometimes the hazards) of operating at a major airport. These fliers are often willing to make use of GA airports located elsewhere in the region if suitable facilities are available.

Diversion of GA traffic from commercial air carrier airports has been taking place for many years. As air carrier traffic grows at a particular location, it almost always tends to displace GA traffic. FAA has encouraged this trend by designating approximately 334 airports as "relievers" to air carrier airports, and earmarking funds especially for developing and upgrading these airports. Many other airports, although not specifically designated as relievers, serve the same function; they provide an alternative operating site for GA aircraft well removed from the main commercial airport of the region.

Not all GA aircraft can make use of reliever airports. Some might be delivering passengers or freight to connect with commercial flights at the air carrier airport. Others might be large business jets that require the longer runways of a major airport. In general, airport authorities do not have the power to exclude GA as a class, although this has been attempted on occasion. For example, in the late 1970s, the airport management and city government of St. Louis attempted to exclude all private aircraft from Lambert Airport. This ordinance was overturned by the courts as discriminatory.

Where they have had any policy on the matter, local airport authorities have attempted to make GA airports attractive to users by offering good facilities or by differential pricing schemes. This approach is most effective where the commercial airport and the principal reliever are operated by the same entity. The state of Maryland, owner of Baltimore-Washington International Airport, operates a separate GA airport, Glenn L. Martin Field, and has a specific policy of encouraging GA traffic to use it rather than the main airport. The master plan for Cleveland Hopkins International Airport depends on the availability of the city-owned Burke Lakefront Airport as a reliever. If that airport should for some reason cease operation as a GA reliever, Hopkins would experience a great increase in traffic, which might necessitate additional construction that is not now planned.

Most local airport authorities, however, do not operate their own GA relievers. Some large airport authorities plan and coordinate activities with nearby reliever airports operated by other municipalities or private individuals, but this has not been the general case. The system of relievers in each region has tended to grow up without any specific planning or coordination on the regional level.

Development of GA relievers is not without problems. These airports are also subject to complaints about noise, and they experience the same difficulties as commercial airports in expanding their facilities or in developing a new airport site. Further, because many GA airports are small and function just on the ragged edge of profitability, problems of noise or competing land use can actually threaten the airport's existence. The number of airports available for public use in the United States has been declining. Although most of the airports that closed were small, privately owned facilities, some industry observers worry that the nation is irrevocably losing many potential reliever airports just as it has become clear that they are vital.

At the largest commercial service airports, GA activity consists primarily of flights by large business and executive aircraft. This type of GA traffic accounts for approximately 10 to 20 percent of the use of major airports, a figure that many consider the "irreducible minimum." The delays that persist at these airports are primarily the result of air carrier demand that can be satisfied only by another commercial service airport. In several metropolitan areas, it is clear that the commercial airports are not used in a balanced manner.

Air carriers, sensitive to public preferences, tend to concentrate their service at the busier airport, where they perceive a larger market. It is in the carrier's economic interest to serve the airport where passengers want to go. The busier airport is a known and viable enterprise, whereas the underutilized alternative airport is a risk. Air carriers are justifiably reluctant to isolate themselves from the major market by moving all of their services to the less popular airport. On the other hand, serving both airports imposes an economic burden that carriers seldom choose to bear, because they would incur the additional expense of setting up and operating duplicate ground services. In addition, splitting their passengers between two airports might make scheduling of flights more complicated and lead to inefficient utilization of aircraft.

These obstacles have sometimes been overcome in locations where airport operators have the authority to encourage a diversion of traffic from one airport to another. For example, in the New York area, the Port Authority of New York and New Jersey operates four air carrier airports, including Stewart International Airport in Newburgh. In theory, this gives the port authority the ability to establish regulatory policies or economic incentives in order to encourage the diversion of some traffic to Stewart, from the three larger and historically more congested airports at Newark, Kennedy, and LaGuardia. In practice, however, measures adopted to promote traffic redistribution have not been fully effective. The recent growth of traffic at Newark has been primarily due to new carriers entering the New York market and not diversion of established carriers.

One technique of administrative management now in use at a few airports is a quota, or slot, system, an administratively established limit on the number of operations per hour. Because delay increases exponentially as demand approaches capacity, a small reduction in the number of hourly operations can have a significant effect on delay. This makes the quota an attractive measure for dealing promptly (and inexpensively) with airport congestion.

Examples of airports with quotas are O'Hare, LaGuardia, JFK, and Ronald Reagan Washington International, airports covered by the FAA high-density rule. The quotas at these airports were established by FAA in 1973 on the basis of estimated limits of the ATC system and airport runways at that time. An example of a locally imposed quota is John Wayne Airport in Orange County, California, which limits scheduled air carrier operations to an annual average of 41 operations per day. This quota is based on noise considerations as well as limitations on the size of the terminal and gate areas.

During busy hours, demand for operational slots typically exceeds the quota. At the airports covered by the high-density rule, the slots are allocated among different user classes. For example, at Ronald Reagan Washington International Airport, where there are 60 slots available per hour, 37 are allotted to air carriers, 11 to commuter carriers, and 12 to general aviation.* During visual meteorological conditions, more than 60 operations can be handled, and aircraft without assigned slots may be accommodated at the discretion of air traffic controllers and the airport manager.

At airports where the quota system is in force, slots may be allocated in various ways: a reservation system, negotiation, or administrative determination. The GA slots are generally distributed through a reservation system—the first user to call for a reservation gets the slot.

However, for commuters and air carriers, the slots at the high-density rule airports are still subject to a great deal of controversy. In 1986, FAA declared the slots the property of the airlines holding them by allowing carriers to sell or lease slots to other airlines. A few available slots were also distributed by lottery.

A systemwide response to alleviate delays at busy airports is redistribution of operations to other, less busy airports in other regions. Some air carriers, especially those with a high proportion of interconnecting flights, might voluntarily move their operations to underutilized airports located some distance from the congested hub. Transfer passengers account for a large percentage of traffic at some large airports. About three-fourths of passengers at Atlanta and nearly half the passengers at Chicago, Denver, and Dallas/Fort Worth arrive at those airports merely to change planes for some other destination. There is an advantage to carriers in choosing a busy airport as a transfer hub—they can offer passengers a wide variety of possible connections; however, when the airport becomes too crowded, the costs of delay might begin to outweigh the advantages of the large airport, and carriers might find it attractive to establish new hubs at smaller, less busy airports.

---

*Since September 11, 2001, general aviation operations have been limited at Ronald Reagan Washington National Airport.

Demand management

Administrative management of airport use, whether by restricted access for certain types of aircraft, by demand balancing among metropolitan area airports, or by imposition of quotas offers the promise of immediate and relatively low-cost relief of airport congestion. As long-term measures, these solutions might not be as attractive. Administrative limits tend to bias the outcome toward maintenance of the status quo when applied over a long period of time. Because the economic value of airport access is not fully considered in setting administrative limits, incumbents cannot be displaced by others who would place a higher value on use of the airport. Further, incumbents and potential new entrants alike have no way to indicate the true economic value they would place on increased capacity. Economists contend that a vital market signal is missing and that airport operators and the federal government cannot obtain a true picture of future capacity needs. Administratively limiting demand, they say, creates an artificial market equilibrium that—over the long term—distorts appreciation of the nature, quality, and costs of air transportation service that the public requires. Some economists, therefore, favor a scheme of allocating airport access by **demand management** that relies on the price mechanism.

At present, price plays a rather weak role in determining airport access or in modulating demand. Access to public-use airports, except for the few large airports where quotas are imposed, is generally unrestricted so long as one is willing to pay landing fees and endure the costs of congestion and delay. Landing fees, most often based solely on aircraft weight and invariant by time of day, make up a very small fraction of operational cost, typically 2 to 3 percent for air carriers and even less for GA. Further, landing fees are not uniform from airport to airport. In many cases, landing fees are set so that, in the aggregate, they make up the difference between the cost of operating the airport and the revenues received from other sources such as concessions, leases, and automobile parking fees.

This leads economists to the conclusion that landing fees are somewhat arbitrary and do not reflect the costs imposed on the airport by an aircraft operation. Economists suggest that by including airport costs and demand as determinants of user fees, delay could be significantly reduced. The two most commonly advocated methods of achieving this are differential pricing and auctioning of landing rights.

Many economists argue that weight-based landing fees are counterproductive because they do not vary with demand, and consequently provide no incentive to utilize airport facilities during off-peak hours. Further, they do not reflect the high capital costs of facilities used only during peak hours. Thus, economists contend, a more effective pricing method would be to charge higher user fees during peak hours and lower fees during off-peak hours. Theoretically, the net effect of such a pricing policy would be a more uniform level of demand.

It is difficult to project accurately the changes in patterns of airport use that might be brought about by peak-hour surcharges. Some analysts estimate that peak-hour surcharges, along with improvement of the ATC system, would reduce anticipated air carrier delays significantly in the future. Others argue that although expansion might be inevitable at many airports, peak-hour surcharges could significantly delay the need for expansion and reduce financial pressure at a number of airports. Another important aspect of peak-hour surcharges noted by the Congressional Budget Office (CBO) is that even if they do not reduce traffic levels at peak hours to the desired levels, they could provide airports with increased revenues to expand facilities and, consequently, to reduce delays.

Some contend that a system of marginal cost pricing should be based on the delay costs that each peak-hour user imposes on other users. For example, during peak hours, airport users would be charged a fee based on the delay costs associated with their operations. This creates a system of user fees where the fees become progressively larger as delays increase. Proponents contend that using marginal delay costs as the basis for pricing airport access provides a stronger incentive for off-peak airport use than a scheme based on marginal facility costs alone.

Implementing a policy of differential pricing, whether based on marginal facility cost, marginal delay cost, or some purely arbitrary scheme, is difficult. It is likely that a significant increase in airport user fees will raise questions of equity. Higher fees might be more burdensome for small airlines and new entrants than for established carriers. There are a number of examples where airport operators have attempted to increase user fees and have been challenged by air carriers and general aviation. In some cases, air carrier landing fees are established in long-term contracts that cannot be easily changed. GA users often contend that differential pricing is discriminatory because it favors those with the ability to pay, and illegal because it denies the right to use a publicly funded facility. Economists rebut this argument by pointing out that time-of-use price is neither discriminatory nor illegal so long as price differences reflect cost differences. They contend that it is fair and just to set prices on the basis of the costs that each user imposes on others and on society generally.

Slot auctions have been advocated as the best method of allocating scarce airport landing rights on the grounds that if airport access must be limited, it should be treated as a scarce resource and priced accordingly. The method to accomplish this is a system whereby the price of airport access is determined by demand. Slot auctions allow peak-hour access only to those users willing to pay a market-determined price. However, as operations increase, there might not be enough extra capacity in the traditional off-peak time periods to accommodate additional operations without significant delays. At this point, slot allocations will only be able to reduce delay by effectively "capping" the total number of operations at the airport. This program can be cumbersome to execute both equitably and efficiently. Its use within this country has been restricted to the four high-density traffic airports, Ronald Reagan Washington International, Chicago O'Hare, New York LaGuardia, and New York Kennedy, where delays have historically affected the performance of the National Airspace System (NAS).

Critics also contend that the current slot sale process gives an advantage to the airlines already operating at the airport and denies access to competitors, providing the existing users with virtual monopolies and a financial windfall. Slot holders know that without a slot, no competitor can enter a market and, consequently, slots represent one of the most significant barriers to entry in the airline business today. Their impact on the industry extends far beyond the few airports where they are imposed because markets critical to many communities either begin or end at one of these facilities.

Concluding remarks

In the recent years before September 11, 2001, the single most pressing issue in the commercial aviation industry was that of airport capacity and delay. In 2001, demands on the system decreased significantly as fears of terrorism, a declining economy, and financial troubles of the major air carriers reduced the numbers of enplaning passengers and aircraft operations. Demand rebounded to the greatest levels in aviation history soon thereafter, only to decline again with the world economic downturn in the last years of the twenty-first century's first decade.

It is generally agreed upon, however, that in the future demands for air travel will soon recover, and in fact, rise to levels greater than in the history of aviation. To be prepared for this growth, airport planners and management, along with private industries and local, regional, and federal governments, should embrace the principles of airport capacity and demand management and seek ways to further improve the system to accommodate the future of air travel. While systemwide improvements continue to develop, airport management should always be aware of their individual environments, particularly when it comes to planning and managing capacity.

Key terms

capacity

throughput capacity

practical capacity

PHOCAP (practical hourly capacity)

PANCAP (practical annual capacity)

LAHSO (land and hold short operations)

ROT (runway occupancy time)

fleet mix

time-space diagram

SIMMOD

TAAM

delay

queuing diagram

OPSNET (Air Traffic Operations Network System)

CODAS (Consolidated Operations and Delay Analysis System)

ASQP (Airline Service Quality Performance)

ETMS (Enhanced Traffic Management System)

demand management

Questions for review and discussion

1. What is the theoretical definition of capacity?
2. What is the difference between throughput capacity and practical capacity?
3. What is PHOCAP? What is PANCAP?
4. What are the various factors that affect capacity and delay?
5. How do runway configurations affect capacity and delay?
6. What is the required lateral separation of parallel runways that allows simultaneous operations under IFR?
7. What is LAHSO? What are the advantages and disadvantages of LAHSO with respect to airport capacity?
8. What is ROT? How does ROT affect airport capacity?
9. How does an aircraft fleet mix affect capacity at an airport?
10. What is a time-space diagram? How can a time-space diagram be used in estimating runway capacity?

11. How are the FAA airport capacity approximation charts used?

12. What is ASV?

13. How is a mix index calculated?

14. What are two of the accepted simulation models that exist to estimate airport capacity?

15. What is the theoretical definition of delay?

16. What is meant by congestive delay?

17. How is queuing theory used to analytically estimate delay?

18. What are the two primary sources used by the FAA to gather aircraft delay data?

19. What are the various approaches to reducing delay?

Suggested readings

Airfield and Airspace Capacity/Delay Policy Analysis, FAA-APO-81-14, Washington, D.C.: FAA, Office of Aviation Policy and Plans, December 1981.

Airport Capacity Enhancement Plan. Washington, D.C.: FAA, October 2002.

Airport Congestion: Background and Some Policy Options. Washington, D.C.: Congressional Research Service, The Library of Congress, May 20, 1994.

Airport System Capacity-Strategic Choices. Washington, D.C.: Transportation Research Board, 1990.

Airport System Development. Washington, D.C.: U.S. Congress, Office of Technology Assessment, August 1984.

de Neufville, Richard. *Airport Systems Planning.* Cambridge, Mass.: MIT Press, 1976.

Newell, Gordon. *Applications of Queuing Theory.* London, England: Chapman-Hall, 1971.

Policy Analysis of the Upgraded Third Generation Air Traffic Control System. Washington, D.C.: Federal Aviation Administration, January 1977.

Report and Recommendations of the Airport Access Task Force. Washington, D.C.: Civil Aeronautics Board, March 1983.

FAA Capacity Analysis Approximation Charts

No.	Runway Configuration Diagram	Mix Index– Percent (C+3D)	Hourly Capacity (operations per hour)		Annual Service Volume (operations per year)
			VFR	IFR	
1.		0 to 20	98	59	230,000
		21 to 50	74	57	195,000
		51 to 80	63	56	205,000
		81 to 120	55	53	210,000
		121 to 180	51	50	240,000
2.	700' to 2,499'	0 to 20	197	59	355,000
		21 to 50	145	57	275,000
		51 to 80	121	56	260,000
		81 to 120	105	59	285,000
		121 to 180	94	60	340,000
3.	2,500' to 3,499'	0 to 20	197	62	355,000
		21 to 50	149	63	285,000
		51 to 80	126	65	275,000
		81 to 120	111	70	300,000
		121 to 180	103	75	365,000
4.	3,500' to 4,299'	0 to 20	197	62	355,000
		21 to 50	149	63	285,000
		51 to 80	126	65	275,000
		81 to 120	111	70	300,000
		121 to 180	103	75	365,000
5.	4,300' or more	0 to 20	197	119	370,000
		21 to 50	149	114	320,000
		51 to 80	126	111	305,000
		81 to 120	111	105	315,000
		121 to 180	103	99	370,000
6.	700' to 2,499' 700' to 2,499'	0 to 20	295	62	385,000
		21 to 50	213	63	305,000
		51 to 80	171	65	285,000
		81 to 120	149	70	310,000
		121 to 180	129	75	375,000

Figure A1-1A. *Preliminary analysis of capacity.*

No.	Runway Configuration Diagram	Mix Index– Percent (C+3D)	Hourly Capacity (operations per hour)		Annual Service Volume (operations per year)
			VFR	IFR	
7.	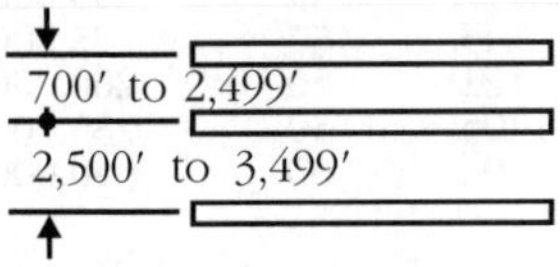 700' to 2,499' / 2,500' to 3,499'	0 to 20 21 to 50 51 to 80 81 to 120 121 to 180	295 219 184 161 146	62 63 65 70 75	385,000 310,000 290,000 315,000 385,000
8.	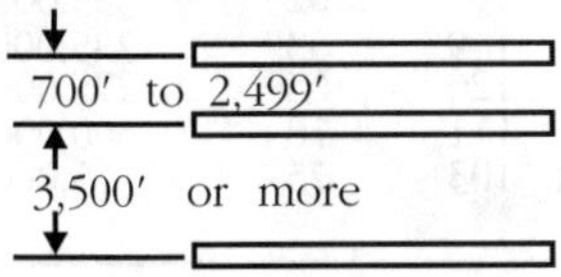 700' to 2,499' / 3,500' or more	0 to 20 21 to 50 51 to 80 81 to 120 121 to 180	295 219 184 161 146	119 114 111 117 120	625,000 475,000 455,000 510,000 645,000
9.	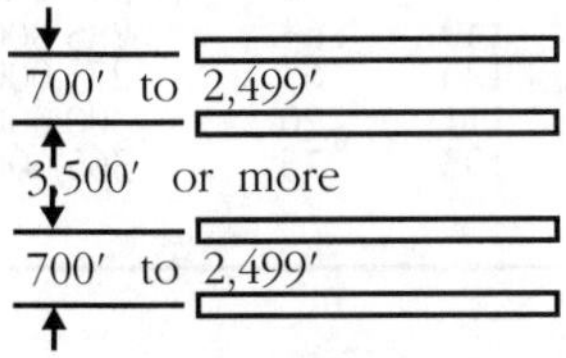 700' to 2,499' / 3,500' or more / 700' to 2,499'	0 to 20 21 to 50 51 to 80 81 to 120 121 to 180	394 290 242 210 189	119 114 111 117 120	715,000 550,000 515,000 565,000 675,000
10.		0 to 20 21 to 50 51 to 80 81 to 120 121 to 180	98 77 77 76 72	59 57 56 59 60	230,000 200,000 215,000 225,000 265,000

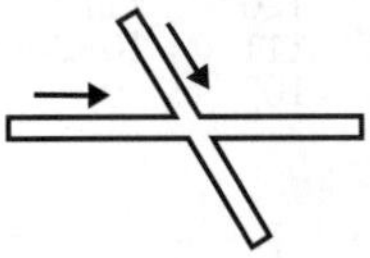

NOTE: ➤   Denotes predominant direction of runway operation.

Figure A1-1B. *Preliminary analysis of capacity (continued).*

No.	Runway Configuration Diagram	Mix Index– Percent (C+3D)	Hourly Capacity (operations per hour)		Annual Service Volume (operations per year)
			VFR	IFR	
11.	700' to 2,499'	0 to 20 21 to 50 51 to 80 81 to 120 121 to 180	197 145 121 105 94	59 57 56 59 60	355,000 275,000 260,000 285,000 340,000
12.	2,500' to 3,499'	0 to 20 21 to 50 51 to 80 81 to 120 121 to 180	197 149 126 111 103	62 63 65 70 75	355,000 285,000 275,000 300,000 365,000
13.	3,500' to 4,299'	0 to 20 21 to 50 51 to 80 81 to 120 121 to 180	197 149 126 111 103	62 63 65 70 75	355,000 285,000 275,000 300,000 365,000
14.	4,300' or more	0 to 20 21 to 50 51 to 80 81 to 120 121 to 180	197 149 126 111 103	119 114 111 105 99	370,000 320,000 305,000 315,000 370,000

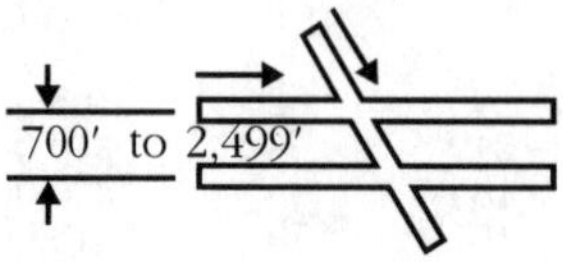

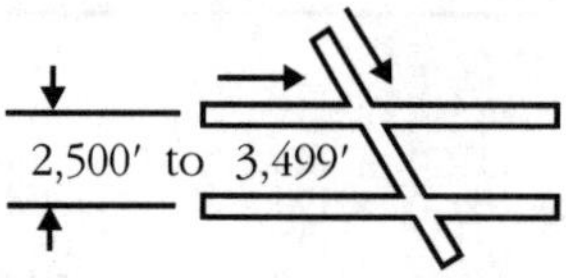

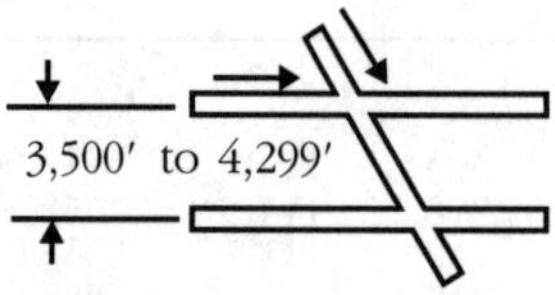

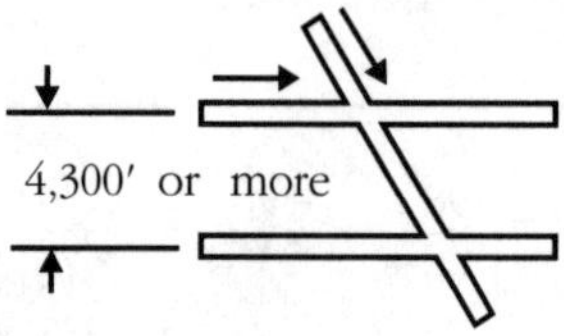

NOTE: →   Denotes predominant direction of runway operation.

Figure A1-1C. *Preliminary analysis of capacity (continued).*

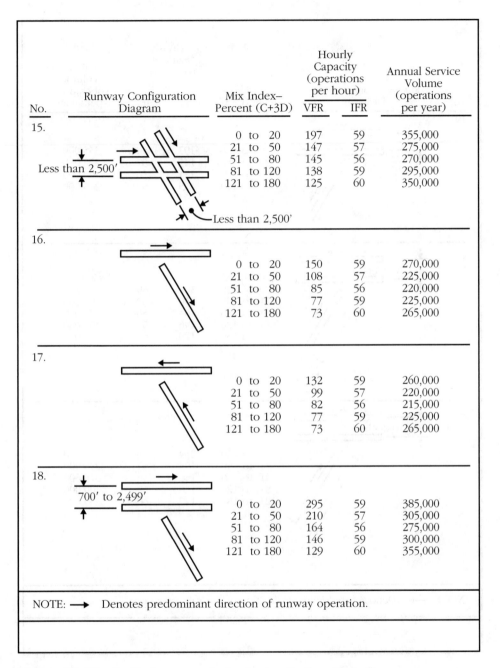

No.	Runway Configuration Diagram	Mix Index– Percent (C+3D)	Hourly Capacity (operations per hour) VFR	IFR	Annual Service Volume (operations per year)
15.	Less than 2,500' / Less than 2,500'	0 to 20	197	59	355,000
		21 to 50	147	57	275,000
		51 to 80	145	56	270,000
		81 to 120	138	59	295,000
		121 to 180	125	60	350,000
16.		0 to 20	150	59	270,000
		21 to 50	108	57	225,000
		51 to 80	85	56	220,000
		81 to 120	77	59	225,000
		121 to 180	73	60	265,000
17.		0 to 20	132	59	260,000
		21 to 50	99	57	220,000
		51 to 80	82	56	215,000
		81 to 120	77	59	225,000
		121 to 180	73	60	265,000
18.	700' to 2,499'	0 to 20	295	59	385,000
		21 to 50	210	57	305,000
		51 to 80	164	56	275,000
		81 to 120	146	59	300,000
		121 to 180	129	60	355,000

NOTE: → Denotes predominant direction of runway operation.

Figure A1-1D. *Preliminary analysis of capacity (continued).*

No.	Runway Configuration Diagram	Mix Index– Percent (C+3D)	Hourly Capacity (operations per hour)		Annual Service Volume (operations per year)
			VFR	IFR	
19.	700' to 2,499'				
		0 to 20	197	59	355,000
		21 to 50	145	57	275,000
		51 to 80	121	56	260,000
		81 to 120	105	59	285,000
		121 to 180	94	60	340,000
20.	700' to 2,499' 700' to 2,499'				
		0 to 20	301	59	385,000
		21 to 50	210	57	305,000
		51 to 80	164	56	275,000
		81 to 120	146	59	300,000
		121 to 180	129	60	355,000
21.	700' to 2,499' 700' to 2,499'				
		0 to 20	264	59	375,000
		21 to 50	193	57	295,000
		51 to 80	158	56	275,000
		81 to 120	148	59	300,000
		121 to 180	129	60	355,000

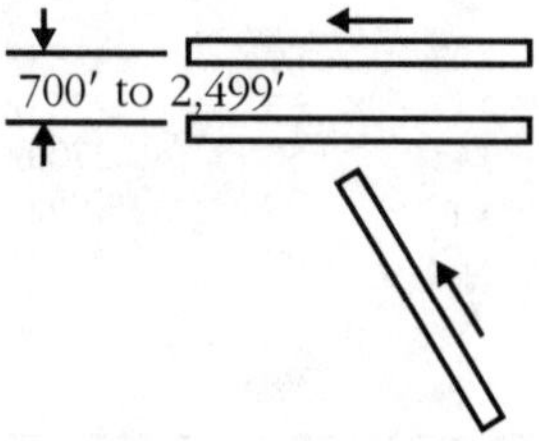

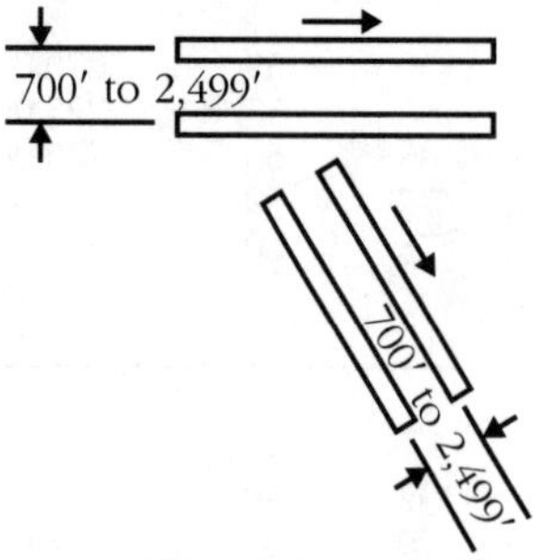

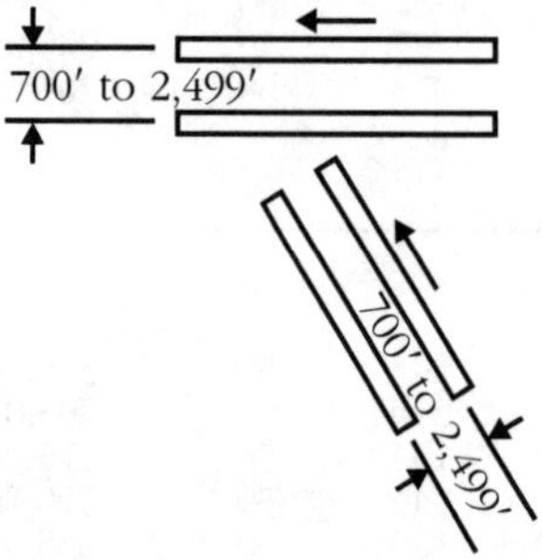

NOTE: ⟶ Denotes predominant direction of runway operation.

Figure A1-1E. *Preliminary analysis of capacity (continued).*

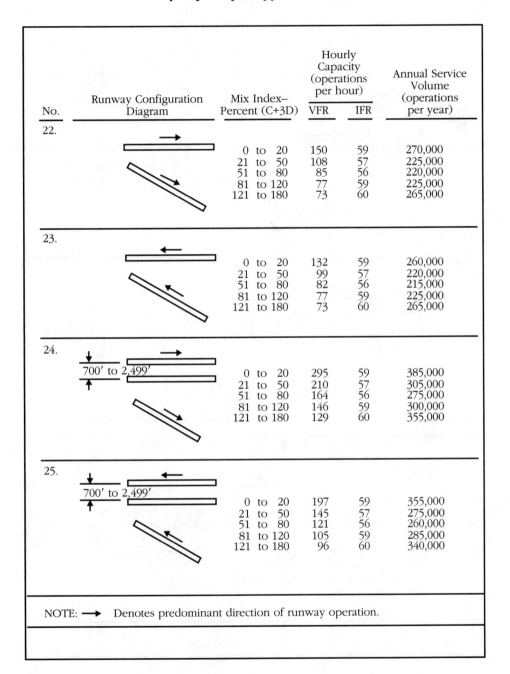

No.	Runway Configuration Diagram	Mix Index– Percent (C+3D)	Hourly Capacity (operations per hour)		Annual Service Volume (operations per year)
			VFR	IFR	
22.		0 to 20	150	59	270,000
		21 to 50	108	57	225,000
		51 to 80	85	56	220,000
		81 to 120	77	59	225,000
		121 to 180	73	60	265,000
23.		0 to 20	132	59	260,000
		21 to 50	99	57	220,000
		51 to 80	82	56	215,000
		81 to 120	77	59	225,000
		121 to 180	73	60	265,000
24.	700' to 2,499'	0 to 20	295	59	385,000
		21 to 50	210	57	305,000
		51 to 80	164	56	275,000
		81 to 120	146	59	300,000
		121 to 180	129	60	355,000
25.	700' to 2,499'	0 to 20	197	59	355,000
		21 to 50	145	57	275,000
		51 to 80	121	56	260,000
		81 to 120	105	59	285,000
		121 to 180	96	60	340,000

NOTE: ➜ Denotes predominant direction of runway operation.

Figure A1-1F. *Preliminary analysis of capacity (continued).*

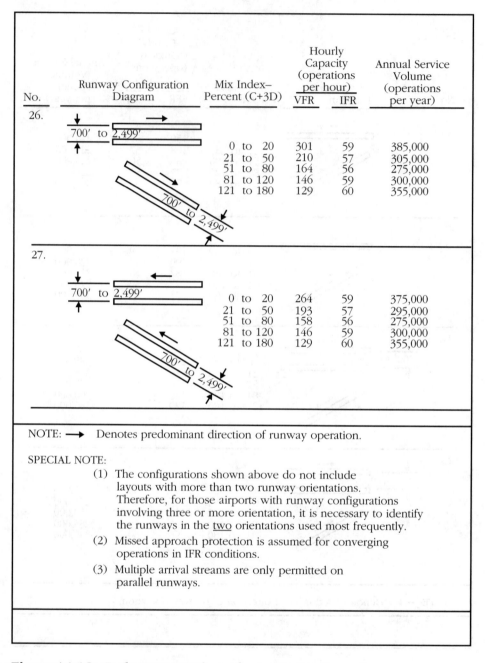

No.	Runway Configuration Diagram	Mix Index– Percent (C+3D)	Hourly Capacity (operations per hour)		Annual Service Volume (operations per year)
			VFR	IFR	
26.	700' to 2,499' 700' to 2,499'	0 to 20	301	59	385,000
		21 to 50	210	57	305,000
		51 to 80	164	56	275,000
		81 to 120	146	59	300,000
		121 to 180	129	60	355,000
27.	700' to 2,499' 700' to 2,499'	0 to 20	264	59	375,000
		21 to 50	193	57	295,000
		51 to 80	158	56	275,000
		81 to 120	146	59	300,000
		121 to 180	129	60	355,000

NOTE: ➡ Denotes predominant direction of runway operation.

SPECIAL NOTE:

(1) The configurations shown above do not include layouts with more than two runway orientations. Therefore, for those airports with runway configurations involving three or more orientation, it is necessary to identify the runways in the <u>two</u> orientations used most frequently.

(2) Missed approach protection is assumed for converging operations in IFR conditions.

(3) Multiple arrival streams are only permitted on parallel runways.

Figure A1-1G. *Preliminary analysis of capacity (continued).*

13

The future of airport management

Outline

- Introduction
- Reviewing previous predictions
  - Restructuring of commercial air carriers
  - New large aircraft, specifically the Airbus A-380
  - Small aircraft transportation systems (SATS)
- The future outlook for airport management
  - Enhanced safety
  - Environmental sustainability
  - FAA reauthorization
  - Future financial and marketing strategies
  - NextGen implementation
  - Globalization
  - The airport cities

Objectives

The objectives of this chapter are to educate the reader with information to:

- Discuss how the events of the early twenty-first century will continue to affect airport management in the future.
- Understand the near-term issues of safety and environmental sustainability with respect to the near-term future of airport management.
- Describe how the implementation of NextGen and new FAA Authorization will affect airport management in the future.
- Discuss the future impacts of increased globalization and potential land-use planning paradigms on airports.

Introduction

The civil aviation system that exists in the early part of the twenty-first century was virtually unimaginable just over 100 years ago, around the time the Wright Brothers made flight using powered, fixed-wing aircraft a reality. With this in mind, there is every reason to offer the consideration that the next 100 years will provide changes to the industry that will render the current civil aviation system obsolete. Airports in the distant future may be completely unrecognizable to their present-day counterparts, and the management of such future facilities may certainly be entirely different from the policies of today.

Although it is impossible to predict precisely what the future of airport management will entail over the next 100 years, it can be said with reasonable certainty that airports and airport management will evolve with changes in technologies, business policies, and governmental regulations. Airport management will further develop in order to address future operational issues, ranging from capacity and delay to safety and security, much the way they have matured over the industry's first 100-plus years.

Reviewing previous predictions

In the 8 years between the previous edition and the current edition of this text, and over the history of aviation, it is clear that, in many ways, the future of airport management is reflected in its past. Recent geopolitical events, economic conditions, and developments in technology have moved the future of airport management through a cycle of maturation.

The previous edition of this text identified the following items of interest for the future of airport management:

- Restructuring of commercial air carriers
- New large aircraft, specifically the Airbus A-380
- Small aircraft transportation systems (SATS)

These issues have and will continue to evolve in the future, perhaps as discussed in the following sections.

Restructuring of commercial air carriers

Commercial air carriers continue to restructure. The world economic downturn beginning in 2008 forced many air carriers out of business and others to consolidate. As a result, airport management has had to, and will continue to, become more flexible in its contractual obligations with air carriers, their aeronautical revenue structures, and increasingly the use of their facilities. Situations where large terminal facilities built for an expanding airline hub suddenly become underutilized because of airline downsizing or consolidation leave airports with little option but to manage the costs of operating such facilities with decreasing aeronautical revenues. In the future, airport management will continue to manage facilities with more flexibility that can robustly accommodate these large swings in demand.

The most recent mergers of Delta and Northwest Airlines and of United and Continental Airlines, and, as of 2010, the potential merger of Southwest Airlines and AirTran Airways perhaps will result in an ultimate consolidation of the industry into four major U.S. carriers in the short term. However, as seen throughout history, changes in aircraft technology, new economic cycles, geopolitical events, and regulatory reform may indeed result in an entirely new expansion of air carrier service to airports, and whether it be through traditional airline models, or small aircraft air taxi services, or even commercial space operations, airport management must, and will, adapt to future commercial air carrier structures.

To accommodate the consolidation of commercial air carriers, some experts foresee airports further focusing on certain segments of the industry. For example, one airport may decide to focus entirely on long-haul and international service while another may focus entirely on short-haul low-fare services. Outside of the scheduled commercial air carrier segment, other airports may focus entirely on cargo, charter, or corporate general aviation services. The "one-airport-fits-all" model may become the exception rather than the rule.

New large aircraft, specifically the Airbus A-380

When the previous edition of this text was published, the Airbus A-380 had yet to be introduced into the commercial air carrier market. As of 2010, more than 30 A-380 aircraft were in service operating under the flags of five air carriers (Air France, Emirates, Lufthansa, Qantas, and Singapore Airlines), none of which were U.S. carriers. Prior to its introduction, the primary concerns for the future of airport management was how to manage the airports runways, taxiways, gate areas, and terminals to accommodate this large and heavy aircraft and the more than 800 passengers per flight. What has resulted has been a relatively seamless, trouble-free adaptation on the part of airports, which reflects the results of thoughtful planning and analysis.

It is expected that more A-380s will be introduced into service at increasing numbers of airports. Those managing larger commercial service airports that do not currently accommodate the A-380 may take the knowledge gained from recent experiences to most quickly, easily, and flexibly adapt to this newest, largest aircraft in the future.

Small aircraft transportation systems (SATS)

The NASA-sponsored small aircraft transportation systems (SATS) program was designed to bring the latest navigation, communication, aircraft systems, and aircraft and engine designs from the research phase to market-ready production, in coordination with the implementation of a next-generation airspace system. In many ways, the program, which came to an end with demonstration events in 2005, successfully achieved its mission. The latest-generation aircraft, particularly in the general aviation sector, are clearly designed for the future of aviation (Fig. 13-1).

As early as 2006, SATS-designed general aviation aircraft had been arriving at the nation's airports. Airport management had thus begun to accommodate these aircraft in several ways, including adapting to new airspace procedures for approaching the airfield, creating new parking and passenger processing facilities, and bringing on new aircraft maintenance and repair services. While the recession, beginning in 2008, significantly reduced the proliferation of these

Figure 13-1. *Artist rendition of an SATS airport. While some elements of the SATS program are already being implemented into the current aviation system, others have yet to emerge.* (Picture courtesy NASA)

aircraft, it can certainly be expected that in the future, with an improvement in the economy, as well as with legislation to accelerate the implementation of the NextGen air transportation system, the proliferation of small aircraft will grow considerably. Moreover, with this proliferation of aircraft, new air transport services, such as air-taxis, will emerge, and airport management must adapt accordingly.

Much of the research performed with respect to automated aircraft control and navigational capabilities in the SATS program may also enhance airside operations at airports in the future, upon the implementation of NextGen. Such operations as described in the previous edition of this text include:

"Visual" approach procedures during IFR conditions: Virtual VFR relates to the concept that despite weather conditions that would force IFR flight plans and approaches, aircraft will be able to navigate similar to that during VFR conditions. Flying virtual VFR may allow for increased capacity because of reduced longitudinal separations for aircraft operations on a single runway, reduced required lateral separations between parallel runways for multiple runway operations, and simply the availability of the airfield under any weather condition without the need for traditional instrument NAVAIDs and procedures.

Multiple instrument approaches at airports: Because of the dynamic nature of enhanced navigation technologies developed via the SATS program, aircraft approaching a given runway may do so simultaneously from varying, and perhaps uniquely defined, approaches. Currently, every aircraft approaching a runway tends to fly the same published approach procedure, which limits the capacity of operations to the airport. With multiple instrument approaches, there is great potential for increasing the capacity.

Permissible land and hold short operations on a broader range of environments, including under IFR conditions: The increase in precision when it comes to tracking of aircraft, along with more highly accurate collision avoidance systems, might allow for reduced restrictions on land and hold short procedures. Allowing multiple operations on converging runways, especially during IFR conditions, would certainly have a positive effect on airfield capacity. Some of these technologies, such as ADS-B and ASDE-X, as described in Chap. 5 of this text, are already being implemented in airports and within the airspace system.

Simultaneous operations on a single runway: Aircraft with high navigation precision on airfields with sufficiently long runways may be able to regularly allow multiple takeoffs and/or landings simultaneously on a single runway. Even though this idea may seem extreme, such operations actually do occur under special-use situations. Formation takeoffs are not uncommon at private and military airfields, where civil aviation regulations are not necessarily enforceable. Furthermore, during periods of high demand, such as air shows, special landing procedures do provide for multiple simultaneous approaches to a single runway (e.g., land short, midfield entry, land long approach procedures). Although much investigation and testing would be necessary to prove that such operations would be safe on a regular basis, the SATS-developed technology associated with navigation may indeed make such operations possible, thereby providing great increases in airfield capacity.

The future outlook for airport management

It is difficult to define the future for an industry as volatile as aviation and airport management. The combination of quickly advancing technologies, volatile world economies, and abrupt air carrier business decisions with an often slower moving regulatory system and the extensive amount of time required to build large capital infrastructure renders many forecast attempts meaningless. However, it is important to at least attempt to forecast the future, so that airport management can have some basis for preparation and planning.

As witnessed in recent history, certain significant external events tend to have the greatest effect on the short-term future of the aviation system, including airports. In the early 2000s, the focus of airports abruptly changed from addressing issues of capacity and delays to enhancing security, in the wake of the events of September 11, 2001, for example. As the industry enters the second decade of the twenty-first century, much of the future for airports still does focus on enhanced security, as terrorist threats and plots continue to be revealed. However, as the events of 2001 fall further into history, other issues that will be addressed by airport management have come to the fore, and will continue to gain more focus. These short-term future issues are thought to be enhanced safety, environmental sustainability, and financial flexibility, along with preparation for a new FAA reauthorization act. In the longer term, adaptation to NextGen, further globalization of the industry, consideration of airport-centric metropolitan areas, and even the prospect of commercial space transportation via our nation's airports are all on the horizon.

Enhanced safety

As discussed earlier in this text, the Federal Aviation Agency (now Administration) was formed for the purpose of maintaining safety in the aviation system. The earliest policies created by the FAA focused on safety within the airspace, and then safety of aircraft and pilots, followed by safety at airports. As a result of these policies, the aviation system is at its safest in history.

One area of safety that continues to be addressed is airfield operational safety. While rare, any incidents that occur on an airfield, particularly a runway incursion, can have catastrophic results. To further mitigate the risk of such events, the FAA is in the process of enhancing safety-related policies that directly affect airport management. The most significant policy is the proposed rule that every airport operating under 14 CFR Part 139 creates and implements a safety management system (SMS), as discussed in Chap. 6 of this text. The implementation of this rule will have a significant effect on airports, and airport management must be prepared to become knowledgeable in SMS over the short-term future to successfully adapt to this rule.

In addition to SMS, the FAA is forecasted to continue to require enhancements to airfield infrastructure, such as signage, lighting, and markings, to enhance safety. There has also been discussion regarding modifications to the dimensions of runway safety areas, object-free zones, and runway protection zones, as well as creation of minimum ARFF personnel staffing requirements. Airport management in the future will have to be very much in communication with the FAA to anticipate, understand, and accommodate any new requirements.

Environmental sustainability

Two recent events in the early 2000s formalized the need for airports to introduce the concept of environmental sustainability. These events were the evidence of global climate change and the economic recession.

Studies of global climate change revealed that the heavy use of fossil fuels may have a significant effect on the atmosphere, resulting in greater climatic extremes, ranging from greater variations in seasonal temperatures to more extreme weather events such as hurricanes, blizzards, or floods. The heating and cooling of large buildings combined with the fuels burned to operate aircraft, ground service vehicles, and the vehicles used to transport passengers to and from airports (including personal automobiles) have resulted in airports becoming centers that extensively use fossil fuels. Thus the burden is on airport management to seek ways of reducing the use of fossil fuels in the future. In addition, with the economic recession, airports, as well as most other industries, have been seeking ways to reduce their overall expenditures and become more efficient. As a result, airport management will spend much of the short-term future focusing on the idea of environmental sustainability, that is, the operation of facilities that minimize the use of fuels and conserve energy.

Examples of environmental sustainability efforts will include the building and management of environment-friendly airport terminals that are well insulated and distribute conditioned air efficiently, airfield operations that minimize the travel of ground vehicles, and the increased use of transit systems that travel to, from, and around the airport that minimize the need for personal automobiles.

Other environmental sustainability issues that will continue to be the focus of airport management over the short term include conserving water and preserving water quality, increasing the use of environmentally sustainable materials and recycling, and of course finding ways to reduce the noise impact of airport operations on surrounding communities.

FAA reauthorization

As of the writing of this chapter in 2010, the Congress had yet to pass a reauthorization bill to replace the expired funding authorization written in the 2003 Vision 100—Century of Aviation Reauthorization Act. Among other issues, debate had ensued in the area of how to fund the Airport and Airway Trust Fund, what the new PFC level should be, and how these funds should be used. While there has been no new act passed, the AIP program has been run under a "continuing resolution," which has limited funding and forced airport management to put many major capital improvement projects on hold. Once a new reauthorization bill is passed, though, airport management must be prepared to move forward with capital projects that will best serve the demands for air service in the future.

Future financial and marketing strategies

Traditionally, airports operated strictly as an arm of municipal government, under a public works model of financial management. Recently, airports have begun to adopt more business-model philosophies for much of their operations, specifically in the areas of financial management and marketing. The future of airport management is expected to see a continuation of these trends.

In the area of revenue generation, traditional revenues generated from long-term airline contracts will continue to be less of a significant component of the overall revenue stream of an airport, in favor of short-term contracts with a more diverse customer base, including concession vendors, nonaeronautical facility rents, and other revenues drawn directly from passengers, such as parking and other services.

In the area of advertising and marketing, airports have begun to, and will continue to, take advantage of social networking platforms and other methods to directly market their services to the wider public at relatively little expense.

NextGen implementation

As described in detail in Chap. 5 of this text, the implementation of NextGen is designed to revolutionize air traffic management. As part of NextGen, airport management can expect newly designed instrument approach and departure procedures, which will in turn affect elements of airport management, ranging from airfield safety and capacity, to land-use planning.

Globalization

Management at many commercial service airports in the United States have had many years of experience with international service. As the range of commercial airliners has increased over the years, international service has also expanded from the traditional European markets directly served by east coast airports, South American markets from southern airports, and Asia-Pacific markets from west coast airports. As we enter the second decade of the twenty-first century, airports located anywhere in the United States can accommodate direct service to points all over the world, including emerging markets in China, India, the Middle East, and South America, provided the airport has the infrastructure to accommodate the aircraft and its passengers and cargo. Such infrastructure includes longer runways, larger passenger processing facilities, and enhanced security and customs processes. In addition to accommodating air service from around the world at airports in the United States, the opportunities for managing airports internationally are continuing to grow.

Look farther out, the longer-term future may bring supersonic or hypersonic air travel, and even low-earth orbit space travel to the public, allowing travel between any two points on earth in a matter of a few hours or less. At this time the requirements to accommodate such service at airports is entirely unknown, but as the years progress, such systems will most certainly develop, literally opening up a whole new world of airport management opportunities and challenges.

The airport cities

It is clear that airports are vital components of metropolitan areas, but often are located miles away from their core commercial and residential centers. This is due to the fact that cities have traditionally developed around a central business district of commercial and industrial activity.

In fact, traditional central business districts were originally developed around major transportation facilities, such as shipping ports or railroad terminals. There is a philosophy developing that perhaps another major transportation center, such as an airport, forms the center of urban development, for it is the airport, and associated intermodal facilities that airports are already beginning to evolve into, that may in fact sustain, and spur, economic development. This philosophy is known as the concept of the "airport city" or "aerotropolis."

Creating urban development with the airport as the core may reveal additional responsibilities to airport management. Airport management may indeed take on greater roles in the land-use planning and economic development activities of their metropolitan areas. This potential certainly provides motivation for airport managers to broaden their education and experience in the area of public administration, land-use planning, and urban economic development.

Concluding remarks

The future of airports as part of the civil aviation system in the United States and internationally is certainly an open book, with no foreseeable conclusion. However, much of the fundamentals that govern airports and airport management will, for the most part, remain a constant. The fundamental physics of flight will never change, nor will the fundamentals that govern air traffic operations. Technologies are sure to change, as are the regulations and funding mechanisms that facilitate the adaptation of technologies. The policies that direct the process of aviation are sure to change as well.

It is the responsibility of airport management and others with interests in airport operations to constantly evolve their knowledge of this ever-changing industry. A constant maturation of airport management from a business, operations, and public relations standpoint, along with a working knowledge of the fundamentals that form the foundation of the industry, will no doubt contribute to civil aviation being a most important part of our world's transportation and social framework.

Key terms

SATS

SMS

Sustainability

Reauthorization

NextGen

Globalization

Aerotropolis

Questions for review and discussion

1. What is the future of the current commercial air carrier industry?
2. How will future changes in air carrier strategies affect airport management?
3. How will the A-380 affect the planning, design, and management of airport facilities?
4. What was SATS? How has SATS affected civil air transportation? How will SATS affect airport management?
5. In your opinion, what is the future of civil air transportation, in general, and airport management, in particular?

6. How will new airport safety management systems impact airport management?

7. In what ways can airports become more environmentally sustainable?

8. How will FAA reauthorization impact airport management?

Suggested readings

Airports in the 21st Century, Washington, D.C.: Transportation Research Circular, April 2000.

Future Flight, A Review of the Small Aircraft Transportation System Concept.

Special Report 263. Washington, D.C.: Transportation Research Board, National Academy Press, 2002.

Kasarda, Lindsay, *Aerotropolis, The Way We Live Next.* Self-published book found at http://www.aerotropolis.com

Airport Sustainability Practices, ACRP Synthesis 10, Airports Cooperative Research Program, FAA Washington, D.C., 2009.

Code of Federal Regulations
14 CFR—Aeronautics and Space, Parts 1 through 199: Federal Aviation Regulations

Subchapter A—Definitions

14 CFR Part 1: Definitions and abbreviations

Subchapter B—Procedural rules

14 CFR Part 11: General rulemaking procedures

14 CFR Part 13: Investigative and enforcement procedures

14 CFR Part 14: Rules implementing the Equal Access to Justice Act of 1980

14 CFR Part 15: Administrative claims under Federal Tort Claims Act

14 CFR Part 16: Rules of practice for federally assisted airport enforcement proceedings

14 CFR Part 17: Procedures for protests and contracts disputes

Subchapter C—Aircraft

14 CFR Part 21: Certification procedures for products and parts

14 CFR Part 23: Airworthiness standards: Normal, utility, acrobatic, and commuter category airplanes

14 CFR Part 25: Airworthiness standards: Transport category airplanes

14 CFR Part 27: Airworthiness standards: Normal category rotorcraft

14 CFR Part 29: Airworthiness standards: Transport category rotorcraft

14 CFR Part 31: Airworthiness standards: Manned free balloons

14 CFR Part 33: Airworthiness standards: Aircraft engines

14 CFR Part 34: Fuel venting and exhaust emission requirements for turbine engine powered airplanes

14 CFR Part 35: Airworthiness standards: Propellers

14 CFR Part 36: Noise standards: Aircraft type and airworthiness certification

14 CFR Part 29: Airworthiness directives

14 CFR Part 43: Maintenance, preventive maintenance, rebuilding, and alteration

14 CFR Part 45: Identification and registration marking

14 CFR Part 47: Aircraft registration

14 CFR Part 49: Recording of aircraft titles and security documents

14 CFR Part 50–59 (reserved)

Subchapter D—Airmen

14 CFR Part 60 (reserved)

14 CFR Part 61: Certification: Pilots, flight instructors, and ground instructors

14 CFR Part 63: Certification: Flight crewmembers other than pilots

14 CFR Part 65: Certification: Airmen other than flight crewmembers

14 CFR Part 67: Medical standards and certification

Subchapter E—Airspace

14 CFR Part 71: Designation of class A, class B, class C, class D, and class E airspace areas; airways; routes; and reporting points

14 CFR Part 73: Special use airspace

14 CFR Part 75 (reserved)

14 CFR Part 77: Objects affecting navigable airspace

Subchapter F—Air traffic and general operating rules

14 CFR Part 91: General operating and flight rules

14 CFR Part 93: Special air traffic rules

14 CFR Part 95: IFR altitudes

14 CFR Part 97: Standard instrument approach procedures

14 CFR Part 99: Security control of air traffic

14 CFR Part 101: Moored balloons, kites, unmanned rockets and unmanned free balloons

14 CFR Part 103: Ultralight vehicles

14 CFR Part 105: Parachute Operations

Subchapter G–Air carriers and operators for compensation or hire: Certification and operations

14 CFR Part 119: Certification: Air carriers and commercial operators

14 CFR Part 121: Operating requirements: Domestic, flag, and supplemental operations

14 CFR Part 125: Certification and operations: Airplanes having a seating capacity of 20 or more passengers or a maximum payload capacity of 6,000 pounds or more; and rules governing persons on board such aircraft

14 CFR Part 129: Operations: Foreign air carriers and foreign operators of U.S.-registered aircraft engaged in common carriage

14 CFR Part 133: Rotorcraft external-load operations

14 CFR Part 135: Operating requirements: Commuter and on demand operations and rules governing persons on board such aircraft

14 CFR Part 136: National parks air tour management

14 CFR Part 137: Agricultural aircraft operations

14 CFR Part 139: Certification of Airports

Subchapter H–Schools and other certificated agencies

14 CFR Part 140 (reserved)

14 CFR Part 141: Pilot schools

14 CFR Part 142: Training centers

14 CFR Part 143 (reserved)

14 CFR Part 145: Repair stations

14 CFR Part 147: Aviation maintenance technician schools

Subchapter I–Airports

14 CFR Part 150: Airport noise compatibility planning

14 CFR Part 151: Federal aid to airports

14 CFR Part 152: Airport aid program

14 CFR Part 155: Release of airport property from surplus property disposal restrictions

14 CFR Part 156: State block grant pilot program

14 CFR Part 157: Notice of construction, alteration, activation, and deactivation of airports

14 CFR Part 158: Passenger facility charges (PFCs)

14 CFR Part 161: Notice and approval of airport noise and access restrictions

14 CFR Part 169: Expenditure of federal funds for nonmilitary airports or air navigation facilities thereon

Subchapter J—Navigational facilities

14 CFR Part 170: Establishment and discontinuance criteria for air traffic control services and navigational facilities

14 CFR Part 171: Non-federal navigation facilities

Subchapter K—Administrative regulations

14 CFR Part 183: Representatives of the Administrator

14 CFR Part 185: Testimony by employees and production of records in legal proceedings, and service of legal process and pleadings

14 CFR Part 187: Fees

14 CFR Part 189: Use of Federal Aviation Administration communications system

14 CFR Part 193: Protection of voluntarily submitted information

Subchapter N—War risk insurance

14 CFR Part 198: Aviation insurance

Code of Federal Regulations
49 CFR—Transportation, 1500 Series:
Transportation Security Regulations

49 CFR Part 1500: Applicability, terms, and abbreviations

49 CFR Part 1502: Organization, functions, and procedures

49 CFR Part 1503: Investigative and enforcement procedures

49 CFR Part 1510: Passenger civil aviation security service fees

49 CFR Part 1511: Aviation security infrastructure fee

49 CFR Part 1520: Protection of sensitive security information

49 CFR Part 1540: Civil aviation security: general rules

49 CFR Part 1542: Airport security

49 CFR Part 1544: Aircraft operator security: air carriers and commercial operators

49 CFR Part 1546: Foreign air carrier security

49 CFR Part 1548: Indirect air carrier security

49 CFR Part 1550: Aircraft security under general operating and flight rules

49 CFR Part 1570: Land transportation security: general rules

49 CFR Part 1572: Credentialing and background checks for maritime and land transportation security

Federal Aviation Administration 150 Series Advisory Circulars

Airport planning

150/5000-5C Designated U.S. International Airports (12-4-96) (AAS-3). Explains the different categories of U.S. airports designated to serve international air traffic and provides a listing of these airports.

150/5000-7 Announcement of Availability Report No. DOT/FAA/PP/ 87-1, Measuring the Regional Economic Significance of Airports.

150/5000-9A Announcement of Availability Report No. DOT/FAA/ PP/92-5, Guidelines for the Sound Insulation of Residences Exposed to Aircraft Operations (7-2-93) (APP-510).

150/5000-10A Announcement of Availability Report No. DOT/FAA/ PP/-92-6, Estimating the Regional Economic Significance of Airports (7-2-93) (APP-400).

150/5000-11 Announcement of Availability: All Cargo Carrier Activity Report (FAA Form 5100-108, Revised) (3-34-93) (APP-400). Provides guidance for the submission of the All Cargo Carrier Activity Report (revised as of March 1993).

150/5000-12 Announcement of Availability: Passenger Facility Charge (PFC) Application (FAA Form 5500-1) (7-15-94) (APP-530). Provides guidance for the submission of the PFC application. View more recent versions of the PFC application.

150/5000-13 Announcement of Availability: RTCA Inc., Document RTCA-221, Guidance and Recommended Requirements for Airport Surface Movement Sensors (9-7-94) (AAS-100).

150/5020-1 Noise Control and Compatibility Planning for Airports (8-5-83) (APP-600). Provides general guidance for noise control and compatibility planning for airports as well as specific guidance for preparation of airport noise exposure maps and airport noise compatibility programs by airport operators for submission under Federal Aviation Regulation Part 150 and the Aviation Safety and Noise Abatement Act of 1979. Contains an expanded Table of Land Uses Normally Compatible with Various Levels of Noise.

150/5050-3B Planning the State Aviation System (1-6-89) (APP-400). Provides general guidance in preparing a state airport system plan. SN 050-007-00813-9.

150/5050-4 Citizen Participation in Airport Planning (9-26-75) (APP-600). Provides guidance for citizen involvement in airport planning. Although not mandatory for airport grant programs, it explains the need for early citizen participation.

150/5050-7 Establishment of Airport Action Groups (6-23-87) (AAS-300). Provides guidance on the establishment of airport action groups.

150/5060-5 Airport Capacity and Delay (9-23-83) (AAS-100) (consolidated reprint includes change 1). Explains how to compute airport capacity and aircraft delay for airport planning and design.

150/5060-5 Change 2 (12-1-95) (AAS-100). Reflects the increased capacity resulting from a change in parallel runway separation criteria, adds a procedure far calculating savings associated with a reduction in aircraft delay, and updates the capacity and delay computer program references.

150/5070-5 Planning the Metropolitan Airport System (5-22-70) (APP-400). Gives guidance in developing airport-system plans for large metropolitan areas. It may be used by metropolitan planning agencies and their consultants in preparing such system plans and by the FAA in reviewing same. SN 050-008-00003-7.

150/5070-6A Airport Master Plans (6-85) (APP-400). Provides guidance for the preparation of airport master plans, pursuant to the provisions of the Airport and Airway Improvement Act of 1982. SN 050-007-00703-5.

Federal-Aid Airport Programs

150/5100-6D Labor Requirements for the Airport Improvement Program (AIP) (10-15-86) (APP-510). Encompasses the basic labor and associated requirements for the Airport Improvement Program (AIP). It is intended for sponsors using program assistance and for contractors and subcontractors working on projects under the program.

150/5100-10A Accounting Records Guide for Airport Aid Program Sponsors (4-13-76) (APP-500). This advisory circular sets forth record-keeping requirements imposed on sponsor of Airport Development Aid Program (ADAP) and the Planning Grant Program (PGP) projects by the Airport and Airway Development Act of 1970, as amended. In addition, the Federal Aviation Regulations (FARs) require a sponsor to establish and maintain a financial management system that meets the standards set forth in FAR 152, Appendix K. This circular provides detailed explanations of these requirements.

150/5100-13 Development of State Standards for General Aviation Airports (3-1-77) (AAS-100). Provides guidelines and programming procedures for the development of state standards for general aviation airports as provided for in the Airport and Airway Development Act Amendments of 1976.

150/5100-13A Development of State Standards for Non-Primary Airports (9-28-99) (AAS-100). This advisory circular provides guidelines for the development of state standards for nonprimary public-use airports as provided for in Title 49 United States Code, Section 47105(c).

150/5100-14C Architectural, Engineering, and Planning Consultant Services for Airport Grant Projects (2-16-94) (AAS-100). Provides guidance for airport sponsors in the selection and employment of architectural, engineering, and planning consultants under Federal Aviation Administration airport grant programs.

150/5100-15A Civil Rights Requirements for the Airport Improvement Program (AIP) (3-31-89) (APP-510). Encompasses the basic civil rights requirements for the Airport Improvement Program (AIP). It is intended for sponsors using program assistance and for contractors and subcontractors working on projects under the program.

150/5100-16A Airport Improvement Program Grant Assurance Number One—General Federal Requirements (10-4-88) (APP-510). Describes the federal requirements contained in Assurance 1 of the Grant Assurances required by the Airport and Airway Improvement Act of 1982, as amended. It is intended for sponsors receiving assistance under the Airport Improvement Program.

150/5100-17 Land Acquisition and Relocation Assistance for Airport Improvement Program Assisted Projects (3-29-96) (APP-600). Provides guidance to sponsors of airport projects developed under the Airport Improvement Program (AIP) to meet the requirements of the Uniform Relocation Assistance and Real Property Acquisition Policies Act of 1970 (Pl 91-646, as amended) and the Regulations of the Office of the Secretary of Transportation, 49 CFR Part 24.

150/5100-18 Guide for Audit Certification by Airport Sponsors (8-31-98) (AAS-400). Superseded and replaced by OMB Circular A-133, Audits of States, Local Governments, and Non-Profit Organizations, revised June 24, 1997, and by the OMB Circular A-133, Compliance Supplement for the Department of Transportation, dated on or after March 2000.

150/5100-19B Change 2 (1-15-03) **Guide for Airport Financial Reports Filed by Airport Sponsors** (AAS-400). Provides airport sponsors with guidance for complying with the airport financial reporting requirements. For additional information and to download Excel versions of the report forms, visit Airport Compliance Program.

Surplus Airport Property Conveyance Programs

150/5150-2B Federal Surplus Personal Property for Public Airport Purposes (10-1-84) (APP-510). Attempts to acquaint public airport owners and other interested parties with the Federal Surplus Personal Property Program for public airports and to outline procedures to be used in applying for and acquiring surplus personal property for this purpose.

Airport Compliance Program

150/5190-1A Minimum Standards for Commercial Aeronautical Activities on Public Airports (12-16-85) (AAS-310). CANCELLED. Replaced by 150/5190-5, Exclusive Rights and Minimum Standards for Commercial Aeronautical Activities. Contact AAS-400 for further information.

150/5190-2A Exclusive Rights at Airports (4-4-72). (Consolidated reprint incorporates change 1) (AAS-300) CANCELLED. Replaced by 150/5190-5, Exclusive Rights and Minimum Standards for Commercial Aeronautical Activities. Contact AAS-400 for further information.

150/5190-4A A Model Zoning Ordinance to Limit Height of Objects Around Airports (12-14-87) (AAS-100) (editorially updated). Provides a model zoning ordinance to be used as a guide to control the height of objects around airports.

150/5190-5 Change 1 (06/10/02), Exclusive Rights and Minimum Standards for Commercial Aeronautical Activities. Contact AAS-400 for further information.

Airport Safety—General

150/5200-12B Fire Department Responsibility in Protecting Evidence at the Scene of an Aircraft Accident (9-3-99) (AAS-300). Furnishes general guidance for airport, employees, airport management, and other personnel responsible for fire fighting and rescue operations, at the scene of an aircraft accident, on the proper presentation of evidence (HTML version).

150/5200-18B Airport Safety Self-Inspection (5-2-88) (AAS-310). Provides information to airport operators on airport self-inspection programs and identifies items that should be included in such a program.

150/5200-18B Autoinspeccion De Seguridad Para Los Aeropuertos (5-2-88). (AAS-310) (Spanish version).

150/5200-28B Notices to Airmen (NOTAMS) for Airport Operators (6-20-96) (AAS-310). Provides guidance for use of the NOTAM system in airport condition reporting.

150/5200-29 Announcement of Availability: Airport Self Inspection Videotape (11-6-87) (AAS-310). Announces the availability of airport self inspection videotape and tells how it can be obtained.

150/5200-30A Airport Winter Safety and Operations (10-1-91) (incorporates Changes 1 & 2) (AAS-100). Provides guidance to assist airport owners/operators in the development of an acceptable airport snow and ice control program and to provide guidance on appropriate field condition reporting procedures.

150/5200-30A Change 2 (3-27-95). Chapters 2 and 3 have been revised to incorporate additional guidance in conducting runway-friction surveys under winter operational conditions, and to provide a generic specification for fluid-based runway deicers/anti-icers that may be used on aircraft operational areas.

AC 150/5200-30A Change 3 (11-30-98) (AAS-100). Provides guidance to assist airport operators in applying sand to runways under winter operational conditions, the use of runway and taxiway edge light markers, and in reporting runway friction measurements taken under winter operational conditions.

AC 150/5200-30A Change 4 (11-15-99) (AAS-100). Airport Winter Safety and Operations. This change modifies guidance provided in reporting runway friction values measured under winter operational conditions.

AC 150/5200-30A Change 5 (1-15-02) (AAS-100). This change updates Appendix 4, FAA-Approved Decelerometers.

AC 150/5200-30A Change 6 (12-16-02) (AAS-100). This change updates manufacturer addresses in Appendix 4, FAA-Approved Decelerometers (HTML version).

150/5200-31A Airport Emergency Plan (9-30-99) (AAS-310). Provides guidance for the preparation and implementation of emergency plans at civil airports.

150/5200-33 Hazardous Wildlife Attractants on or Near Airports (5-1-97) (AAS-310). Provides guidance on locating certain land uses having the potential to attract hazardous wildlife to or in the vicinity of public-use airports.

150/5200-32 Announcement of Availability: Bird Strike Incident/ Ingestion Report (2-14-90) (AAS-310). Explains the nature of the revision of FAA Form 5200-7, Bird Strike Incident/Ingestion Report and how it can be obtained.

150/5200-34 Construction or Establishment of Landfills near Public Airports (8-26-2000) (AAS-300). Contains guidance on complying with new federal statutory requirements regarding the construction or establishment of landfills near public airports.

150/5210-2A Airport Emergency Medical Facilities and Services (11-27-84) (AAS-300). Provides information and advice so that airports may take specific voluntary preplanning actions to assure at least minimum first-aid and medical readiness appropriate to the size of the airport in terms of permanent and transient personnel.

150/5210-5B Painting, Marking, and Lighting of Vehicles Used on an Airport (7-11-86) (AAS-100). Provides guidance, specifications, and standards, in the interest of airport personnel safety and operational efficiency, for painting, marking, and lighting of vehicles operating in the airport air operations areas.

150/5210-6C Aircraft Fire and Rescue Facilities and Extinguishing Agents (1-28-85) (AAS-100). Outlines scales of protection considered as the recommended level compared with the minimum level in Federal Aviation Regulation Part 139.49 and tells how these levels were established from test and experience data.

150/5210-7C Aircraft Rescue and Firefighting Communications (7-1-99) (AAS-300). Provides guidance for planning and implementing the airport Aircraft Rescue and Firefighting (ARFF) communications systems.

150/5210-13A Water Rescue Plans, Facilities, and Equipment (5-31-91) (AAS-100). Provides guidance to assist airport operators in preparing for water rescue operations.

150/5210-14A Airport Fire and Rescue Personnel Protective Clothing (7-13-95) (AAS-100). Developed to assist airport management in the development of local procurement specifications for an acceptable, cost-effective proximity suit for use in aircraft rescue and firefighting operations.

150/5210-15 Airport Rescue and Firefighting Station Building Design (7-30-87) (AAS-100). Provides standards and guidance for planning, designing, and constructing and airport rescue and firefighting station.

150/5210-17 Programs for Training of Aircraft Rescue and Firefighting Personnel (3-9-94) (AAS-100). Provides information on courses and reference materials for training of aircraft and firefighting (ARFF) personnel.

150/5210-17 Change 1 (4-6-95). Changed to reflect a new source for the FAA Standard Basic Aircraft Rescue and Firefighting Curriculum, and to update other sources of training programs.

150/5210-18 Systems for Interactive Training of Airport Personnel (4-13-94) (AAS-100). Provides guidance in the design of systems for interactive training of airport personnel.

150/5210-19 Driver's Enhanced Vision System (DEVS) (12-23-96) (AAS-100). Contains performance standards, specifications, and recommendations for DEVS.

150/5210-20 Ground Vehicle Operations on Airports (6-21-02) (AAS-300). Contains guidance to airport operators developing ground vehicle operation training programs.

150/5220-4B Water Supply Systems for Aircraft Fire and Rescue Protection (7-29-92) (AAS-100). Provides guidance for the selection of a water source and standards for the design of a distribution system to support aircraft rescue and firefighting (ARFF) service operations on airports.

150/5220-9 Aircraft Arresting System for Joint Civil/Military Airports (4-6-70) (AAS-300). Updates existing policy and describes and illustrates the various types of military aircraft emergency arresting systems that are now installed at various joint civil/military airports. It also informs users of criteria concerning installations of such systems at joint civil/military airports.

150/5220-10C Guide Specification for Water/Foam Aircraft Rescue and Firefighting Vehicles (2-18-02) (AAS-100). Contains performance standards, specifications, and recommendations for the design, construction, and testing of a family of aircraft rescue and firefighting (ARFF) vehicles (HTML version).

150/5220-13B Runway Surface Condition Sensor Specification Guide (3-27-91) (AAS-100). Provides guidance to assist airport operators, consultants, and design engineers in the preparation of procurement specifications for sensor systems which monitor and report runway surface conditions.

150/5220-16A Automated Weather Observing Systems (AWOS) for Non-Federal Applications (6-12-90) (ANW-140). Contains the FAA standard for non-Federal Automated Weather Observing Systems.

150/5220-17A Design Standards for an Aircraft Rescue and Firefighting Training Facility (1-31-92) (AAS-100). Contains standards, specifications, and recommendations for the design of an aircraft rescue and fire-fighting training facility utilizing either propane or a flammable liquid hydrocarbon (FLH) as the fuel.

150/5220-17A Change 1 (11-24-98)**. Change 1 to Design Standards for an Aircraft Rescue and Firefighting (ARFF) Training Facility.**

150/5220-18 Buildings for Storage and Maintenance of Airport Snow and Ice Control Equipment and Materials (10-15-92) (AAS-100). Provides guidance for site selection, design, and construction of buildings used to store and maintain airport snow and ice control equipment and materials.

150/5220-19 Guide Specification for Small Agent Aircraft Rescue and Fire Fighting Vehicles (12-7-93) (AAS-100). Contains performance standards, specifications, and recommendations for the design, construction, and testing of a family of small, dual agent aircraft rescue and fire fighting (ARFF) vehicles.

150/5220-20 Airport Snow and Ice Control Equipment (6-30-92) (AAS-100). Provides guidance to assist airport operators in the procurement of snow and ice control equipment for airport use.

150/5220-20 Change 1 (3-1-94). This change provides guidance to airport operators involved in the procurement of snow sweepers to control ice and snow at airports during inclement weather.

150/5220-21B Guide Specification for Devices Used to Board Airline Passengers with Mobility Impairments (3-17-00) (AAS-100). Contains performance standards, specifications, and recommendations for the design, construction, and testing of devices used to assist in the boarding of airline passengers with mobility impairments.

150/5220-22 Engineered Materials Arresting Systems (EMAS) for Aircraft Overruns. (8-21-98) (AAS-100). Contains standards for the planning, design, and installation of EMAS in runway safety areas.

150/5220-22 Change 1 (10/06/00). Guidance in installing EMAS where the area available is no longer than required, based on stopping the designed aircraft exiting the runway at 70 knots, has been added to paragraph 6.b.

150/5230-4 Aircraft Fuel Storage, Handling, and Dispensing on Airports (8-27-82) (AAS-300) (consolidated reprint includes changes 1 and 2). Provides information on aviation fuel deliveries to airport storage and the handling, cleaning, and dispensing of fuel into aircraft.

Design construction and maintenance–General

150/5300-7B FAA Policy on Facility Relocations Occasioned by Airport Improvements or Changes (11-8-72) (ABU-10). Reaffirms the aviation community of the FAA policy governing responsibility for funding relocation, replacement and modification to air traffic control and air navigation facilities that are made necessary by improvements or changes to the airport.

150/5300-9A Predesign, Prebid, and Preconstruction Conferences for Airport Grant Projects (5-1-85) (AAS-100). Provides guidance for conducting predesign, prebid, and preconstruction conferences for projects funded under the FAA airport grant program.

150/5300-13 Airport Design (9-29-89) (AAS-110) (consolidated reprint includes changes 1 through 5). Contains the FAA's standards and recommendations for airport design. SN 050-007-01046-0.

150/5300-13 Change 5 (2-14-97). Provides guidance to assist airport sponsors in their evaluation and preparation of the airport landing surface to support instrument approach procedures and incorporates change 4 criteria into the airport layout plan preparation guidance.

150/5300-13 Change 6 (9-30-00). Provides expanded guidance for new approach procedures and incorporates new flight standards requirements.

150/5300-13 Change 7 (10-1-02). Provides new guidance consistent with Runway Safety Area Program requirements, clarifies nighttime threshold siting requirements, and revises new instrument approach procedures and requirements for preparing airport layout plans.

150/5300-14 Design of Aircraft Deicing Facilities (8-23-93) (AAS-100). Provides standards, specifications, and guidance for designing aircraft deicing facilities.

150/5300-14 Change 1 (8-13-99). Design of Aircraft Deicing Facilities, Change 1. This change updates the definitions of aircraft deicing facilities and holdover times of fluids, design criteria for aircraft de/anti-icing fluid storage and transfer systems, information concerning recycling of glycols, and references.

150/5300-14 Change 2 (8-31-00) (AAS-100). Design of Aircraft Deicing Facilities, Change 2. This change provides standards and recommendations to build infrared aircraft deicing facilities and adds anaerobic bioremediation as an alternative method to mitigate the runoff effects of de/anti-icing products.

150/5300-15 Use of Value Engineering for Engineering and Design of Airport Grant Projects (9-9-93) (AAS-100). Provides guidance for the use of value engineering (VE) in airport projects funded under the Airport Grant Program.

150/5320-5B Airport Drainage (7-1-70) (AAS-100). Provides guidance for engineers, airport managers, and the public in the design and maintenance of airport drainage systems.

150/5320-6D Airport Pavement Design and Evaluation (1-30-96) (AAS-100). Provides guidance to the public for the design and evaluation of pavements at civil airports.

150/5320-6D Change 1 (1-30-96). Corrects errors in the graph for Fig. 2-4, Effect of Subbase on Modulus of Subgrade Reaction, and changes a typographical error in paragraph 339b.

150/5320-6D Change 2 (6-3-02). Incorporates recent changes and corrections.

150/5320-12C Measurement, Construction, and Maintenance of Skid Resistant Airport Pavement Surfaces (3-18-97) (AAS-100). Contains guidelines and procedures for the design and construction of skid-resistant pavement; pavement evaluation, without or with friction equipment; and maintenance of high skid-resistant pavements.

150/5320-14 Airport Landscaping for Noise Control Purposes (1-31-78) (APP-400). Provides guidance to airport planners and operators in the use of tree and vegetation screens in and around airports.

150/5320-15 Management of Airport Industrial Waste (2-11-91) (AAS-100). Provides basic information on the characteristics, management, and regulations of industrial wastes generated at airports.

150/5320-15 Change 1 (4-22-97). This change provides guidance on best management practices to eliminate, prevent, or reduce pollutants in storm water runoff associated with airport industrial activities.

150/5320-16 Airport Pavement Design for the Boeing 777 Airplane. Provides thickness design standards for pavements intended to serve the Boeing 777 airplane. Download Pavement Design Computer Program.

150/5325-4A Runway Length Requirements for Airport Design (1-29-90) (AAS-110) (consolidated reprint includes Change 1 dated 3-11-91). Provides design standards and guidelines for determining recommended runway lengths.

150/5335-5 Standardized Method of Reporting Airport Pavement Strength PCN (6-15-83) (AAS-100). Provides guidance for using the standardized International Civil Aviation Organization (ICAO) method to report airport pavement strength. The standardized method is known as the ACN/PCN method.

150/5335-5 Change 1 (3-6-87).

150/5340-1H Standards for Airport Markings (8-31-99) (AAS-300). Describes the standards for markings used on airport runways, taxiways, and aprons.

150/5340-1H Change 1 Standards for Airport Markings (12-01-00) (AAS-300). Revised to increase the size of various holding position markings to improve conspicuity and, thereby, reduce surface incidents, including runway incursions.

150/5340-4C Installation Details for Runway Centerline Touchdown Zone Lighting Systems (5-6-75) (AAS-100) (consolidated reprint includes Changes 1 and 2). Describes standards for the design and installation of runway centerline and touchdown zone lighting systems.

150/5340-5B Segmented Circle Airport Marker System (12-21-84) (AAS-100) (consolidated reprint includes change 1). Sets forth standards for a system of airport marking consisting of certain pilot aids and traffic control devices.

150/5340-14B Economy Approach Lighting Aids (6-19-70) (AAS-100) (consolidated reprint includes Changes 1 and 2). Describes standards for the design selection, siting, and maintenance of economy approach lighting aids.

150/5340-14B Change 3. Updates AC 150/5340-14B.

150/5340-17B Standby Power for Non-FAA Airport Lighting Systems (1-6-86) (AAS-100). Describes standards for the design, installation, and maintenance of standby power for non–agency owned airport visual aids associated with the National Airspace System (NAS) and the national system of airports.

150/5340-18C Standards for Airport Sign Systems (7-31-91) (AAS-4). Contains the Federal Aviation Administration standards for the siting and installation of signs on airport runways and taxiways.

150/5340-18C Change 1 (11-13-91).

150/5340-21 Airport Miscellaneous Lighting Visual Aids (3-25-71) (AAS-100). Describes standards for the system design, installation, inspection, testing, and maintenance of airport miscellaneous visual aids; i.e., airport beacons, beacon towers, wind cones, wind tees, and obstruction lights.

150/5340-23B Supplemental Wind Cones (5-11-90) (AAS-100). Describes criteria for the location and performance of supplemental wind cones.

150/5340-24 Runway and Taxiway Edge Lighting System (9-3-75) (AAS-100) (consolidated reprint includes change 1). Describes standards for the design, installation, and maintenance of runway and taxiway edge lighting.

150/5340-26 Maintenance of Airport Visual Aid Facilities (8-26-82) (Spanish Language version available) (AAS-100). Provides recommended guidelines for maintenance of airport visual aid facilities.

150/5340-27A Air-to-Ground Radio Control of Airport Lighting Systems (3-4-86) (AAS-100). Contains the FAA standard operating configurations for air-to-ground radio control of airport lighting systems.

150/5340-28 [Appendix 2 (Figs. 1–15)] [Appendix 2 (Figs. 16–30)] Low Visibility Taxiway Lighting Systems (9-1-98) (AAS-100). Describes the standards for design, installation, and maintenance of low visibility taxiway lighting systems, including taxiway centerline lights, stop bars, runway guard lights, and clearance bars.

150/5340-29 [Appendix 2 (figures)] Installation Details for Land and Hold Short Lighting Systems (12-30-99) (AAS-100). Describes standards for the design, installation, and maintenance of land and hold short lighting systems.

150/5345-1V Approved Airport Equipment (6-6-97) (AAS-100). AC 150/5345-1U is canceled. Any reference to AC 150/5345-1 shall be replaced with AC 150/5345-53.

150/5345-3E Specification for L-821 Panels for Control of Airport Lighting (9-1-98) (AAS-100). Provides the specified manufacturing requirements for panels used for remote control of airport lighting and auxiliary systems.

150/5345-5A Circuit Selector Switch (3-23-82) (AAS-100). This advisory circular contains a specification for a circuit selector switch for use in airport lighting circuits.

150/5345-7E Specification for L-824 Underground Electrical Cable for Airport Lighting Circuits (8-2-01) (AAS-100). Describes the specification requirements for L-824 electrical cables.

150/5345-10E Specification for Constant Current Regulators Regulator Monitors (10-16-84) (AAS-100). Contains a specification for constant current regulators used on airport lighting circuits, and for a monitor that reports on the status of the regulator. Airport Technical Advisory on Electromagnetic Interference (EMI).

150/5345-12C Specification for Airport and Heliport Beacon (1-9-84) (AAS-100). Contains equipment specifications for light beacons that are used to locate and identify civil airports, seaplane bases, and heliports.

150/5345-13A Specification for L-841 Auxiliary Relay Cabinet Assembly for Pilot Control of Airport Lighting Circuits (8-8-86) (AAS-100). Contains the specification requirements for a relay cabinet used to control airfield lighting circuits. The L-841 consists of an enclosure containing a DC power supply, control circuit protection and 20 pilot relays.

150/5345-26C FAA Specification for L-823 Plug and Receptacle, Cable Connectors (4-17-00) (AAS-100). Describes the subject specification requirements.

150/5345-27C Specification for Wind Cone Assemblies (7-19-85) (AAS-100). This advisory circular contains a specification for wind cone assemblies to be used to provide wind information to pilots of aircraft.

150/5345-28D Precision Approach Path Indicator (PAPI) Systems (5-23-85) (reprint incorporates Change 1) (AAS-100). Contains the Federal Aviation Administration (FAA) standards for Precision Approach Path Indicator (PAPI) systems, which provide pilots with visual glideslope guidance during approach for landing.

150/5345-28D Change 1 (11-1-91) (AAS-250).

150/5345-39B FAA Specification L-853, Runway and Taxiway Center-line Retroreflective Markers (12-9-80) (AAS-100) (consolidated reprint includes changes 1). Describes specification requirements for L-853 Runway and Taxiway Retroreflective markers, for the guidance of the public.

150/5345-42C Specification for Airport Light Bases, Transformer Houses, Junction Boxes and Accessories (6-8-89) (AAS-100). Contains specification for containers designed to serve as airport light bases, transformer housings, junction boxes, and related accessories (incorporates Change 1).

150/5345-42C Change 1 (10-29-91) (AAS-100). Corrects, clarifies, and revises portions of the specification and corrects drafting errors.

150/5345-43E Specification for Obstruction Lighting Equipment (10-19-95) (AAS-100). Contains the FAA specification for obstruction lighting equipment.

150/5345-44F Specification for Taxiway and Runway Signs (1-5-94) (consolidated reprint incorporates change 1) (AAS-100). Contains a specification for lighted and unlighted signs to be used on taxiways and runways.

150/5345-44F Change 1 (8-23-94). Adds a requirement to the qualification tests and deletes the prohibition on unlighted swinging signs.

150/5345-45A Lightweight Approach Light Structure (12-9-87) (AAS-100). Presents the specifications for lightweight structures for supporting lights as used in visual navigational aid systems.

150/5345-46B Specification for Runway and Taxiway Light Fixtures (9-1-98) (AAS-100). Contains FAA specifications for light fixtures to be used on airport runways and taxiways.

150/5345-47A Isolation Transformers for Airport Lighting Systems (12-9-87) (AAS-100). Contains the specifications requirements for series-to-series isolation transformers for use in airport lighting systems.

150/5345-49A Specification L-854, Radio Control Equipment (8-8-86) (AAS-100). Contains the specification for radio control equipment to be used for controlling airport lighting facilities.

150/5345-50 Specification for Portable Runway Lights (10-16-78) (AAS-100) (consolidated reprint includes change 1). Creates a standard and specification for a battery operated light unit to be used to outline a runway area temporarily.

150/5345-51 Specification for Discharge-Type Flasher Equipment (8-14-81) (AAS-100) (consolidated reprint includes change 1). Contains the specifications for discharge-type flashing light equipment to be used for runway end identifier lights (REIL) and for an omnidirectional approach lighting system (ODALS).

150/5345-52 Generic Visual Glideslope Indicators (CVGI) (6-21-88) (AAS-100). Contains the FAA standards for generic visual glideslope indicators systems. GVGI systems provide pilots with visual glideslope guidance during approaches for landing at general aviation airports.

150/5345-53B (Including Change 1) Airport Lighting Equipment Certification Program (10-23-98) (AAS-100). Describes the Airport Lighting Equipment Certification Program (ALECP).

150/5345-53B Addendum (7-15-03) (AAS-100). Airport Lighting Equipment Certification Program. Addendums to Appendices 3 and 4 of AC 150/5345-53B (HTML version).

150/5345-53B Change 1 (09-08-00). This change reinstates L-824, Underground Electrical Cable for Airport Lighting Circuits (AC 150/5345-7D) and adds L-884, Power and Control Unit for Land and Hold Short Lighting Systems (AC 150/5345-54) to the Airport Lighting Equipment Certification Program.

150/5345-54A (08-09-00). Specification for L-884 Power and Control Unit for Land and Hold Short Lighting Systems. A consolidated version of 150/5345-54A (including change 1) will soon be available online. For more information, contact Richard Smith at 202-267-9529.

150/5345-54A Change 1 (06-29-01) (AAS-100). Provides a correction to paragraph 3.3.1.4, Painting and Finishing.

150/5345-55 Lighted Visual Aid to Indicate Temporary Runway Closure (7-14-03) (AAS-100). Provides guidance in the design of a lighted visual aid to indicate temporary runway closure.

150/5360-8B Announcement of Availability of Information on Foreign Airport Planning, Design, Construction, and Trade Opportunities (6-26-97) (AAS-100). Provides information on trade opportunity programs and publications related to foreign airport planning, design, and construction.

150/5360-9 Planning and Design of Airport Terminal Facilities at Non-Hub Locations (4-4-80) (AAS-100). This advisory circular provides guidance material for the planning and design of airport terminal buildings at nonhub locations.

150/5360-10 Announcement of Availability—Airport Landside Simulation Model (ALSIM) (4-24-84) (AAS-3). Announces the availability and describes some of the features of a dynamic simulation computer model for use by airport planners, designers, and operators in evaluating and comparing the effectiveness of alternative airport terminal and landside configurations and facilities.

150/5360-11 Energy Conservation for Airport Buildings (5-31-84) (AAS-100). Provides guidance for promoting energy conservation in the design and operation of airport buildings; for initiating energy conservation programs; and for conducting airport building energy assessments.

150/5360-12D Airport Signing and Graphics (7-1-03) (AAS-100). Provides guidance on airport related signs and graphics.

150/5360-13 Planning and Design Guidelines for Airport Terminal Facilities (4-22-88) (AAS-100). Provides guidelines for the planning and design of airport terminal buildings and related access facilities (incorporates Change 1).

150/5360-14 Access to Airports by Individuals With Disabilities (6-30-99) (AAS-100). Designed to assist airports in complying with the current laws and regulations regarding individuals with disabilities by (1) identifying the relevant statutes and regulations which impact upon airports, (2) presenting in a single document the main features of each of the statutes and regulations, (3) providing legal citations to facilitate research, (4) listing sources of assistance or additional information, and (5) identifying Final Rules. It presents and reconciles the federal accessibility regulations implementing the Americans with Disabilities Act of 1990 (ADA), the Air Carrier Access Act of 1986 (ACAA), the Rehabilitation Act of 1973, as amended (RA), and the Architectural Barriers Act of 1968, as amended (ABA) which affect the architectural or program accessibility of airports in the U.S. transportation system and employment opportunities on these airports for individuals with disabilities.

150/5370-2E Operational Safety on Airports During Construction (1-17-03) (AAS-300). Concerning operational safety on airports with special emphasis on safety during periods of construction activity, to assist airport operators in complying with part 139 (Word version).

150/5370-6B Construction Progress and Inspection Report—Airports Grant Program (8-7-89) (AAS-100). Provides guidance for submission of a report on construction progress of Airport Development Aid Program (ADAP) projects.

150/5370-10A Standards for Specifying Construction of Airports (2-17-89) (with Changes 1–8) (AAS-100). Provides standards for construction of airports. Items covered include earthwork, drainage, paving, turfing, lighting, and incidental construction.

150/5370-10A Change 1 (6-15-90). Revised Item P-501, Portland Cement Concrete Pavement, to reflect the use of cementitous materials and chemical admixtures.

150/5370-10A Change 2 150/5370-10A Change 2 (11-2-90). Revised Items P-155, D-701, D-705, D-751, and F-162, have been revised to reflect changes in material specifications and new materials.

150/5370-10A Change 3 (1-9-91). Revised Item P-625, Coal-Tar Emulsion Sealcoat, is withdrawn.

150/5370-10A Change 4 (7-7-92) (AAS-100). Revised Item P-401, Plant Mix Bituminous Pavements to incorporate quality assurance provisions.

150/5370-10A Change 5 (4-2-93) (AAS-100). Revises Table 2 and deletes requirement for voids filled with asphalt.

150/5370-10A Change 6 (1-25-94) (AAS-100). Adds Sections 100 and 110 to the General Provisions of the AC. The Notice to Users, Notice, and Item P-401 have been revised.

150/5370-10A Change 8 (7-6-94) (AAS-100). Revised Item F-162, Chain Link Fences, to incorporate new materials.

150/5370-10A Change 9 (9-10-96) (AAS-100). Revised Item P-620, Runway and Taxiway Painting.

150/5370-10A Change 10 (3-11-98) (AAS-100). General provisions Section 110—Method of Estimating Percentage of Material Within Specification Limits (PWL), Item P-401—Plant Mix Bituminous Pavements, and Item P-501—Portland Cement Concrete Pavement, have been revised to clarify PWL concepts, incorporate new acceptance criteria, test procedures, and pay adjustment schedules.

150/5370-10A Change 11 (5-20-98) (AAS-100). Item P-626—Emulsified Asphalt Slurry Seal Surface Treatment, Item F-162—Chain-link Fences, Item D-701—Pipe for Storm Drains and Culverts, Item D-705—Pipe Underdrains for Airports, and Item D-751—Manholes, Catch Basins, Inlets, and Inspection Holes have been revised. Item D-702—Slotted Drains has been added.

150/5370-10A Change 12 (2-22-99) (AAS-100). Revised Item P-620—Runway and Taxiway Painting.

150/5370-10A, Item P-401, thru Change 12 (3-15-00) (AAS-100). Item P-401—Plant Mix Bituminous Pavements specification with all changes through Change 12.

150/5370-10A, Item P-501, thru Change 12. (3-15-00) (AAS-100). Item P-501, Portland Cement Concrete Pavement specification with all changes through Change 12.

150/5370-10A, Change 13 (1-30-01) (AAS-100). Updates Item D-701—Pipe for Storm Drains and Culverts, and Item D-705—Pipe Underdrains for Airports.

150/5370-11 Use of Nondestructive Testing Devices in the Evaluation of Airport Pavements (6-4-76) (AAS-100) (consolidated reprint includes change 1). Provides guidance to the public on the use of nondestructive testing devices as aids in the evaluation of the load-carrying capacity of airport pavements.

150/5370-12 Quality Control of Construction for Airport Grant Projects (9/6/85) (AAS-100). Provides information to ensure the quality of construction accomplished under the FAA Airports Grant Program.

150/5370-13 Offpeak Construction of Airport Pavements Using Hot-Mix Asphalt (8-27-90) (AAS-100). Provides guidance for the planning, coordination, management, design, testing, inspection, and execution of offpeak construction of airport pavements using hot mix asphalt paving materials.

150/5380-5B Debris Hazards at Civil Airports (7-5-96) (AAS-100). Discusses problems of debris at airports, gives information on foreign objects, and tells how to eliminate such objects from operational areas.

150/5380-6A Guidelines and Procedures for Maintenance of Airport Pavements (7-14-03) (AAS-100). Provides guidelines and procedures for maintenance of rigid and flexible airport pavements (Word version).

150/5380-7 Pavement Management System (9-28-88) (AAS-100). Presents the concepts of a Pavement Management System, discusses the essential components of such a system, and outlines how it can be used in making cost-effective decisions regarding pavement maintenance and rehabilitation.

150/5390-2A Heliport Design (1-20-94) (AAS-100). Provides recommendations and standards for heliport and helistop design.

150/5390-3 Vertiport Design (5-31-91) (AAS-100). Provides guidance to planners and communities interested in developing a civil vertiport or vertistop.

150/5395-1 Seaplane Bases (6-29-94) (AAS-100). Provides guidance to assist operators in planning, designing, and constructing seaplane base facilities.

139/201-1 Airport Certification Manual (ACM) & Airport Certification Specifications (ACS) (7-15-88) (AAS-300). Provides methods for meeting the certification requirements specified in Part 39 of the Federal Aviation Regulations.

Phonetic alphabet

A—Alpha
B—Bravo
C—Charlie
D—Delta
E—Echo
F—Foxtrot
G—Golf
H—Hotel
I—India
J—Juliet
K—Kilo
L—Lima
M—Mike

N—November
O—Oscar
P—Papa
Q—Quebec
R—Romeo
S—Sierra
T—Tango
U—Uniform
V—Victor
W—Whiskey
X—X-ray
Y—Yankee
Z—Zulu

Abbreviations

AAS	Airport Advisory Service
AC	advisory circular
ACM	Airport Certification Manual
ACS	Airport Certification Specifications
ADAP	Airport Development Aid Program
ADC	Air Defense Command
ADF	automatic direction finder
ADS	Automated Dependent Surveillance
ADS-B	Automated Dependent Surveillance—Broadcast
AERA	advanced en route automation
AFFF	aqueous film-forming foam
AGL	above ground level
AID	airport information desk
AIM	*Airman's Information Manual*
AIP	Airport Improvement Program
AIREP	air report
AIRMET	airmen's meteorological information
ALNOT	alert notice
ALP	airport layout plan
ALS	approach lighting system
AMASS	Airport Movement Area Safety System
AMIS	aircraft movement information service
AOA	air operations area
AOCNet	Airline Operations Center Network
ARC	Airport Reference Code
ARFF	aircraft rescue and firefighting
ARP	airport reference point
ARSR	air route surveillance radar
ARTCC	Air Route Traffic Control Center
ARTS	Automated Radar Terminal System
ASDE	airport surface detection equipment
ASDI	aircraft situation display to industry

ASQP	airline service quality performance
ASOS	Automated Surface Observing Systems
ASP	Airport Security Plan
ASR	airport surveillance radar
ATC	air traffic control
ATCRBS	Air Traffic Control Radar Beacon System
ATCSCC	ATC Systems Command Center
ATCT	air traffic control tower
ATIS	Automatic Terminal Information Service
ATOMS	Air Traffic Operations Management System
ATSA	Aviation and Transportation Security Act of 2001
AWOS	Automated Weather Observing Systems
BOT	build, operate, and transfer contract
CAA	Civil Aeronautics Administration
CAB	Civil Aeronautics Board
CAT	clear-air turbulence
CBD	Central Business District
CDM	Collaborative Decision Making
CFR	Code of Federal Regulations
CIP	Capital Improvement Plan
CIS	Cockpit Information System
CNEL	community noise equivalent level
CODAS	Consolidated Operations and Delay Analysis System
CPDLC	Controller-to-Pilot Data Link Communications
CTAS	Center Terminal Radar Approach Control Automation System
CUSS	common-use self-service kiosks
CUTE	common-use terminal equipment
DABS	Discrete Address Beacon System
dB	decibel
DF	direction finder
DH	decision height
DLAND	Development of Landing Areas for National Defense
DME	distance measuring equipment
DOT	Department of Transportation
DP	Departure Procedure

DSR	display system replacement
EAS	Essential Air Service Program
EDS	explosive detection systems
EFAS	en route flight advisory service
EIR	Environmental Impact Review
EIS	Environmental Impact Statement
EMAS	Engineered Materials Arresting System
EPNL	effective perceived noise level
ETD	explosive trace detection
ETMS	enhanced traffic management system
F&E	facilities and equipment program
FAA	Federal Aviation Administration
FAAP	Federal-Aid Airport Program
FAR	Federal Aviation Regulation
FAWS	Flight Advisory Weather Service
FIS	Federal Inspection Services
FMA	Final Monitor Aid
FOD	foreign object debris
FONSI	Finding of No Significant Impact
FSD	Federal Security Director
FSM	Flight Schedule Monitor
FSS	flight service station
GARB	general airport revenue bond
GCA	ground-controlled approach
GDL	guidance light facility
GOB	general obligation bond
GPS	Global Positioning System
GS	glide slope
HAT	height above touchdown
ICAO	International Civil Aviation Organization
ICP	Initial Conflict Probe
IFR	instrument flight rules
ILS	Instrument Landing System
IM	inner marker
IMC	instrument meteorological conditions

ITWS	Integrated Terminal Weather System
JAWOS	Joint Automated Weather Observation System
JPDO	Joint Planning and Development Office
LAAS	Local Area Augmentation System
LBO	lease, build, and operate agreement
LDIN	lead-in light facility
Ldn	day/night noise level
LF	low frequency
LOC	ILS localizer
LOI	federal letter of intent
MAP	Military Airport Program
MDA	minimum descent altitude (FAR Part I)
MEA	minimum en route IFR altitude
MLS	Microwave Landing System
MM	middle marker
MOCA	minimum obstruction clearance altitude
MPO	Metropolitan Planning Organization
MSL	elevation above mean sea level
NAFEC	National Aviation Facilities Experimental Center
NAP	National Airport Plan
NAR	National Airspace Review
NAS	National Airspace System
NASP	National Airport System Plan
NAVAID	navigational aid
NDB	Non-Directional Radio Beacon
NDT	nondestructive testing
NEXCOM	Next-Generation Air-to-Ground Communications
NEXRAD	next generation weather radar
NextGen	Next-Generation Air Transportation System
NOTAM	notice to airmen
NPIAS	National Plan of Integrated Airport Systems
NRP	National Route Program
NTSB	National Transportation Safety Board
NWS	National Weather Service
OM	outer marker

OPSNET	Air Traffic Operations Network System
PANCAP	practical annual capacity
PAPI	Precision Approach Path Indicator
PAR	precision approach radar
PATCO	Professional Air Traffic Controllers Organization
pFAST	passive final approach spacing tool
PFC	passenger facility charge
PGP	Planning Grant Program
PHOCAP	practical hourly capacity
PIREP	pilot report
PNL	perceived noise level
PPBM	positive passenger baggage matching
PRM	Precision Runway Monitor
RAIL	runway alignment indicator light
RAPCON	Radar Approach Control
RATCF	radar air traffic control facility
REIL	runway end identifier lights
RHSM	reduced horizontal separation minima
RNAV	radio (area) navigation
RNP	Required Navigation Performance
RON	remain overnight
ROT	runway occupancy time
RVR	runway visual range
RVSM	reduced vertical separation minima
SALS	short approach light system
SASP	state aviation system plans
SATS	Small Aircraft Transportation Systems
SCIA	simultaneous converging instrument approaches
SECRA	secondary radar
SFL	sequenced flashing lights
SIDA	security identification display area
SIGMET	significant meteorological information
SMA	Surface Movement Advisor
SMS	Safety Management Systems
SMSA	standard metropolitan statistical area

STARS	Standard Terminal Automation Replacement System
STOL	short takeoff and landing
TACAN	tactical air navigation
TCAS	Traffic Alert and Collision Avoidance System
TCH	threshold crossing height
TDWR	terminal Doppler weather radar
TERPs	terminal instrument procedures
TMA	Traffic Management Advisor
TRACON	Terminal Radar Approach Control
TSA	Transportation Security Administration
TSR	Transportation Security Regulations
UHF	ultrahigh frequency
VAS	Vortex Advisory System
VASI	visual approach slope indicator
VFR	visual flight rules (FAR Part 91)
VMC	visual meteorological conditions
VOR	very-high-frequency omnidirectional range
VTOL	vertical takeoff and landing
WAAS	Wide Area Augmentation System
WARP	Weather and Radar Processor
Z	Zulu time

Glossary

This glossary includes key terms appearing at the ends of the chapters, as well as many other terms used in the text and others of significance in airport planning and management. The definitions are meant to be brief and straightforward, rather than technically precise and all-inclusive.

abandoned airport: An airport permanently closed to aircraft operations, which may be marked in accordance with current FAA standards for marking and lighting of deceptive, closed, and hazardous areas on airports.

above ground level (AGL): The altitude above terrain at any given location.

access/egress link: As used in the passenger handling system, the link that includes all of the ground transportation facilities, vehicles, and other modal transfer facilities required to move the passenger to and from the airport.

access/processing interface: As used in the passenger handling system, the link in which the passenger makes the transition from the vehicular mode of transportation to pedestrian movement into the passenger processing activities.

access taxiway: A taxiway that provides access to a particular location or area.

active based aircraft: Aircraft that have a current Airworthiness Certificate and are based at an airport.

actual runway length: The length of a full-width usable runway from end to end of full-strength pavement where those runways are paved.

administrative building: A building or buildings accommodating airport administration activity and public facilities for itinerant and local flying, usually associated with general aviation fixed-based operations.

administrative management: A method of controlling airport access by setting quotas on passenger enplanements or on the number and type of aircraft operations that will be accommodated during a specific period.

advanced en route automation (AERA): A traffic management system that enables ATC personnel to detect and resolve problems concerning an aircraft's flight path on an approach to an airport. AERA will assist controllers in finding the open route closest to the preferred one if the latter is unavailable.

advisory circulars (AC): Documents published by the Federal Aviation Administration to assist airports and other components of the aviation system in operations and planning.

aiming point: A distinctive mark placed on the runway to serve as a point for judging and establishing a glide angle for landing aircraft. It is usually 1,000 feet from the landing threshold. Also known as a touchdown zone marking.

AIP Temporary Extension Act of 1994: Authorized the extension of AIP funding through 1994. Amended the percentage of AIP funds that must be set aside for reliever, commercial service, nonprimary, and system planning projects.

Air Cargo Deregulation Act of 1976: This act deregulated the nation's air cargo industry, allowing air cargo carriers to freely enter and exit markets and set rates without government regulation or approval.

air carrier: A person who undertakes directly by lease, or other arrangement, to engage in air transportation. (FAR Part 1)

air carrier airport: An airport (or runway) designated by design and/or use for air carrier operations.

air carrier—certificated route: An air carrier holding a Certificate of Public Convenience and Necessity issued by the Civil Aeronautics Board to conduct scheduled services over specified routes and a limited amount of nonscheduled operations.

air carrier—commuter: An air taxi operator that (1) performs at least five round-trips per week between two or more points and publishes flight schedules that specify the times, days of the week, and places between which such flights are performed; or (2) transports mail by air pursuant to a current contract with the U.S. Postal Service.

Air Commerce Act of 1926: Designed to promote the development of and stabilize commercial aviation. Included the first licensing of aircraft, pilots, and mechanics and established the first rules and regulations for operating aircraft in the airway system.

aircraft: A device that is used or intended to be used for flight in the air. (FAR Part 1)

aircraft capacity: The rate of aircraft movements on the runway/taxiway system that results in a given level of delay.

aircraft fleet mix: The types of categories of aircraft that are to be accommodated at the airport.

Aircraft Noise and Capacity Act of 1990: Establishes a national aviation noise policy, including a general prohibition against the operation of stage 2 aircraft of more than 75,000 pounds after December 31, 1999.

aircraft rescue and firefighting (ARFF): Services used to perform firefighting and rescue operations at airports.

aircraft rescue, fire fighting chief: Develops procedures and implements aircraft rescue, firefighting, and disaster plan.

Aircraft Situation Display to Industry (ASDI): System that provides near-real-time position and other relevant flight data for every aircraft operating that is subject to traffic flow management planning.

aircraft tiedowns: Positions on the ground surface that are available for securing aircraft.

airfield: The component of an airport that includes all the facilities located on the physical property of the airport to facilitate aircraft operations, including runways and taxiways, navigational aids, lighting, signage, and marking.

Airline Deregulation Act of 1978: Marked the beginning of the end of economic regulation of the certificated air carriers by the CAB. The act called for the gradual phaseout of the CAB with its termination on December 31, 1984. All remaining essential functions were transferred to DOT and other agencies.

air operations area (AOA): The area on the airport used or intended to be used for landing, takeoff, or surface maneuvering of aircraft.

Airline Operations Center Network (AOCNet): A private intranet that provides an enhanced capability for the FAA and airline operations control centers to rapidly exchange and share a single integrated source of CDM-related aeronautical information concerning delays and constraints in the NAS.

air marker: An alphanumeric or graphic symbol on ground or building surfaces designed to give guidance to pilots in flight.

air navigation facility (NAVAID): Any facility used as, available for use as, or designed for use as an aid to air navigation, including landing areas, lights, any apparatus or equipment for disseminating weather information; for signaling; for radio direction finding; or for radio or other electronic communication; and any other structure or mechanism having a similar purpose for guiding and controlling flight in the air or the landing or take-off of aircraft.

Air Piracy: The act of hijacking an aircraft.

airport: An area of land or water that is used or intended to be used for the landing and takeoff of aircraft, and includes its buildings and facilities, if any. (FAR Part 1)

airport access plans: Proposed routing of airport access to the central business district and to points of connection with existing or planned ground transportation arteries.

airport accounting: Involves the accumulation, communication, and interpretation of economic data relating to the financial position of an airport and the results of its operations for decision-making purposes.

Airport and Airway Development Act Amendments of 1976: Extended the 1970 Act for 5 years and included a number of amendments including the types of airport development projects eligible for ADAP funding, increased the federal share for ADAP and PGP grants, and initiated a number of studies concerning the National Airport System Plan (NASP).

Airport and Airway Development Act of 1970: A federal aid to airports program administered by the FAA for the 10-year period ending in 1980. Over $4.1 billion was invested in the airport system during this period.

Airport and Airway Improvement Act of 1982: Reestablished the operation of the Airport and Airway Trust Fund with a slightly revised schedule of user taxes.

Airport and Airway Revenue Act of 1970: Created an airport and airway trust fund to generate revenues for airport aid. Taxes included an 8 percent surcharge on domestic passenger fares, a $3 surcharge on international passenger tickets, a 7 cent surcharge on fuel, a 5 percent surcharge on airfreight waybills, and an annual registration fee of $25 on all civil aircraft.

Airport and Airway Safety and Capacity Expansion Act of 1987: Extended the Airport and Airway Improvement Act for 5 years. Also provided that 10 percent of funding be available for disadvantaged small businesses.

Airport and Airway Safety, Capacity, Noise Improvement, and Intermodal Transportation Act of 1992: Authorized the extension of AIP funding through 1993. Expanded eligibility under the Military Airport Program and State Block Grant Program.

Airport and Airway Trust Fund: A federal fund, originally established in 1970 and funded by levies on aviation users, to be used toward airport capital improvement projects.

airport authority: Similar to a port authority but with the single purpose of setting policy and management direction for airports within its jurisdiction.

airport beacon: A visual navigation aid displaying alternating white and green flashes to indicate a lighted airport or white flashes only for an unlighted airport.

Airport Certification Manual (ACM): A comprehensive list of airport operational procedures required for full compliance with FAR Part 139.

airport closed to the public: An airport not available to the public without permission from the owner.

airport configuration: The relative layout of component parts of an airport such as the runway-taxiway-terminal arrangement.

Airport Development Aid Program (ADAP): A federal aid to airports program established under the Airport and Airway Development Act of 1970 for the development of airport facilities.

airport director: Sometimes referred to as airport manager or supervisor, the person responsible for the overall day-to-day operation of the airport.

Airport District/Development Office (ADO): Regional FAA offices that keep in contact with airports within their respective regions to ensure compliance with federal regulations and to assist airport management in safe and efficient airport operations and planning.

airport elevation: The highest point of an airport's usable runways measured in feet from mean sea level.

airport facilities requirements: Part of airport master plan that specifies new or expanded facilities that will be needed during the planning period.

This involves cataloging existing facilities and forecasting future traffic demand. The planner compares the capacity of existing facilities with future demand, identifying where demand will exceed capacity and what new facilities will be necessary.

airport geographical position: The designated geographical center of the airport (latitude and longitude) that is used as a reference point for the designation of airspace regulations.

airport identification beacon: Coded lighted beacon used to indicate the location of an airport where the airport beacon is more than 5,000 feet from the landing area.

airport imaginary surfaces: Imaginary surfaces established at an airport for obstruction determination purposes and consisting of primary, approach-departure, horizontal, vertical, conical, and transitional surfaces. (FAR Part 77)

Airport Improvement Program (AIP): A federal aid to airports program similar to ADAP covering the period from 1983 to 2007 (awaiting reauthorization as of 2010).

airport layout: The major portion of the airport layout plan drawing including existing and ultimate airport development and land uses drawn to scale.

airport layout plan (ALP): A plan for an airport showing boundaries and proposed additions to all areas owned or controlled by the sponsor for airport purposes, the location and nature of existing and proposed airport facilities and structures, and the location on the airport of existing and proposed nonaviation areas and improvements thereon.

airport layout plan drawing: Includes the airport layout, location map, vicinity map, basic data table, and wind information.

airport master plan: Presents the planner's conception of the ultimate development of a specific airport. It presents the research and logic from which the plan was evolved and displays the plan in a graphic and written report.

Airport Movement Area Safety System (AMASS): Enhances the function of the ASDE:3 radar by providing automated alerts and warnings of potential runway incursions and other hazards.

Airport Noise and Capacity Act of 1990: Sets a deadline date of December 31, 1999, for the elimination of stage 2 aircraft weighing more than 75,000 pounds in the contiguous United States.

airport of entry: See international airport.

airport open to the public: An airport open to the public without prior permission and without restrictions within the physical capacities of available facilities.

airport premises liability policy: Designed to protect the airport operator for losses arising out of legal liability for all activities carried on at the airport.

airport reference code (ARC): A code determined by the wingspan and approach speed of the design aircraft for a particular airport project.

airport safety self-inspection: Provides a safety inspection checklist primarily designed for GA airports. (Advisory Circular 150/5200-18)

Airport Security Plan (ASP): A required set of procedures for adhering to federal airport security regulations.

airport sponsor: A public agency or tax-supported organization, such as an airport authority, that is authorized to own and operate the airport, to obtain property interests, to obtain funds, and to be legally, financially, and otherwise able to meet all applicable requirements of current laws and regulations.

airport surface detection equipment (ASDE): Radar equipment specifically designed to detect all principal features on the surface of an airport, including aircraft vehicular traffic, and to present the entire picture on a radar indicator console in the control tower.

Airport Surface Detection Equipment (ASDE-3): A high-resolution ground-mapping radar that provides surveillance of taxiing aircraft and service vehicles at high-activity airports.

airport surveillance radar (ASR): Radar providing position of aircraft by azimuth and range data. It does not provide elevation data. It is designed for range coverage up to 60 nautical miles and is used by terminal area air traffic control.

airport system planning: Airport plans as part of a system that includes national, regional, state, and local transportation planning.

airport-to-airport distance: The great-circle distance, measured in statute miles, between airports as listed in the Civil Aeronautics Board's official airline route and mileage manual.

airport traffic area: Unless otherwise specifically designated in FAR Part 93, that airspace within a horizontal radius of 5 statute miles from the geographical center of any airport at which a control tower is operating, extending from the surface up to, but not including, an altitude of 3,000 feet above the elevation of the airport. (FAR Part 1)

airport use agreement: Legal contracts for the air carriers' use of the airport and leases for use of terminal facilities.

air route: Navigable airspace between two points that are identifiable.

Air route surveillance radar (ARSR): A radar facility remotely connected to an air route traffic control center used to detect and display the azimuth and range of en route aircraft operating between terminal areas, enabling an ATC controller to provide air traffic service in the air route traffic control system.

Air Route Traffic Control Center (ARTCC): A facility established to provide air traffic control service to aircraft operating on an IFR flight plan within controlled airspace and principally during the en route phase of flight.

airside facilities: The airfield on which aircraft operations are carried out, including runways and taxiways.

airside-landside concept: Terminal concept that emphasizes a physical separation of facilities that handles passengers and ground vehicles and those that deal primarily with aircraft handling.

airspace: Space in the air above the surface of the earth or a particular portion of such space, usually defined by the boundaries of an area on the surface projected upward.

air taxi aircraft: Aircraft operated by the holder of an Air Taxi Operating Certificate, which authorizes the carriage of passengers, mail, or cargo for revenue in accordance with FAR Parts 135 and 121.

air taxi operator: An operator providing either scheduled or unscheduled air taxi service or mail service.

air traffic: Aircraft operating in the air or on an airport surface, exclusive of loading ramps and parking areas. (FAR Part 1)

air traffic clearance: An authorization by air traffic control, for the purpose of preventing collision between known aircraft, for an aircraft to proceed under specified traffic conditions within controlled airspace. (FAR Part 1)

air traffic control (ATC): A service operated by appropriate authority to promote the safe, orderly, and expeditious flow of air traffic. (FAR Part 1)

Air Traffic Control Beacon Interrogator (ATCBI): That part of the ATCRBS system located on the ground that interrogates the airborne transponder and receives the reply.

Air Traffic Control Radar Beacon System (ATCRBS): A radar system in which the object to be detected is fitted with cooperative equipment in the form of a radio receiver/transmitter (transponder). Radio pulses transmitted from the searching transmitter/receiver (interrogator) site are received in the cooperative equipment and used to trigger a distinctive transmission from the transponder. This latter transmission rather than a reflected signal is then received back at the transmitter/receiver site.

air traffic control tower (ATCT): A central operations facility in the terminal air traffic control system, consisting of a tower cab structure, including an associated IFR room if radar equipped, using air/ground communications and/or radar, visual signaling, and other devices to provide safe and expeditious movement of terminal air traffic.

airport traffic control service: Air traffic control service provided by an airport traffic control tower or aircraft operating on the movement area and in the vicinity of an airport.

Air Traffic Operations Management System (ATOMS): FAA personnel record aircraft that are delayed 15 or more minutes by specific cause (weather, terminal volume, center volume, closed runways or taxiways, and NAS equipment interruptions).

air transportation: Interstate, overseas, or foreign air transportation, or the transportation of mail by aircraft. (FAR Part 1)

airway: A path through the navigable airspace designated by appropriate authority within which air traffic service is provided.

Airways Modernization Act of 1957: Passed by Congress in response to several serious aircraft accidents and the coming of jet equipment; it was designed to provide for the development and modernization of the national system of navigation and traffic control facilities. Expiration was planned for June 30, 1960.

alert areas: Areas of airspace that contain a high volume of pilot training or an unusual type of aerial activity.

alphanumeric display: Use of letters of the alphabet and numerals to show altitude, beacon code, and other information about a target on a radar display.

alternate airport: An airport at which an aircraft may land if a landing at the intended airport becomes inadvisable. (FAR Part 1)

ancillary processing facilities: Facilities located at airport terminals that are not essential for passenger processing, but are often provided to improve the overall travel experience.

AOPA Airport Watch: A security program, developed by the Aircraft Owners and Pilots Association, that advocates self-reporting of observed suspicious activity by users of general aviation.

apportionment funds: Represent the largest funding category, making up approximately half of all AIP funding. For example, apportionment funds to primary airports are based on those airports' annual enplanements.

approach and clear zone layout: A graphic presentation to scale of the imaginary surfaces defined in FAR Part 77.

approach area: The defined area the dimensions of which are measured horizontally beyond the threshold over which the landing and takeoff operations are made.

approach clearance: Authorization issued by air traffic control to the pilot of an aircraft for an approach for landing under instrument flight rules.

approach control facility: A terminal air traffic control facility (TRACON, RAPCON, RATCF) providing approach control service.

approach control service: Air traffic control service provided by an approach control facility for arriving and departing VFR/IFR aircraft.

approach fix: The fix from or over which final approach (IFR) to an airport is executed.

approach gate: That point on the final approach course which is 1 mile from the approach fix on the side away from the airport or 5 miles from the landing threshold, whichever is farther from the landing threshold.

approach light beacon: An aeronautical beacon placed on the extended centerline of a runway at a fixed distance from the threshold.

approach light contact height: The height on the glide path of an instrument landing system from which a pilot making an approach can expect to see the high-intensity approach lights.

approach lighting system (ALS): An airport lighting facility that provides visual guidance to landing aircraft by radiating light beams in a directional pattern by which the pilot aligns the aircraft with the extended centerline of the runway on the final approach and landing.

approach path: A specific flight course laid out in the vicinity of an airport and designed to bring aircraft in to safe landings; usually delineated by suitable navigational aids.

approach sequence: The order in which aircraft are positioned while awaiting approach clearance or while on approach.

approach slope ratio: The ratio of horizontal-to-vertical distance indicating the degree of inclination of the approach surface.

approach surface: An imaginary surface longitudinally centered on the extended centerline of the runway, beginning at the end of the primary surface and rising outward and upward to a specified height above the established airport elevation.

appropriation by activity: A form of budget where appropriate expenses are planned according to major work area or activity with no further detailed breakdown.

apron: A defined area, on a land airport, intended to accommodate aircraft for purposes of loading or unloading passengers or cargo, refueling, parking, or maintenance.

aqueous film-forming foam (AFFF): Materials used as part of aircraft rescue and firefighting services.

arriving passenger: A passenger who has deplaned an aircraft and entered the terminal from the flight interface with the intention of leaving the airport terminal for his or her final destination through the access/processing interface.

assistant director—finance and administration: Responsible for overall matters concerning finance, personnel, purchasing, facilities management, and office management.

assistant director—maintenance: Responsible for planning, coordinating, directing, and reviewing the maintenance of buildings, facilities, vehicles, and utilities.

assistant director—operations: Responsible for all airside and landside operations including security and crash, fire, and rescue operations.

assistant director—planning and engineering: Provides technical assistance to all airport organizations, and ensures the engineering integrity of construction, alteration, and installation programs.

ATC data transfer and display equipment: Equipment for ATC facilities intended to provide a symbolic display of the data necessary for the control function by automatic means and for certain computation, storage, and recall of display data.

ATC Systems Command Center (ATCSCC): A facility responsible for the operation of four distinct but integrated functions: Central Flow Control Function (CFCF), Central Altitude Reservations Function (CARF), Airport Reservation Position, and the Air Traffic Service Contingency Command Post (ATSCCP).

Automated Dependent Surveillance (ADS): An onboard system that will replace verbal aircraft position reports, thereby enhancing surveillance coverage and accuracy in flight and on the airport surface.

Automated Dependent Surveillance—Broadcast (ADS-B): A technology that leverages GPS navigation and digital communications to enhance air traffic management.

Automated Radar Terminal System (ARTS): Computer-aided radar display subsystems capable of associating alphanumeric data with radar returns. Systems of varying functional capability, determined by the type of automation equipment and software, are denoted by a number/letter suffix following the name abbreviation.

Automated Weather Observing System (AWOS): Gathers weather data from unmanned sensors, automatically formulates weather reports, and distributes them to airport control towers.

Automatic Terminal Information Service (ATIS): The continuous broadcast of recorded noncontrol information in selected high-activity terminal areas. Its purpose is to improve controller effectiveness and to relieve frequency congestion by automating the repetitive transmission of essential but routine information.

Aviation and Transportation Security Act of 2001 (ATSA): Legislation passed to address immediate needs of aviation security in the wake of the events of September 11, 2001.

Aviation Safety and Capacity Expansion Act of 1990: This act authorized a passenger facility charge (PFC) program to provide funds to finance airport-related projects and a military airport program to finance the transition of selected military airfields to civil use.

Aviation Safety and Noise Abatement Act of 1979: Provides assistance to airport operators to prepare and carry out noise compatibility programs. Authorizes the FAA to help airport operators develop noise abatement programs and makes them eligible for grants under ADAP.

avigation easement: A grant of a property interest in land over which a right of unobstructed flight in the airspace is established.

baggage claim: Facilities at which arriving passengers claim checked luggage.
baggage claim carousels: Equipment on which checked luggage is loaded for presentation to passengers at baggage claim.

baggage handling: Services that include a number of activities involving the collection, sorting, and distribution of baggage.

balanced runway concept: A runway length design concept wherein length of prepared runway is such that the accelerate-stop distance is equal to the takeoff distance for the aircraft for which the runway is designed.

based aircraft: The total number of active general aviation aircraft that use or may be expected to use an airport as a home base.

basic data table: Shown on the airport layout plan drawing, it includes the airport elevation, runway identification and gradient, percent of wind coverage by principal runway, ILS runway when designated, normal or mean maximum daily temperature of the hottest month, pavement strength of each runway, plan for obstruction removal, and relocation of facilities.

basic runway length: Runway length resulting when actual length is corrected to mean sea level and standard atmospheric and no-gradient conditions.

basic transport airport (or runway): An airport (or runway) that accommodates turbojet-powered aircraft up to 60,000 pounds gross weight.

basic utility (BU) airport: Accommodates most single-engine and many of the small twin-engine aircraft.

biometrics: Term used to describe technologies that measure and analyze human body characteristics for identification and authentication purposes.

blast fence: A barrier that is used to divert or dissipate jet or propeller blast.

blast pad: A specially prepared surface placed adjacent to the ends of runways to eliminate the erosive effect of the high wind forces produced by airplanes at the beginning of their takeoff rolls.

bond rating: A rating, reflecting the reliability of an airport's ability to pay back the bond, based on financial indicators of the airport's ownership.

boundary markers: Markers indicating the boundary of the surface usable for landing and takeoff of aircraft.

break-even need: The annual revenue amount required to cover cost of capital investment and costs of administration, operation, and maintenance.

budgets: The planned dollar amounts needed to operate and maintain the airport during a definite period of time such as a year. There are capital budgets for major capital expenditures (such as runway resurfacing) and operating budgets to meet daily expenses.

building area: An area on an airport to be used, considered, or intended to be used, for airport buildings or other airport facilities or rights-of-way, together with all airport buildings and facilities located thereon.

building restriction line: A line shown on the airport layout plan beyond which airport buildings must not be positioned in order to limit their proximity to aircraft movement areas.

buildings and facilities chief: Responsible for assuring that buildings are adequately maintained with a minimum of cost.

build, operate, and transfer contract (BOT): A contract in which private investment is used to construct and operate a facility for a period defined in the terms of the contract. At the end of the contract period, the ownership of the facility is transferred to the airport owner.

Bureau of Air Commerce: Established in 1934 as a separately constituted bureau of the Department of Commerce to promote and regulate aeronautics. The bureau consisted of two divisions, the division of air navigation and the division of air regulation.

business travel: A type of trip purpose that describes a passenger traveling primarily for business purposes.

busy-hour operations: The total number of aircraft operations expected to occur at an airport at its busiest hour, computed by averaging two adjacent busiest hours of a typically high-activity day.

capacity: The ability of an airport to handle a given volume of traffic (demand). It is a limit that cannot be exceeded without incurring an operational penalty.

capital improvement expenses: Very large, periodic expenses that contribute to significant airport infrastructure improvement or expansion.

Capital Improvement Plan (CIP): An FAA program, established in 1991, which outlined the further enhancement of the air traffic control system.

Category II operation: With respect to the operation of aircraft, means a straight-in ILS approach to the runway of an airport under a Category II ILS instrument approach procedure issued by the administrator or other appropriate authority. (FAR Part I)

causal models: Highly sophisticated mathematical models that are developed and tested using historical data. The model is built on a statistical relationship between the forecasted (dependent) variable and one or more explanatory (independent) variables.

Center Terminal Radar Approach Control Automation System (CTAS): Will provide users with airspace capacity improvement, delay reductions, and fuel savings by introducing computer automation to assist controllers in efficiently descending, sequencing, and spacing arriving aircraft.

Central Business District (CBD): The "downtown" of a metropolitan area.

centralized passenger processing: Facilities for ticketing, baggage check-in, security, customs, and immigration—all done in one building and used for processing all passengers using the building.

certificated route air carrier: One of a class of air carriers holding Certificates of Public Convenience and Necessity issued by the Civil Aeronautics Board. These carriers are authorized to perform scheduled air transportation over specified routes and a limited amount of nonscheduled operations.

chief accountant: Responsible for financial planning, budgeting, accounting, payroll, and auditing.

chief—airside operations: Responsible for all airfield operations.

chief—landside operations: Responsible for all landside operations.

chief purchasing agent: Directs the procurement of materials and services to support the airport; prepares, negotiates, interprets, and administers contracts with vendors.

Civil Aeronautics Act of 1938: Created one administrative agency responsible for the regulation of aviation and air transportation. Under reorganization in 1940, two separate agencies were created: the Civil Aeronautics Board, primarily concerned with economic regulation of the air carriers; and the Civil Aeronautics Administration, responsible for the safe operation of the airway system.

Civil Aeronautics Administration (CAA): Forerunner to the FAA, responsible for supervising the construction, maintenance, and operation of the airway system including enforcement of safety regulations.

Civil Aeronautics Board (CAB): Responsible for the economic regulation of the certificated air carriers during the period from 1940 to 1985.

civil airport user categories: As used by airport planners, refers to the four major types of airports: certificated air carrier, commuter, general aviation, and military.

Class A airspace—positive control airspace: Airspace located throughout the United States beginning at an altitude of 18,000 feet MSL up to 60,000 MSL.

Class B airspace—terminal radar service areas: Airspace surrounding the nation's busiest airports.

Class C airspace—airport radar service areas: Airspace surrounding airports of moderately high levels of IFR operations or passenger enplanements.

Class D airspace—control zones: Airspace surrounding those airports not in Class B or Class C airspace but do have an air traffic control tower in operation.

Class E airspace—general controlled airspace: Airspace that generally exists in the absence of Class A, B, C, or D airspace extending from the surface to 18,000 feet MSL within 5 miles or airports without control towers but with instrument approach procedures.

Class G airspace—uncontrolled airspace: Airspace in the absence of Class A, B, C, D, or E airspace.

clear-air turbulence (CAT): Turbulence encountered in air where no clouds are present; more popularly applied to high-level turbulence associated with wind shear; often encountered in the vicinity of the jet stream.

clear zone: Areas constituting the innermost portions of the runway approach areas as defined in FAR Part 77.

closed airport: An airport temporarily closed to aircraft operations for maintenance, construction, or some other purpose while the operator is still in business.

closed field marking: Panels placed in the center of the segmented circle, or in the center of the field, in the form of a cross which will signify that the field is closed to all traffic.

closed runway marking: Panels placed on the ends of the runway and at regular intervals in the form of a cross, signifying that a runway is closed to all traffic.

Cockpit Information System (CIS): Will process and display Flight Information Service (FIS) information and integrate it with navigation, surveillance, terrain, and other data available in the cockpit.

Code of Federal Regulations (CFR): The published federal rules and regulations that are used to govern national policies.

Collaborative Decision Making (CDM): A joint FAA/industry initiative designed to improve traffic flow management through increased interaction and collaboration between airspace users and the FAA.

combined unit terminal: A unit terminal configuration where two or more airlines would share a common building but have separate passenger and baggage processing facilities.

commercial service airports: Public-use commercial airports receiving scheduled passenger service and enplaning at least 2,500 passengers annually.

common-use self-service kiosks (CUSS): Automated facilities that provide self-service check-in for multiple air carriers.

Common-use terminal equipment (CUTE): A computer-based system that can accommodate the operating systems of any air carrier that shares a common ticketing facility.

community noise equivalent level (CNEL): A method that considers the sensitivity of a community toward aircraft noise levels in estimating the degree of airport-generated noise impacts on a region.

compensatory cost approach: A financial management approach under which the airport operator assumes the major financial risk of running the airport and charges the airlines fees and rental rates set so as to recover the actual costs of the facilities and services that they use.

Computer-Assisted Passenger Pre-screening System (CAPPS II): A passenger profiling system that uses passenger information to verify identity and then determine the security risk of a ticketed air carrier passenger.

computerized aircraft manifest: Produces aircraft load sheets, passenger manifests, and automatic telex reservations.

computerized baggage sorting equipment: A new technique to sort baggage through the use of machine-readable tags.

computerized ticket systems: Provide passengers advance reservation and sales, preassignment of seats, and automatic tagging of baggage.

concession agreement: An agreement between the airport and a concession regarding the conduct of business on airport property.

concessions: Ancillary processing facilities that generate revenue for the airport through the sale of products and services to passengers.

concourse: A passageway for passengers and public between the principal terminal building waiting area and the fingers and/or aircraft landing positions.

conical surface: A surface extending from the periphery of the horizontal surface outward and upward at a slope of 20 to 1 for the horizontal distances and to the elevations above the airport elevation as prescribed by FAR Part 77.

Consolidated Operations and Delay Analysis System (CODAS): An improved aircraft delay reporting system. Using the FAA's Enhanced Traffic Management System and Aeronautical Radio Incorporated data it will calculate delay by phase of flight and will include weather data from National Oceanic and Atmospheric Administration.

consumer market survey: A qualitative forecasting method that seeks the opinions of the consumer base of the airport.

Continuing Appropriations Act of 1982: An amendment to the Airport and Airway Improvement Act of 1982 that added a section providing authority to issue discretionary grants in lieu of unused apportioned funds under certain circumstances.

controlled access: Measures used around airports to prevent or control the movement of persons and vehicles to and from security-sensitive areas of the airport property.

controlled airspace: Airspace within which some or all aircraft may be subject to air traffic control. (FAR Part 1)

Controller-to-Pilot Data Link Communications (CPDLC): A data link service that will replace sets of controller/pilot voice messages with data messages displayed in the cockpit.

controlled firings areas: Areas of airspace that contain civilian and military activities that could be hazardous to nonparticipating aircraft, such as rocket testing, ordinance disposal, and blasting.

controlling obstruction: The highest obstruction relative to a prescribed plane within a specific area.

criteria for design and development: Specific performance measures used by airport planners in designing terminal building space requirements.

criteria for inclusion in the NPIAS: The principal criteria are (1) that the airport has (or is forecast to have within 5 years) at least 10 based aircraft (or engines), (2) that it be at least a 30-minute drive from the nearest existing or proposed airport currently in the NPIAS, and (3) that there is an eligible sponsor willing to undertake ownership and development of the airport.

crosswind: A wind blowing across the line of flight of an aircraft.

crosswind component: A wind component that is at a right angle to the longitudinal axis of the runway or the flight path of the aircraft.

crosswind leg: A flight path at right angles to the landing runway off its upwind leg.

crosswind runway: A runway additional to the primary runway to provide for wind coverage not adequately provided by the primary runway.

curbside check-in: Designed to speed passenger movement by separating baggage handling from other ticket counter and gate activities and thereby disencumbering those locations, allowing baggage to be consolidated and moved to aircraft more directly.

daylight beacon operation: Operation of an airport rotating beacon during the hours of daylight means that the reported ground visibility in the control zone is less than 3 miles and/or the reported ceiling is less than 1,000 feet and that the ATC clearance is required for landing, takeoffs, and flight in the traffic pattern.

day/night average noise level (Ldn): A method that considers the impact of nighttime operations in estimating the degree of airport-generated noise impacts on a region.

debt service coverage: The requirement that the airport's revenue, net of operating and maintenance expenses, be equal to a specified percentage in excess of the annual debt service (principal and interest payments) for revenue bond issues.

decentralized passenger processing: The passenger handling facilities are provided in smaller units and repeated in one or more buildings.

decibel (dB): A unit of noise level representing a relative quantity. This reference value is a sound pressure of 20 micronewtons per square meter.

decision height (DH): With respect to the operation of aircraft, means the height at which a decision must be made during the ILS or PAR instrument approach, to either continue the approach or to execute a missed approach. (FAR Part 1)

defederalization: Refers to a proposal to withdraw assistance for major air carrier airports.

delay: The difference between the time an operation actually takes and the time it would have taken under uncongested conditions without interference from other aircraft.

Delphi method: A qualitative forecasting method that involves an iterative process of interviewing experts in the field of interest, responding to initial answers, and revising or giving further arguments into support of their answers.

demand management: A method of controlling airport access by promoting more effective or economically efficient use of existing facilities. The two most prevalent methods are differential pricing and auctioning of landing rights.

departing passenger: A passenger entering the terminal from the ground access system through the access/processing interface.

Department of Transportation (DOT): Established in 1967 to promote coordination of existing federal programs and to act as a focal point for future research and development efforts in transportation.

deplanements (or deplaned passengers): The total number of revenue passengers deplaning an aircraft at an airport.

depreciable investment: The annual cost of capital invested in plant and equipment.

Development of Landing Areas for National Defense (DLAND): A program approved by Congress in 1940 that appropriated $40 million to be spent by the CAA for 250 airports necessary for the national defense.

destination sign: Airfield sign, marked by a yellow background with black inscription, indicating a destination on the airfield, followed by an arrow showing the direction of the taxiing route to the destination.

Development of Landing Areas for National Defense (DLAND): Program passed in 1940 that authorized the appropriation of $40 million for the construction of up to 250 airports.

directional marker: An airway marker located on the ground and used to give visual direction to an aircraft; consists of an arrow indicating true north and arrows indicating names and states of the nearest towns.

direction sign: Airfield sign that identifies the designations of intersecting taxiways leading out of the intersection at which an aircraft is located.

Discrete Address Beacon System (DABS): A sophisticated air traffic control surveillance system capable of interrogating each airborne DABS transponder in an "all-call" mode or with a discrete address signal encoded for each specific aircraft operating in the system. The data acquired upon response from each transponder are then processed to provide range, azimuth, altitude, and identity of each aircraft in the system on an individual basis but in sequence on a programmed interroschedule. Because aircraft are addressed individually in DABS, the surveillance system automatically provides a natural vehicle for a data link between ground and aircraft that can be used for ATC control purposes including the proposed intermittent positive control (IPC) concepts.

discretionary funds: Grants that go to projects that address goals established by the Congress, such as enhancing capacity, safety, and security, or mitigating noise at all types of airports.

displaced threshold: A runway marking that defines a threshold that is located at a point on the runway other than the beginning of the runway pavement, where aircraft are permitted to taxi and take off, but not land.

display system replacement (DSR): Part of a joint FAA and Department of Defense program to replace automated radar terminal systems and other older technology systems at air traffic control facilities.

downwind leg: A flight path in the traffic pattern parallel to the landing runway in the direction opposite to landing. It extends to the intersection of the base leg.

dynamic hydroplaning: A condition where landing gear tires ride up on a cushioning film of water on the runway surfaces.

effective perceived noise level (EPNL): Time-integrated perceived noise level calculated with adjustments for irregularities in the sound spectrum, such as that caused by discrete-frequency components (tone correction). The unit of effective perceived noise level is the decibel, with identifying prefix for clarification, EPNdB.

effective runway length: (a) Effective runway length for takeoff means the distance from the end of the runway at which the takeoff is started to the point at which the obstruction clearance plane associated with the other end of the runway intersects the runway centerline. (b) Effective runway length for landing means the distance from the point at which the obstruction clearance plane associated with the approach end of the runway intersects the centerline of the runway to the far end thereof. (FAR Part 121)

elevation mean sea level (MSL): Term used to describe elevation of a location or the altitude of an aircraft with respect to sea level.

enplaned passengers: The total number of revenue passengers boarding aircraft, including originating, stopover, and transfer passengers, in scheduled and nonscheduled services.

enplaning air cargo: Includes the total tonnage of priority, nonpriority, and foreign mail, express, and freight (property other than baggage accompanying passengers) departing on aircraft at an airport, including origination, stopover, and transfer cargo.

en route air traffic control service: Air traffic control service provided to aircraft on an IFR flight plan, generally by centers, when these aircraft are operating between departure and destination terminal areas.

en route flight advisory service (EFAS): A specialized system providing near-real-time weather service to pilots in flight.

entrance taxiway: A taxiway that provides entrance for aircraft to the takeoff end of the runway.

Environmental Impact Review (EIR): A document reviewing the potential environmental impacts of a moderate to major airport planning project.

Environmental Impact Statement (EIS): A document comprehensively assessing the magnitude of any environmental impacts that will exist as a result of an airport planning project, and identifying strategies for which to mitigate the impact.

essential air service: Guarantees air carrier service to selected small cities and provided subsidies (through 1988) if needed so as to prevent these cities from losing service.

Essential Air Service Program (EAS): A program established as part of the Airline Deregulation Act of 1978 that provided subsidies to the last remaining carrier in a market so as to prevent selected cities from losing air service altogether.

essential processing facilities: Passenger processing facilities that must be present at airport terminals to ensure appropriate processing for passengers traveling on each itinerary segment.

exclusive area: Any portion of a secured area, AOA, or SIDA, for which an aircraft operator has assumed responsibility for the security of its area.

Exclusive-Use Gate Usage Agreement: A gate usage agreement in which an air carrier retains sole authority to use a particular gate or set of gates at an airport terminal.

exclusive-use ticket counters: Ticketing facilities typically configured with information systems, computers, and other equipment specific to one air carrier.

executive aircraft operator: A corporation, company, or individual that operates owned or leased aircraft, flown by pilots whose primary duties involve pilotage of aircraft, as a means of transportation of personnel or cargo in the conduct of company business.

exit taxiway: A taxiway used as an exit from a runway to another runway, apron, or other aircraft operating area.

Explosive Detection Systems (EDS): Equipment using computed tomography technology, to detect and identify metal and trace explosives that may be hidden in checked baggage.

explosive trace detection (ETD): Equipment that uses molecular spectrometry to detect and identify trace explosives that may be hidden in checked or carry-on baggage.

externalities: Environmental impacts of activities that occur as a result of operations from other sources, as an indirect result of an airport's presence.

FAA high-density rule: Quotas imposed at selected airports on the basis of estimated limits of the air traffic control (ATC) system and airport runway capacity.

facilities and equipment (F&E) program: Provides funding for airports for the installation of navigational aids and control towers, as necessary.

facilities chief: Establishes criteria and procedures for the administration of all airport property. Responsible for inventory control of all equipment and facilities.

fan noise: General term for the noise generated within the fan stage of a turbofan engine; includes both discrete frequencies and random noise.

FAR: Federal Aviation Regulation.

Federal-Aid Airport Program (FAAP): A program, established in 1946, that provided federal funding assistance to local municipalities for moderate to major airport construction projects.

Federal Airport Act of 1946: A federal aid to airports program administered by the CAA (later the FAA) to give the United States a comprehensive system of airports. Over $1.2 billion in airport development aid funds were disbursed by the federal government during the act's 24-year history.

Federal Aviation Act of 1958: Created the Federal Aviation Agency (FAA) with an administrator responsible to the president. The law retained the CAB as an independent agency and transferred the safety-rule-making powers to the FAA along with the functions of the CAA and the Airways Modernization Board.

Federal Aviation Administration (FAA): Created by the act that established the Department of Transportation. Assumed all of the responsibilities of the former Federal Aviation Agency.

Federal Aviation Administration Authorization Act of 1994: Authorized the extension of AIP funding through 1996. Increased the number of airports eligible for Military Airport Program funding, universal access control, and explosives detection security.

Federal Aviation Agency (FAA): Established in 1958 to regulate, promote, and develop air commerce in a safe manner. FAA was also given the responsibility of operating the airway system and consolidating all research and development of air navigation facilities.

Federal Aviation Administration Reauthorization Act of 1996: Authorized the extension of AIP funding through 1998. Various changes were made to the formula computation of primary and cargo entitlements, state apportionment, and discretionary set-asides.

Federal Inspection Services (FIS): Conducts customs and immigration services including passport inspection, inspection of baggage, and collection of duties on certain imported items, and sometimes inspection for agricultural materials, illegal drugs, or other restricted items.

federal letters of intent (LOI): Issued by the FAA for projects that will significantly enhance systemwide airport capacity.

Federal Security Director (FSD): Representative of the Transportation Security Administration charged with overseeing airport security at one or more commercial service airports.

final approach areas: Areas of defined dimensions protected for aircraft executing instrument approaches.

final approach (IFR): The flight path of an aircraft which is inbound to the airport on an approved final instrument approach course, beginning at the final approach fix or point and extending to the airport or the point where circling for landing or missed approach is executed.

final approach (VFR): A flight path, in the traffic pattern, of a landing aircraft in the direction of landing along the extended runway centerline from the base leg to the runway.

final controller: That controller providing final approach guidance utilizing radar equipment.

Final Monitor Aid (FMA): The FMA is a high-resolution color display that is equipped with the controller-alert hardware and software that is used in the PRM system.

financial plan: An economic evaluation of the entire master plan development including revenues and expenditures.

Finding of No Significant Impact (FONSI): A finding determined as a result of an environmental impact review that reveals no significant environmental impacts to be caused by a given airport planning project.

finger: A roofed structure, with or without sidewalls, extending from the main terminal building or its concourse to the aircraft loading positions.

flareout: That portion of a landing maneuver in which the rate of descent is reduced to lessen the impact of landing.

fleet mix: The percentage of aircraft, by type, operating at an airport.

flexible pavement: A pavement structure consisting of a bituminous surface course, such as asphalt, a base course, and in most cases, a subbase course.

flight advisory service: Advice and information provided by a facility to assist pilots in the safe conduct of flight and aircraft movement.

Flight Advisory Weather Service (FAWS): Flight advisory and aviation forecast service provided by the National Weather Service.

Federal Information Service (FIS): A ground-based data server and data links to provide a variety of nonoperational control information to the cockpit such as weather and traffic information, special-use airspace status, notices to airment, and obstruction updates.

flight interface: As used in the passenger handling system, the link between the passenger processing activities and the flight.

flight plan: Specified information relating to the intended flight of an aircraft that is filed orally or in writing with air traffic control. (FAR Part 1)

Flight Schedule Monitor (FSM): A primary component of the collaborative decision-making (CDM) system which collects and displays arrival information, retrieves real-time demand and schedule information, and monitors ground delay performance.

flight service station (FSS): A central operations facility in the national flight advisory system utilizing data interchange facilities for the collection and dissemination of NOTAM, weather, and administrative data and providing preflight and inflight advisory service and other services to pilots via air/ground communication facilities.

flight time: The time from the moment the aircraft first moves under its own power for the purpose of flight until the moment it comes to rest at the next point of landing ("block-to-block" time). (FAR Part 1)

flow control: Restriction applied by ATC to the flow of air traffic to keep elements of the common system, such as airports or airways, from becoming saturated.

foreign object debris (FOD): Debris located on a runway or taxiway that may cause damage to passing aircraft.

free flight: A concept for safe and efficient flight operating capability under instrument flight rules (IFR) in which the operators have the freedom to select path and speed in real time.

fuel flowage fees: Fees levied by the airport operator per gallon of aviation gasoline and jet fuel sold at the airport.

gate arrival: A centralized terminal building layout that is aimed at reducing the walking distance by bringing the automobile as close as possible to the aircraft.

gate position: A designated space or position on an apron for an aircraft to remain parked while loading or unloading passengers and cargo.

gate usage agreement: A formal contract between an airport and an air carrier as to the lease of gates at the airport terminal.

general airport revenue bond (GARB): A bond secured solely by revenue generated by operations of the airport and is not backed by any additional governmental subsidy or tax levy.

general aviation: That portion of civil aviation that encompasses all facets of aviation except air carriers holding a Certificate of Convenience and Necessity from the Civil Aeronautics Board, and large aircraft commercial operators.

general aviation (GA) airports: Those airports with fewer than 2,500 annual enplaned passengers and those used exclusively by private and business aircraft not providing common carrier passenger service.

general aviation itinerant operations: Takeoffs and landings of civil aircraft (exclusive of air carrier) operating on other than local flights.

general concept evaluation: A set of general considerations that an airport planner uses to evaluate and select among alternative concepts in a preliminary fashion prior to any detailed design and development.

general obligation bonds (GOBs): Bonds that are issued by states, municipalities, and other general-purpose governments and backed by the full faith, credit, and taxing power of the issuing government agency.

general utility (GU) airport: Accommodates all general aviation aircraft.

glide slope transmitter: An ILS navigation facility in the terminal area electronic navigation system, providing vertical guidance for aircraft during approach and landing by radiating a directional pattern of VHF radio waves modulated by two signals that, when received with equal intensity, are displayed by compatible airborne equipment as an "on-path" indication.

Global Positioning System (GPS): A satellite-based navigation system that will enhance user-preferred routing, reduce separation standards, and increase access to airports under instrument meteorological conditions (IMC) through more precision approaches.

grant programs: Federal and state programs from which owners of public-use airports could acquire funds, provided without responsibility to paying any monies back, for airport development.

ground access systems: Existing and planned highway and mass transit systems in the area of the airport.

ground-controlled approach (GCA): A radar landing system operated from the ground by air traffic control personnel transmitting instructions to the pilot by radio. Approach may be conducted with surveillance radar only or with both surveillance and precision approach radar.

grounds chief: Responsible for assuring that the grounds are maintained in good repair and that the landscape is adequately maintained.

guidance light facility (GDL): A lighting facility in the terminal area navigation system located in the vicinity of an airport consisting of one or more high-intensity lights to guide a pilot into the takeoff or approach corridor, away from populated areas for safety and noise abatement.

handoff: Passing of control of an aircraft from one controller to another.

height and hazard zoning: Protects the airport and its approaches from obstructions to aviation while restricting certain elements of community growth.

heliport: An area of land, water, or structure used or intended to be used for the landing and takeoff of helicopters. (FAR Part 1)

high-intensity light: A runway or threshold light whose main beam provides a minimum intensity of 12,000 candlepower in white light through a vertical angle of 3 degrees and a horizontal angle of 6 degrees.

holding areas: Areas located at or very near the ends of runways for pilots to make final checks and await final clearance for takeoff.

holding bay: An area where aircraft can be held, or bypassed, to facilitate efficient ground traffic movement.

horizontal surface: A specified portion of a horizontal plane located 150 feet above the established airport elevation which establishes the height above which an object is determined to be an obstruction to air navigation.

hub: A city or a standard metropolitan statistical area requiring aviation services and classified by each community's percentage of the total enplaned passengers in scheduled service of certain domestic certificated route air carriers.

hydroplaning: The condition in which moving aircraft tires are separated from a pavement surface by a water or liquid rubber film or by steam, resulting in a derogation of mechanical braking effectiveness.

IFR airport: An airport with an authorized instrument approach procedure.

IFR conditions: Weather conditions below the minimum for flight under visual flight rules. (FAR Part 1)

ILS Category I: An ILS that provides acceptable guidance information from the coverage limits of the ILS to the point at which the localizer course line intersects the glide path at a height of 100 feet above the horizontal plane containing the runway threshold. A Category I ILS supports landing minima as low as 200 feet HAT and 1,800 RVR.

ILS Category II: An ILS that provides acceptable guidance information from the coverage limits of the ILS to the point at which the localizer course line intersects the glide path at a height of 50 feet above the horizontal plane containing the runway threshold. A Category II ILS supports landing minima as low as 100 feet HAT and 1,200 RVR.

ILS Category III: An ILS that provides acceptable guidance information from the coverage limits of the ILS with no decision height specified above the horizontal plane containing the runway threshold. See ILS CAT III A, B, C operations.

ILS CAT IIIA operation: Operation, with no decision height limitation, to and along the surface of the runway with a runway visual range not less than 700 feet.

ILS CAT IIIB operation: Operation, with no decision height limitation, to and along the surface of the runway without reliance on external visual reference; and, subsequently, taxiing with external visual reference with a runway visual range not less than 150 feet.

ILS CAT IIIC operation: Operation, with no decision height limitation, to and along the surface of the runway and taxiways without reliance on external visual reference.

inactive airport: An airport where all flying activities have ceased yet has remained in an acceptable state of repair for civil use and is identifiable from the air as an airport.

Initial Conflict Probe (ICP): Provides controllers with the ability to identify potential separation conflicts up to 20 minutes in advance, and to do this with greater precision and accuracy.

inner marker (IM): An ILS navigational facility in the terminal area navigation system located between the middle marker and the end of the ILS runway, transmitting a 75-megahertz fan-shaped radiation pattern modulated at 3,000 hertz, keyed at six dots per second and received by compatible airborne equipment indicating to the pilot, both aurally and visually, that the aircraft is directly over the facility at an altitude of 100 feet on the final ILS approach, providing the pilot is on the glide path.

in-runway lighting: A lighting system consisting of flush or semiflush lights placed in the runway pavement in specified patterns.

instrument approach: An approach to an airport, with intent to land, by an aircraft flying in accordance with an IFR flight plan.

instrument approach runway: A runway served by electronic aid providing at least directional guidance adequate for a straight-in approach.

instrument flight rules (IFR): FAR rules that govern the procedures for conducting instrument flight. (FAR Part 91)

Instrument Landing System (ILS): A system that provides, in the aircraft, the lateral, longitudinal, and vertical guidance necessary for a landing.

instrument meteorological conditions (IMC): Meteorological conditions expressed in terms of visibility and ceiling less than the minimum specified for visual meteorological conditions.

instrument runway: A runway equipped with electronic and visual navigation aids and for which a straight-in (precision or nonprecision) approach procedure has been approved.

integrated airport system planning: As defined in the Airport and Airway Improvement Act of 1982, "the initial as well as continuing development for planning purposes of information and guidance to determine the extent, type, nature, location, and timing of airport development needed in a specific area to establish a viable, balanced, and integrated system of public use airports."

integrated noise model (INM): Computer software used to estimate the noise impacts of airport operations on a surrounding region.

Integrated Terminal Weather System (ITWS): A fully automated weather-prediction system installed at ARTCCs that will give both air traffic personnel and pilots better information on near-term weather hazards in the airspace within 60 nanometers of an airport.

intermodalism: To improve the speed, reliability, and cost-effectiveness of the country's overall transportation system by integrating transportation strategy to promote intermodal exchanges among highway, railway, waterway, and air transportation.

international airport: (1) An airport of entry that has been designated by the secretary of treasury or commissioner of customs as an international airport for customs service. (2) A landing rights airport at which specific permission to land must be obtained from customs authorities in advance of contemplated use. (3) Airports designated under the Convention of International Civil Aviation as an airport for use by international commercial air transport and/or international general aviation. (4) As pertaining to ICAO facilitation, any airport designated by the contracting state in whose territory it is situated as an airport of entry and departure for international air traffic, where the formalities incident to customs, immigration, public health, animal and plant quarantine, and similar procedures are carried out.

International Civil Aviation Organization (ICAO): A membership-based organization comprised of 188 contracting states that span the world, which publishes a series of recommended policies and regulations to be applied by individual states in the management of their airports and civil aviation systems.

intersecting runways: Two or more runways that cross or meet within their lengths.

itinerant operations: All aircraft arrivals and departures other than local operations.

jet noise: The noise generated externally to a jet engine in the turbulent jet exhaust.

Joint Automated Weather Observation System (JAWOS): Automatically gathers local weather data and distributes it to other air traffic control facilities and to the National Weather Service.

Joint Planning and Development Office (JPDO): The intergovernmental organization established in 2003 to provide leadership in the implementation of NextGen.

joint-use airport: An airport owned by the military, a public body, or both, where an agreement exists for joint civil-military, fixed-based aviation operations.

judgmental forecasts: Forecasts based on intuition and subjective evaluations by an individual who is closely acquainted with the factors related to the variable being forecast.

jury of executive opinion method: A qualitative forecasting method that seeks the predictions of management and administration of the airport and the airport's tenants.

Kelly Act of 1925: Authorized the postmaster general to enter into contracts with private persons or companies for the transportation of the mail by air.

land and hold short operations (LAHSO): Operations conducted simultaneously in intersecting runways under a policy that requires landing aircraft to hold short of intersecting runways upon landing.

landing area: Any locality, either on land or water, including airports, heliports, and STOLports, that is used or intended to be used for the landing and takeoff or surface maneuvering of aircraft, whether or not facilities are provided for the shelter, servicing, or repair of aircraft, or for receiving or discharging of passengers or cargo.

landing rights airport: See international airport.

landing roll: The distance from the point of touchdown to the point where the aircraft can be brought to a stop, or exit the runway.

landing strip: A term formerly used to designate (1) the graded area upon which the runway was symmetrically located and (2) the graded area suitable for the takeoff and landing of airplanes where a paved runway was not provided.

landing strip lighting: Lines or rows of lights located along the edges of the designated landing and takeoff path within the strip. See landing strip.

landside operations: Those parts of the airport designed to serve passengers, including the terminal buildings, vehicular circular drive, and parking facilities.

land use plan: Shows on-airport land uses as developed by the airport sponsor under the master plan effort and off-airport land uses as developed by surrounding communities.

land use zoning: Zoning by cities, towns, or counties restricting the use of land to specific commercial or noncommercial activities.

large aircraft: Aircraft of more than 12,500 pounds maximum certificated take-off weight. (FAR Part 1)

large hubs: Those airports that account for at least 1 percent of the total annual U.S. passenger enplanements.

lead-in light facility (LDIN): A facility in the terminal area navigation system providing special light guidance to aircraft in approach patterns or landing procedures. Facility configuration consists of any number of flashers so located as to visually guide an aircraft through an approach corridor, bypassing high-density residential, commercial, or obstruction areas.

lease, build, and operate agreement (LBO): An agreement in which the airport owner allows a private sector company to build and manage an airport facility, while leasing the property and facility from the airport.

leisure travel: A type of trip purpose that describes a passenger traveling primarily for leisure purposes.

lighted airport: An airport where runway and associated obstruction lighting are available from sunset to sunrise or during periods of reduced visibility or on request of the pilot.

linear or curvilinear terminal: A type of simple terminal layout that is repeated in a linear extension to provide additional apron frontage, more gates, and more room within the terminal for passenger processing.

line-item budget: The most detailed form of budget used quite extensively at the large commercial airports. Budgets are established for each item and often adjusted to take into consideration changes in volume of activity.

Local Area Augmentation System (LAAS): A differential GPS system that provides localized measurement correction signals to basic GPS signals to improve navigation accuracy, integrity, continuity, and availability.

localizer beacon: An ILS navigation facility in the terminal area electronic navigation system, providing horizontal guidance to the runway centerline for aircraft during approach and landing by radiating a directional pattern of VHF radio waves modulated by two signals that, when received with equal intensity, are displayed by compatible airborne equipment as an "on-course" indication, and when received in unequal intensity are displayed as an "off-course" indication.

local operations: Pertains to air traffic operations, aircraft operating in the local traffic pattern or within sight of the tower; aircraft known to be departing for, or arriving from, flight in local practice areas located within a

20-mile radius of the control tower; aircraft executing simulated instrument approaches or low passes at the airport.

local traffic: Aircraft operating in the local traffic pattern or within sight of the tower, or aircraft known to be departing for or arriving from flight in local practice areas, or aircraft executing simulated instrument approaches at the airport.

local VFR flight plan: Specific information provided to air traffic service units, relative to the intended flight of an aircraft under visual flight rules within a specific local area.

location map: Shown on the airport layout plan drawing, it depicts the airport, cities, railroads, major highways, and roads within 20 to 50 miles of the airport.

location sign: Airfield sign used to identify either a taxiway or a runway on which an aircraft is located.

long-term parking: Designed for travelers who leave their vehicles at the airport while they travel.

low-intensity light: A runway or threshold light from which the light distribution through 360 degrees of azimuth and a selected 6 degrees in the vertical is not less than 10 candlepower in white light.

Low-Level Wind Shear Alert System (LLWAS): Provides the air traffic control tower with information on wind conditions near the runway. It consists of an array of anemometers that read wind velocity and direction around the airport and signal the sudden changes that indicate wind shear.

lump sum appropriation: The simplest form of budget and generally only used at small GA airports. There are no specific restrictions as to how the money should be spent.

magnetometer: A device used at passenger screening checkpoints to detect the presence of metal objects on, or carried by, the person being screened.

majority-in-interest clauses: Found in some airport use agreements that give the airlines accounting for a majority of traffic at an airport the opportunity to review and approve or veto capital projects that would entail significant increases in the rates and fees they pay for the use of airport facilities.

management contract: An agreement under which a firm is hired to operate a particular service on behalf of the airport.

manager of public relations: Responsible for all public relations activities including the development of advertising and publicity concerning the airport.

mandatory instruction sign: Airfield sign marked with a red background and white inscription used to denote an entrance to a runway or critical area and areas where an aircraft is prohibited from entering without proper authorization.

marking: On airports, a pattern of contrasting colors placed on the pavement, turf, or other usable surface by paint or other means to provide specific information to aircraft pilots, and sometimes to operators of ground vehicles, on the movement areas.

medium hubs: Those airports that account for between 0.25 and 1 percent of the total passenger enplanements.

metering: Regulating the arrival time of aircraft in the terminal area so as not to exceed a given acceptance rate.

Metropolitan Planning Organization (MPO): A regional, state, or local transportation planning body, designed to develop comprehensive transportation plans for metropolitan or regional areas as a whole.

Microwave Landing Systems (MLS): An instrument approach and landing system that operated in the microwave frequencies (5.0–5.25 GHz/15.4–15.7 GHz) that provided precision guidance in azimuth, elevation, and distance measurement.

middle marker (MM): An ILS navigation facility in the terminal area navigation system located approximately 3,500 feet from the runway edge on the extended centerline, transmitting a 75-MHz fan-shaped radiation pattern, modulated at 1,300 Hz, keyed alternately dot and dash, and received by compatible airborne equipment, indicating to the pilot both aurally and visually, that he or she is passing over the facility.

Military Airport Program (MAP): A program established as a funding set aside under the Airport Improvement Program (AIP) to provide money for airport master planning and capital development for military airfields transitioning to civilian airports.

military operations areas (MOA): Areas of airspace that contain certain military activities.

missed approach procedure: A procedure performed by pilots in the event that there is insufficient visibility or cloud clearance to complete a landing while performing an instrument approach to a runway.

mobile lounge or transporter: Used to transport passengers to and from the terminal building to aircraft parked on the apron.

mode S data link: An addition to the ATCRBS transponder that permits direct, automatic exchange of digitally encoded information between the ground controller and individual aircraft.

moving target detection: An electronic device that will permit radar scope presentation only from targets that are in motion.

multiple unit terminals: Unit terminals built as separate buildings for each airline, each building behaving as its own unit terminal.

multiplier effect: Revenues generated by the airport are channeled throughout the community.

municipally operated airport: An airport owned by a city and run as a department of the city, with policy direction by the city council and, in some cases, by a separate airport commission or advisory board.

National Airport Plan (NAP): The first organized airport system planning effort in the United States, established in 1944, which called attention to the private airport deficiencies of inadequate distribution and inadequate facilities, and formed the basis for airport system planning and federal funding programs for airports in the United States.

National Airspace System (NAS): The current organization of airports, airspace, and air traffic control that make up the civil aviation system in the United States.

National Airport System Plan (NASP): A plan specifying in terms of general location and type of development the projects considered by the administrator to be necessary to provide a system of public airports adequate to anticipate and meet the needs of civil aeronautics. Replaced by the NPIAS. See criteria for inclusion in the NPIAS.

National Airspace Redesign (NAR): A large-scale analysis of the national airspace structure that began by identifying problems in the congested airspace of New York and New Jersey. The goal is to ensure that the design and management of the national airspace system is prepared as the system evolves toward free flight.

National Environmental Policy Act of 1969: Requires the preparation of detailed environmental statements for all major federal airport development actions significantly affecting the quality of the environment.

National Plan of Integrated Airport Systems (NPIAS): The Airport and Airway Improvement Act of 1982 required the FAA to develop the NPIAS by September 1984. The legislation called for the identification of national airport system needs, including development costs in the short and long run.

National Plan of Integrated Airport Systems (NPIAS) levels of need: The NPIAS relates the airport system improvements to three levels of need: level I: maintain the airport system in its current condition, level II: bring the system up to current design standards, and level III: expand the system.

National Route Program (NRP): Gives airlines and pilots increased flexibility in choosing their routes. This flexibility allows airlines to plan and fly the most cost-effective routes and increases the efficiency of the aviation system.

national system of airports: The inventory of selected civil airports that are highly correlated with those aviation demands most consistent with the national interest.

National Transportation Safety Board (NTSB): Created by the act that established the Department of Transportation to determine the cause of transportation accidents and review on appeal the suspension or revocation of any certificates or licenses issued by the secretary of transportation.

National Weather Service (NWS): U.S. government agency concerned with the prediction and dissemination of weather information.

NAVAID: Any facility used in, available for use in, or designated for use in aid of air navigation, including lights; any apparatus or equipment for disseminating weather information, for signaling, for radio direction finding, or for radio or other electronic communication; and any other structure or mechanism having a similar purpose for guiding or controlling flight in the air or the landing or takeoff of aircraft.

navigable airspace: Airspace at and above the minimum flight altitudes prescribed in the FARs, including airspace needed for safe takeoff and landing. (FAR Part 1)

Next-Generation Air-to-Ground Communications (NEXCOM): A digital radio system designed to alleviate the problems associated with the current analog-based communication system.

Next-Generation Air Transportation System (NextGen): A program established in 2003 and coordinated by the JPDO to modernize the nation's air traffic management system.

next-generation weather radar (NEXRAD): Advanced radar systems designed to observe significant weather conditions, such as thunderstorms and hurricanes.

noise compatibility programs: Outlines measures to improve airport land use compatibility.

noise exposure forecast (NEF): A method developed to predict the degree of community annoyance from aircraft noise (and airports) on the basis of various acoustical and operational data.

noise exposure maps: Identify noise contours and land use incompatibilities and are useful in evaluating noise impacts and discouraging incompatible development.

nondepreciable investment items: Those assets, such as the cost of land acquisition, that have a permanent value even if the airport site is converted to other uses.

nondestructive testing (NDT): Techniques used to test the strength of pavements without physically destroying existing pavement.

nonhub primary airports: Those airports that enplane less than 0.05 percent of all commercial passenger enplanements but at least 10,000 annually.

non-directional radio beacon (NDB): A radio-based navigational aid that emits low- or medium-frequency radio signals whereby the pilot of an aircraft properly equipped with an automatic direction finder (ADF) can determine bearings and "home in" on the station.

noninstrument runway: A runway intended for the operation of aircraft using visual approach procedures. See visual runway.

nonprecision approach procedure: A standard instrument approach procedure in which no electronic glide slope is provided. (FAR Part 1)

nonprecision instrument runway: A runway having an existing instrument approach procedure utilizing air navigation facilities with only horizontal guidance for which a straight-in nonprecision instrument approach procedure has been approved.

nose-in, angled nose-in, angled nose-out, and parallel parking: Aircraft parking positions at various angles with respect to the terminal building. The nose-in parking position is the most frequently used at major airports.

notices to airmen (NOTAM): Notices containing information (not known sufficiently in advance to publicize by other means) concerning the establishment, condition, or change in any component (facility, service, or procedure) of, or hazard in, the National Airspace System, the timely knowledge of which is essential to personnel concerned with flight operations.

objective of the airport master plan: To provide guidelines for future development of the airport which will satisfy aviation demand and be compatible with the environment, community development, other modes of transportation, and other airports.

obstruction light: A light, or one of a group of lights, usually red, mounted on a surface structure or natural terrain to warn pilots of the presence of a flight hazard; either an incandescent lamp with a red globe or a strobe light.

obstruction marking/lighting: Distinctive marking and lighting to provide a uniform means for indicating the presence of obstructions.

Office of Management and Budget Circular A-95: Prior to July 1982, required that designated regional agencies review airport projects before federal grants were given.

O'Hare Agreement: An agreement established in the 1950s between the city of Chicago and the airlines that established a precedent in revenue bond financing which pledged the airlines to meet any shortfall in income needed to pay off the principal and interest on the bonds.

open-V runways: Two intersecting runways whose extended centerlines intersect beyond their respective thresholds.

operating statement: Records an airport's revenues and expenses over a particular time period (quarterly and annually).

operational activity forecasts: Includes forecasts of operations by major user categories (air carrier, commuter, general aviation, and military).

operation and maintenance costs (O&M): Those expenses that occur on a regular basis and are required to maintain the current operations at the airport.

organization chart: Shows the formal authority relationships between superiors and subordinates at various levels, as well as the formal channels of communication within the organization.

other commercial service airports: Commercial service airports enplaning 2,500 to 10,000 passengers annually.

outer fix: A fix in the destination terminal area, other than the approach fix, to which aircraft are normally cleared by an air route traffic control center or an approach control facility, and from which aircraft are cleared to the approach fix or final approach course.

outer marker (OM): An ILS navigation facility in the terminal area navigation system located 4 to 7 miles from the runway edge on the extended centerline transmitting a 75-megahertz fan-shaped radiation pattern, modulated at 400 hertz, keyed at two dashes per second, and received by compatible airborne equipment indicating to the pilot, both aurally and visually, that the aircraft is passing over the facility and can begin its final approach.

overrun: To run off the end of the runway after touching down on the runway.

overrun area: An area beyond the end of the designated runway with a stabilized surface of the same width as the runway and centered on the extended runway centerline. Also known as a stopway.

parallel runways: Two or more runways at the same airport whose centerlines are parallel.

parallel taxiways: Taxiways that are parallel to an adjacent runway.

parking apron: An apron intended to accommodate parked aircraft.

Part 150 of the Federal Aviation Regulations: Established a system for measuring aviation noise in the community and for providing information about land uses that are normally compatible with various levels of noise exposure.

passenger screening: The inspection of passengers for prohibited items at security checkpoints in airport terminals.

passenger facility charges (PFCs): The Airway Safety and Capacity Expansion Act of 1990 authorized the imposition of PFCs at commercial service airports. The airport operator may propose collecting $1, $2, or $3, $4, or $4.50 per enplaned passenger, domestic or foreign, to fund approved airport capital projects.

passenger handling system: A series of links or processes that a passenger goes through in transferring from one mode of transportation to another.

passenger movers: Designed to speed passenger movement through the terminal. Includes buses, mobile lounges, moving sidewalks, and automated guideway systems.

passenger processing link: As used in the passenger handling system, the link that accomplishes the major processing activities required to prepare the passenger for using air transportation.

passive final approach spacing tool (pFAST): An air traffic control tool that helps controllers select the most efficient arrival runway and arrival sequence within 50 nautical miles of an airport.

pavement grooving: The mechanical serration of a pavement surface to provide escape paths for water and slush in order to promote improved aircraft mechanical braking effectiveness.

pavement maintenance: Any regular or recurring work necessary on a continuing basis to preserve existing airport facilities in good condition, any work involved in the care or cleaning of existing airport facilities, and any incidental or minor repair work on existing airport facilities.

pavement management system: Evaluates the present condition of a pavement and predicts its future condition through the use of a pavement condition index.

pavement rehabilitation: Work required to preserve, repair, or restore the physical integrity of the pavement; for example, a structural overlay (laying more asphalt on the runway surface).

pavement structure: The combination of runway base and subbase courses and surface course that transmits the traffic load to the subgrade.

pavement subgrade: The upper part of the soil, natural or constructed, that supports the loads transmitted by the runway pavement structure.

pavement surface course: The top course of a pavement, usually portland cement concrete or bituminous concrete, that supports the traffic load.

perimeter fencing: Physical method of creating a barrier in otherwise easily accessible areas of an airport's secured area boundary.

personnel manager: Responsible for administering the airport personnel program.

pier finger terminal: A type of terminal layout evolving in the 1950s when gate concourses (fingers) were added to simple terminal buildings.

pier satellite terminal: Terminals with concourses extending as piers ending in a round atrium or satellite area.

Planning Grant Program (PGP): A federal aid to airports program established under the Airport and Airway Development Act of 1970 for approved airport planning and development project costs.

port authorities: Legally chartered institutions with the status of public corporations that operate a variety of publicly owned facilities, such as harbors, airports, toll roads, and bridges.

positive passenger baggage matching (PPBM): The act of reconciling boarded passengers with their checked-in baggage on a given aircraft.

practical capacity: The number of operations (takeoffs and landings) that can be accommodated with no more than a given amount of delay, usually expressed in terms of maximum acceptable average delay.

precision approach: A standard instrument approach using a precision approach procedure. See precision approach procedure.

precision approach path indicator (PAPI): A visual glide path indicator that uses light units in a single row of two to four units, to identify the location of landing aircraft with respect to a safe glide path.

precision approach procedure: A standard instrument approach procedure in which an electronic glide slope is provided, such as ILS and PAR. (FAR Part 1)

precision approach radar (PAR): A radar facility in the terminal air traffic control system used to detect and display, with a high degree of accuracy, azimuth, range, and elevation of an aircraft on the final approach to a runway.

precision instrument runway: A runway having an existing instrument approach procedure utilizing an instrument landing system (ILS) or precision approach radar (PAR).

Precision Runway Monitor (PRM): The PRM system consists of an improved monopulse antenna system that provides high azimuth and range accuracy and higher data rates than the current terminal ASR systems. It will improve the accuracy of monitoring simultaneous approaches to parallel runways.

preferential use gate usage agreement: A gate usage agreement in which one air carrier has preferential use of the gate. Should that air carrier not be using the gate during some period of the day, other air carriers subscribing to the agreement may use the gate, as long as its use does not interfere with upcoming operations from the preferential carrier.

primary airports: Public-use commercial airports enplaning at least 10,000 annual enplanements.

primary commercial service airport: A public-use airport that serves at least 10,000 enplaned passengers annually.

primary radar: *See* search radar.

primary surface: A rectangular surface longitudinally centered about a runway. Its width is a variable dimension and it usually extends 200 feet beyond each end of the runway. The elevation of any point on this surface coincides with the elevation of its nearest point on the runway centerline or extended runway centerline.

primary taxiway system: Taxiways that provide aircraft access from runways to aprons and the service areas.

privatization: Shifting of government functions and responsibilities, in whole or in part, to the private sector.

Private Charter Program: A program that mandates all aircraft used for private charter operations with a maximum certified takeoff weight of 45,000 kilograms, or with passenger seating configuration of 61 or more must ensure that all passengers and their carry-on baggage are screened prior to aircraft boarding.

Professional Air Traffic Controllers Organization (PATCO): Labor organization that represented federal air traffic controllers that led a strike of air traffic controllers in 1981.

prohibited areas: Areas of airspace over security-sensitive ground facilities. All aircraft are prohibited from flight operations within a prohibited area unless specific prior approval is obtained.

public airport: An airport for public use, publicly owned and under control of a public agency.

public relations: The management function that attempts to create goodwill for an organization and its products, services, or ideals, with groups of people who can affect its present and future welfare.

public-use airport: An airport open to the public without prior permission and without restrictions within the physical capacities of available facilities.

quadradar: Ground radar equipment named for its four presentations: (1) surveillance, (2) airport surface detection, (3) height finding, and (4) precision approach.

qualitative forecasting: Forecasting methods that rely primarily on the judgment of forecasters based on their expertise and experience with the airport and surrounding environment.

quantitative forecasting: Forecasting methods that use numerical data and mathematical models to derive numerical forecasts.

queuing diagram: A graphical analytic tool used to estimate queuing and delays within the airport.

Radar Approach Control (RAPCON): A joint-use air traffic control facility, located at a U.S. Air Force base, utilizing surveillance and precision approach radar equipment in conjunction with air/ground communication equipment, providing for the safe and expeditious movement of air traffic within the controlled airspace of that facility.

radar beacon system: A radar system in which the object to be detected is fitted with cooperative equipment in the form of a radio receiver/transmitter (transponder). Radio pulses transmitted from the searching transmitter/ receiver (interrogator) site are received in the cooperative equipment and used to trigger a distinctive transmission from the transponder. This latter transmission, rather than a reflected signal, is then received back at the transmitter/receiver site.

radar (radio detection and ranging): A device that, by measuring the time interval between transmission and reception of radio pulses and correlating the angular orientation of the radiated antenna beam or beams in azimuth and/or elevation, provides information on range, azimuth, and/or elevation of objects in the path of the transmitted pulses.

ramp: A defined area, on a land airport, intended to accommodate aircraft for purposes of loading or unloading passengers or cargo, refueling, parking, or maintenance.

reduced horizontal separation minima (RHSM): Programs to reduce the lateral separation of aircraft over oceanic waters.

reduced vertical separation minima (RVSM): Air traffic control program to reduce the vertical separation of aircraft above flight level 290 (29,000 feet MSL) from the current 2000-foot minimum to 1,000-foot minimum.

regional airport planning: Air transportation planning for the region as a whole including all airports in the region, both large and small.

regression analysis: The application of specific mathematical formulas to estimate forecast equations, which may then be used to forecast future activity.

reliever airports: A subset of general aviation airports that has the function of relieving congestion at primary commercial airports and providing more access for general aviation to the overall community.

relocated threshold: An area preceding the runway arrows unusable for takeoff or landing.

remain overnight (RON): Term used to describe a commercial aircraft that stays overnight at an airport terminal prior to an early morning departure.

remote parking: Consists of long-term parking lots located away from the airport terminal buildings. Buses or vans are available to transport passengers to the terminal.

residual cost approach: The airlines collectively assume significant financial risk by agreeing to pay any costs of running the airport that are not allocated to other users or covered by nonairline sources of revenue.

restricted areas: Areas of airspace where ongoing or intermittent activities that create unusual hazards to aircraft.

revenue bonds: Bonds that are payable solely from the revenues derived from the operation of a facility that was constructed or acquired with the proceeds of the bonds.

rigid pavement: A pavement structure consisting of portland cement concrete that may or may not include a subbase course.

RNAV: A generic term that refers to any instrument navigation performed outside conventional routes defined by ground-based navigational aids or by intersections formed by two navigational aids.

rotorcraft: A heavier-than-air aircraft that depends principally for its support in flight on the lift generated by one or more rotors. (FAR Part 1)

rolling hubs: An airline operating strategy to more uniformly distribute the arrivals and departures of aircraft at hub airports.

runway: A defined rectangular area on a land airport prepared for the landing and takeoff run of aircraft along its length.

runway aiming points: Two rectangular markings consisting of a broad wide stripe located on each side of the runway centerline and approximately 1,000 feet from the landing threshold, serving as visual aiming points for landing aircraft.

runway alignment indicator light (RAIL): This airport lighting facility in the terminal area consists of five or more sequenced flashing lights installed on the extended centerline of the runway. The maximum spacing between lights is 200 feet, extending out from 1,600 feet to 3,000 feet from the runway threshold. Even when collocated with ALS, RAIL will be identified as a separate facility.

runway bearing: The magnetic or true bearing of the runway centerline as measured from magnetic or true north.

runway capacity: The maximum number of aircraft operations that can be accommodated by a runway over a given period of time.

runway centerline: A series of uniformly spaced stripes and gaps that run along the longitudinal center of the runway.

runway centerline lighting system (RCLS): The runway centerline lighting system consists of single lights installed at uniform intervals along the runway centerline so as to provide a continuous lighting reference from threshold to threshold.

runway clear zone: An area at ground level whose perimeter conforms to the runway's innermost approach surface projected vertically. It begins at the end of the primary surface and it terminates directly below the point or points where the approach surface reaches a height of 50 feet above the elevation of the runway end.

runway configuration: Layout or design of a runway or runways, where operations on the particular runway or runways being used at a given time are mutually dependent. A large airport can have two or more runway configurations operating simultaneously.

runway contamination: Deposition or presence of dirt, grease, rubber, or other materials on runway surfaces that adversely affect normal aircraft operation or that chemically attack the pavement surface.

runway designator: A number identifying a runway at an airport, defined by the direction (in degrees with respect to magnetic north, divided by 10, and rounded to the nearest integer) on which aircraft operate on the runway.

runway identifier: A whole number to the nearest one-tenth of the magnetic bearing of the runway and measured in degrees clockwise from magnetic north.

runway end identification lights (REIL): An airport lighting facility in the terminal area navigation system consisting of one flashing white high-intensity light installed at each approach end corner of a runway and directed toward the approach zone, which enables the pilot to identify the threshold of a usable runway.

runway environment: The runway threshold or approach lighting aids or other markings identifiable with the runway.

runway gradient (effective): The average gradient consisting of the difference in elevation of the two ends of the runway divided by the runway length may be used provided that no intervening point on the runway profile lies more than 5 feet above or below a straight line joining the two ends of the runway. In excess of 5 feet, the runway profile will be segmented and aircraft data will be applied for each segment separately.

runway grooving: One-quarter-inch grooves spaced approximately 11/4 inches apart in the runway surface designed to provide better drainage and furnish escape routes for water under the tire footprint in order to prevent hydroplaning.

runway length—landing: The measured length from the threshold to the end of the runway.

runway length—physical: The actual measured length of the runway.

runway length—takeoff: The measured length from where the takeoff is designated to begin to the end of the runway.

runway lights: Lights having a prescribed angle of emission used to define the lateral limits of a runway. Runway light intensity may be controllable or preset. Lights are uniformly spaced at intervals of approximately 200 feet.

runway markings: (1) *Basic marking:* markings on runways used for operations under visual flight rules, consisting of centerline marking and runway direction numbers, and if required, letters. (2) *Instrument marking:* markings on runways served by nonvisual navigation aids and intended for landings under instrument weather conditions, consisting of basic marking plus threshold marking. (3) *All-weather marking:* markings on runways served by nonvisual precision approach aids and on runways having special operational requirements, consisting of instrument markings plus landing zone marking and side strips.

runway occupancy time (ROT): The time from when an approaching aircraft crosses the threshold until it turns off the runway or from when a departing aircraft takes the active runway until it clears the departure end.

runway orientation: The magnetic bearing of the centerline of the runway.

runway safety area: Cleared, drained, graded, and usually turfed area abutting the edges of the usable runway and symmetrically located about the runway. It extends 200 feet beyond each runway end. The width varies according to the type of runway. (Formerly called "landing strip.")

runway strength: The assumed ability of a runway to support aircraft of a designated gross weight for each of single-wheel, dual-wheel, and dual–tandem-wheel gear types.

runway surface lighting: Also referred to as "in-runway lighting," consists essentially of touchdown zone (narrow gauge) lights, runway centerline lights, and exit taxiway turnoff lights installed in the pavement.

runway threshold marking: Markings so placed as to indicate the longitudinal limits of that portion of the runway usable for landing.

runway visibility: Visible distance associated with the instrument runways or by an observer stationed at the approach end of a runway.

runway visual range (RVR): An instrumentally derived value that represents the horizontal distance a pilot can see down the runway from the approach end; it is based on the sighting of either high-intensity runway lights or on the visual contrast of other targets, whichever yields the greater visual range.

sales force composite method: A qualitative forecasting method that seeks the judgment of airport employees, and the employees of those firms that do business at the airport for their predictions of future activity.

satellite terminals: A type of terminal layout in which all passenger processing is done in a single terminal that is connected by concourses to one or more satellite structures. The satellite generally has a common waiting room that serves a number of gate positions.

scheduled service: Transport service operated over routes based on published flight schedules, including extra sections and related nonrevenue flights.

search radar: A radar system in which a minute portion of a radio pulse transmitted from a site is reflected off an object and then received back at that site.

secondary radar: See radar beacon system.

secondary runway: A runway that provides additional wind coverage or capacity to expedite traffic handling.

secondary taxiway system: Taxiways that provide aircraft access from runways to hangars and tiedown areas not commonly associated with itinerant and service areas.

secure area: The area at the airport where commercial air carriers conduct the loading and unloading of passengers and baggage between their aircraft and the terminal building.

security chief: Enforces interior security, traffic, and safety rules and regulations and participates in law enforcement activities at the airport.

security identification display area (SIDA): The portion of an airport in which only persons displaying proper identification may have access.

security lighting: Lighting systems that provide a means of continuing, during the hours of darkness, a degree of protection approaching that which is maintained during daylight hours.

segmented circle: A basic marking device used to aid pilots in locating airports, and that provides a central location for such indicators and signal devices as may be required.

self-liquidating general obligation bonds: Like general obligation bonds, these bonds are backed by the full faith, credit, and taxing power of the issuing government body; however, there is enough cash flow from the operation of the facility to cover the debt service and other costs of operation so that the debt is not legally considered part of the community's debt limitation.

semiflush light: A light mounted in pavement capable of rollover by aircraft.

sequencing: Specifying the exact order in which aircraft will take off or land.

set-aside funds: Available to any eligible airport sponsor and allocated according to congressionally mandated requirements for a number of different set-aside subcategories such as minimum allocations to all 50 states, the District of Columbia, and insular areas on the basis of land area and population.

shared-use gate usage agreement: A gate usage agreement in which air carriers and other aircraft schedule use of gates in coordination with airport management and other air carriers serving the airport.

short takeoff and landing (STOL) aircraft: An aircraft that, at some weight within its approved range of STOL operating weight, is capable of operating from an STOL runway in compliance with the applicable STOL characteristics, airworthiness, operations, noise, and pollution standards.

short-term parking: Usually located close to terminal buildings for motorists picking up or dropping off travelers. These motorists generally remain at the airport less than 3 hours.

shoulder: As it pertains to airports, an area adjacent to the edge of a paved surface so prepared to provide a transition between the pavement and the adjacent surface for aircraft running off the pavement, for drainage, and sometimes for blast protection.

SIMMOD: A software program used to simulate airport operations, in part for the purposes of analyzing system capacity.

simple-unit terminal: A type of gate arrival terminal layout that consists of a common waiting and ticketing area with several exits onto a small aircraft parking apron.

simultaneous converging instrument approaches (SCIA): The use of new air traffic control procedures to allow the simultaneous use of instrument approaches on converging runways at an airport.

simultaneous offset instrument approaches: An attempt to increase airport capacity and reduce delay at airports with closely spaced parallel runways by allowing pilots to fly a straight-but-angled instrument (and possibly autopilot) approach until descending below the cloud cover.

single runway: An airport having one runway.

site selection: Part of the airport master plan that evaluates airspace, environmental factors, community growth, airport ground access, availability of utilities, land costs, and site development costs.

slot: A block of time allocated to an airport user to perform an aircraft operation (takeoff or landing).

small aircraft: Aircraft of 12,500 pounds or less maximum certificated takeoff weight. (FAR Part 1)

Small Aircraft Transportation System (SATS): NASA-sponsored program to develop enhanced aviation, communication, and navigation technologies for smaller general aviation aircraft.

small hubs: Those airports that enplane 0.05 percent to 0.25 percent of the total passenger enplanements.

spacing: Establishing and maintaining the appropriate interval between successive aircraft, as dictated by considerations of safety, uniformity of traffic flow, and efficiencies of runway use.

spalling: Fractured edges in and around the joint area of concrete that are due to the tremendous pressures generated during expansion and contraction of the slabs.

special facilities bond: A bond secured by the revenue from the indebted facility, such as a terminal, hangar, or maintenance facility, rather than the general revenue of the airport.

special-use airspace: Airspace controlled by the Department of Defense.

standard metropolitan statistical area (SMSA): Regions defined by local metropolitan planning organizations to comprise the urban and suburban regions surrounding a major city.

Standard Terminal Automation Replacement System (STARS): Will replace outdated air traffic control computers with twenty-first century systems at nine large consolidated TRACONs and approximately 173 FAA and 60 DOD terminal radar approach control sites across the country.

state aviation system plans (SASP): Plan for the development of airports within a state.

State Block Grant Program: Under this program, selected states are given responsibility of AIP grants at other than primary airports. Each state is responsible for determining which locations will receive funds within the state.

state-operated airports: Airports generally managed by the state's Department of Transportation.

sterile area: The part of the airport to which passenger access must be gained through TSA passenger screening checkpoints.

STOLport: An airport specifically designed for STOL aircraft, separate from conventional airport facilities.

straight-in approach (IFR): An instrument approach wherein final approach is commenced without first having executed a procedure turn. (Not necessarily completed with a straight-in landing.)

straight-in approach (VFR): Entry into the traffic pattern by interception of the extended runway centerline without executing any other portion of the traffic pattern.

stub taxiway: A short connecting taxiway to an airport facility that serves as the only connection with the remaining airport complex.

subdivision regulations: Provisions prohibiting residential construction in intense noise exposure areas.

subgrade: The ground underlying airfield pavements.

Surface Movement Advisor (SMA): A system developed by the FAA and NASA to promote the sharing of dynamic information among airlines, airport operators, and air traffic controllers in order to control the efficient flow of aircraft and vehicles on the airport surface.

Surface Transportation Assistance Act of 1983: An amendment to the Airport and Airway Improvement Act of 1982 that increased the annual authorizations for the AIP for fiscal years 1983 to 1985.

TAAM: A software program used to simulate airport operations, in part for the purposes of analyzing system capacity.

taxiway: A defined path, usually paved, over which aircraft can taxi from one part of an airport to another.

taxiway centerline lighting: A system of green flush or semiflush in-pavement lights indicating the taxiway centerline.

taxiway safety area: A cleared, drained, and graded area, symmetrically located about the extended taxiway centerline and adjacent to the end of the taxiway safety area.

taxiway turnoff lighting: Single lights installed in the pavement at uniform intervals to define the path of aircraft travel from the runway centerline to a point on the taxiway.

taxiway turnoff markings: Signs or lights along the runways, taxiways, and ramp surfaces of an airport used to assist a pilot in finding his or her way.

technological improvements: Refers to new devices and equipment as well as operational concepts and procedures designed to relieve congestion, increase capacity, or reduce delay.

temporary flight restrictions (TFR): Established in the wake of September 11, 2001, areas of airspace that are identified as restricted areas for a period of time for reasons of national security.

terminal apron: An area provided for parking and positioning of aircraft in the vicinity of the terminal building for loading and unloading.

terminal area: The area used or intended to be used for such facilities as terminal and cargo buildings, gates, hangars, shops, and other service buildings; automobile parking, airport motels and restaurants, and garages and vehicle service facilities used in connection with the airport; and entrance and service roads used by the public within the boundaries of the airport.

terminal area capacity: The ability of the terminal area to accept the passengers, cargo, and aircraft that the airfield accommodates.

terminal building: A building or buildings designed to accommodate the enplaning and deplaning activities of air carrier passengers.

terminal Doppler weather radar (TDWR): Radar systems designed to observe significant weather conditions, such as thunderstorms and hurricanes.

terminal facilities: The airport facilities providing services for air carrier operations which serve as a center for the transfer of passengers and baggage between surface and air transportation.

terminal finger: An extension of the terminal building to provide direct access to a large number of airport terminal apron gate positions.

terminal instrument procedures (TERPs): Procedures used for conducting independent instrument approaches to converging runways under instrument meteorological conditions.

Terminal Radar Approach Control (TRACON): Regional air traffic control centers that control the movement of air traffic in busy areas at altitudes under 18,000 MSL.

terrorism: The systematic use of terror or unpredictable violence against governments, publics, or individuals to attain a political objective.

tetrahedron: A device with four triangular sides that indicates wind direction and that may be used as a landing direction indicator.

T hangar: An aircraft hangar in which aircraft are parked alternately tail to tail, each in the T-shaped space left by the other row of aircraft or aircraft compartments.

3-D UPT Flight Trials Project: An attempt to quantify the savings associated with unrestricted flight.

threshold: The designated beginning of the runway that is available and suitable for the landing of airplanes.

threshold crossing height (TCH): The height of the straight-line extension of the visual or electronic glide slope above the runway threshold.

threshold lights: Lighting arranged symmetrically about the extended centerline of the runway identifying the runway threshold. They emit a fixed green light.

throughput capacity: The rate at which aircraft can be brought into or out of the airfield, without regard to any delay they might experience.

ticketing: Facilities staffed by air carrier personnel or infrastructure that provides passengers with tickets and boarding passes for scheduled departures.

time-series analysis or trend extension: The oldest and in many cases the most widely used method of forecasting air transportation demand. It consists of interpreting the historical sequence of data and applying the interpretation to the immediate future. Historical data are plotted on a graph, and a trend line is drawn.

time-space diagram: A graphical analytic tool used to estimate runway capacity.

total operations: All arrivals and departures performed by military, general aviation, and air carrier aircraft.

touchdown: (1) The point at which an aircraft first makes contact with the landing surface. (2) In a precision radar approach, the point on the landing surface toward which the controller issues guidance instructions.

touchdown zone: The area of a runway near the approach end where airplanes normally alight.

touchdown zone lighting (TDZL): This system in the runway touchdown zone area presents, in plain view, two rows of transverse light bars located symmetrically about the runway centerline. The basic system extends 3,000 feet along the runway.

touchdown zone markings: Groups of rectangular bars, symmetrically arranged in pairs about the runway centerline, serving to identify the touchdown zone for landing operations.

tower: See airport traffic control tower.

Tower Automated Ground Surveillance System (TAGS): Intended to be used in conjunction with airport surface detection equipment at major airports, it will provide, for transponder-equipped aircraft, a flight identification label alongside the position indicator on the ASDE display.

Traffic Alert and Collision Avoidance System (TCAS): Technology placed in aircraft cockpits that shows the relative positions and velocities of aircraft in the vicinity, up to 40 miles away.

Traffic Management Advisor (TMA): Air traffic control technology that provides en-route controllers the capacity to manage the flow of traffic from a single center into selected major airports.

Traffic Management System (TMS): A new software that will perform several important functions to increase the efficiency of airport and airspace utilization.

traffic pattern: The traffic flow that is prescribed for aircraft landing at, taxiing on, and taking off from an airport (FAR Part 1). The usual components of a traffic pattern are upwind leg, crosswind leg, downwind leg, base leg, and final approach.

transfer passengers: Passengers at an airport transferring from one aircraft to another as part of their itineraries.

transitional surface: A surface that extends outward and upward from the sides of the primary and approach surfaces normal to the runway centerline that identifies the height limitations on an object before it becomes an obstruction to air navigation.

transition area: Controlled airspace extending upward from 700 feet or more above the surface of the earth when designated in conjunction with an airport for which an instrument approach procedure has been prescribed; or from 1,200 feet or higher above the surface of the earth when designated in conjunction with airway route structures or segments. Unless otherwise limited, transition areas terminate at the base of the overlying controlled airspace.

Transportation Security Administration (TSA): Established November 2001 to address issues concerning airport security in the wake of the terrorist attacks on New York City and Washington, D.C., on September 11, 2001.

Transportation Security Regulations (TSR): Regulations found in Part 1500 of Title 49 in the Code of Federal Regulations that describe national regulations governing aviation and transportation security.

turnaround: A taxiway adjacent to the runway ends that aircraft use to change direction, hold, or bypass other aircraft.

turning radius: The radius of the arc described by an aircraft in making a self-powered turn, usually given as a minimum.

turnoff taxiway: A taxiway specifically designed to provide aircraft with a means to expedite clearing a runway.

Twelve-Five Program: A program that mandates all aircraft with a maximum certified takeoff weight of 12,500 pounds or more must be thoroughly searched before departure and all passengers, crew members, and other persons and their accessible property must be screened before boarding the aircraft.

typical peak-hour passenger volume (design volume): The peak hour of an average day in the peak month that is used as the hourly design volume for terminal space.

undershoot: To touch down short of the point of intended landing.

unicom: Frequencies authorized for aeronautical advisory services to private aircraft. Services available are advisory in nature, primarily concerning the airport services and airport utilization.

upwind leg: A flight path parallel to the landing runway in the direction of landing.

utility airport (or runway): An airport (or runway) that accommodates small aircraft excluding turbojet-powered aircraft.

variance: The differences between actual expenses and the budgeted amount.

vehicle chief: Responsible for the maintenance of all vehicles utilized by the airport.

vertical takeoff and landing (VTOL): Aircraft that have the capability of vertical takeoff and landing. VTOL aircraft are not limited to helicopters.

VFR airport: An airport without an authorized or planned instrument approach procedure; also, a former airport design category indicating an airport serving small aircraft only and not designed to satisfy the requirements of instrument landing operations.

VFR tower: An airport traffic control tower that does not provide approach control service.

VHF omnidirectional range (VOR): A radio transmitter facility in the navigation system radiating a VHF radio wave modulated by two signals, the relative phases of which are compared, resolved, and displayed by a compatible airborne receiver to give the pilot a direct indication of bearing relative to the facility.

vibratory (or dynamic) testing: A technique used to measure the strength of a composite pavement system by subjecting it to vibratory load and measuring the amount the pavement responds or deflects under this known load.

vicinity map: Shown on the airport layout plan drawing, it depicts the relationship of the airport to the city or cities, nearby airports, roads, railroads, and built-up areas.

viscous hydroplaning: Occurs when a thin film of oil, dirt, or rubber particles mixes with water and prevents tires from making sure contact with the pavement.

Vision 100—Century of Aviation Reauthorization Act: An act made effective in 2003 that reauthorized the FAA Airport Improvement Program, raised allowable PFCs to $4.50 per segment, and created the JPDO to oversee NextGen implementation.

visual approach: An approach wherein an aircraft on an IFR flight plan, operating in VFR conditions under the control of a radar facility and having an air traffic control authorization, may deviate from the prescribed instrument approach procedure and proceed to the airport of destination, served by an operational control tower, by visual reference to the surface.

visual approach slope indicator (VASI): An airport lighting facility in the terminal area navigation system used primarily under VFR conditions. It provides vertical visual guidance to aircraft during approach and landing by radiating a directional pattern of high-intensity red and white focused light beams that indicate to the pilot that the aircraft is "on path" if the pilot sees red/white, "above path" if white/white, and "below path" if red/red.

visual flight rules (VFR): Rules that govern the procedures for conducting flight under visual conditions. (FAR Part 91)

visual meteorological conditions (VMC): Meteorological conditions expressed in terms of visibility and ceiling equal to or better than specified minima.

visual runway: A runway intended solely for the operation of aircraft using visual approach procedures, with no straight-in instrument approach procedure and no instrument designation indicated on an FAA-approved airport layout plan or a military service–approved military airport layout plan, or by a planning document submitted to the FAA by competent authority. (FAR Part 77)

vortices: As pertaining to aircraft, circular patterns of air created by the movement of an airfoil through the atmosphere. As an airfoil moves through the atmosphere in sustained flight, an area of high pressure is created beneath it and an area of low pressure is created above it. The air flowing from the high-pressure area to the low-pressure area around and about the tips of the airfoil tends to roll up into two rapidly rotating vortices, cylindrical in shape. These vortices are the most predominant parts of aircraft wake turbulence and their rotational force is dependent upon the wing loading, gross weight, and speed of the generating aircraft.

wake vortex: A phenomenon resulting from the passage of an aircraft through the atmosphere. It is an aerodynamic disturbance that originates at the wingtips and trails in corkscrew fashion behind the aircraft. When used by ATC it includes vortices, thrust stream turbulence, jet wash, propeller wash, and rotor wash.

warning areas: Areas of airspace that contain the same kind of hazardous flight activity as restricted areas, but are located over domestic and international waters, beginning 3 miles offshore.

Weather and Radar Processor (WARP): Will collect and process weather data from Low-Level Windshear Systems (LLWAS), Next-Generation Weather Radar (NEXRAD), Terminal Doppler Weather Radar (TDWR), and surveillance radar, and will disseminate this data to controllers, traffic management specialists, pilots, and meteorologists.

Wendell H. Ford Aviation Investment and Reform Act for the 21st Century (AIR-21): Legislation which increased annual levels of funding for aviation investments by $10 billion, with most of the funding appropriated toward air traffic control modernizations and airport construction and improvement projects.

Wide Area Augmentation System (WAAS): An augmentation of GPS that includes integrity broadcasts, differential corrections, and additional ranging signals; its primary objective is to provide accuracy, integrity, availability, and continuity required to support all phases of flight.

wind cone: A free-rotating fabric truncated cone that, when subjected to air movement, indicates wind direction and wind force.

wind rose: A diagram for a given location showing relative frequency and velocity of wind from all compass directions.

wind shear: Variation of wind speed and wind direction with respect to a horizontal or vertical plane. Low-level shear in the terminal area is a factor in the safe and expeditious landing of aircraft.

wind sock: A hollow flaglike object located on an airfield that depicts approximate wind direction and speed.

wind tee: A T-shaped free-rotating device to indicate wind direction. Sometimes capable of being secured for use as a landing direction indicator.

zero-based budget: Derives from the idea that each program or departmental budget should be prepared from the ground up, or base zero. By calculating the budget from a zero base, all costs are newly developed and reviewed entirely to determine their necessity.

Zulu time (Z): Time at the prime meridian in Greenwich, England.

Index